SCIENCE AND HUMANISM

Historically and conceptually, influential traditions of thought and practice associated with humanism and science have been deeply connected. This book explores some of the most pivotal relations of humanistic and scientific engagement with the world to inspire a reconsideration of them in the present. Collectively, its contributions illuminate a fundamental but contested feature of a broadly humanist worldview: the hope that science may help to improve the human condition, as well as the myriad relationships of humanity to the natural and social worlds in which we live. Arguably, these relationships are now more profoundly interwoven with our sciences and technologies than ever before. Addressing scientific and other forms of inquiry, approaches to integrating humanism with science, and cases in which science has failed, succeeded, and could do more to promote our collective welfare, this book enjoins us to articulate a compelling, humanist conception of the sciences for our times. This title is also available as Open Access on Cambridge Core.

ANJAN CHAKRAVARTTY is the Appignani Foundation Professor of Philosophy at the University of Miami. He is the author of *A Metaphysics for Scientific Realism: Knowing the Unobservable* (2007) and *Scientific Ontology: Integrating Naturalized Metaphysics and Voluntarist Epistemology* (2017).

SCIENCE AND HUMANISM

Knowledge, Values, and the Common Good

EDITED BY

ANJAN CHAKRAVARTTY

University of Miami

CAMBRIDGE
UNIVERSITY PRESS

CAMBRIDGE
UNIVERSITY PRESS

Shaftesbury Road, Cambridge CB2 8EA, United Kingdom

One Liberty Plaza, 20th Floor, New York, NY 10006, USA

477 Williamstown Road, Port Melbourne, VIC 3207, Australia

314–321, 3rd Floor, Plot 3, Splendor Forum, Jasola District Centre, New Delhi – 110025, India

103 Penang Road, #05-06/07, Visioncrest Commercial, Singapore 238467

Cambridge University Press is part of Cambridge University Press & Assessment, a department of the University of Cambridge.

We share the University's mission to contribute to society through the pursuit of education, learning and research at the highest international levels of excellence.

www.cambridge.org
Information on this title: www.cambridge.org/9781009626866

DOI: 10.1017/9781009626880

First published 2025

Cover image: Andy Potts, andy-potts.com; represented by agoodson.com.

A catalogue record for this publication is available from the British Library

A Cataloging-in-Publication data record for this book is available from the Library of Congress

ISBN 978-1-009-62686-6 Hardback

Contents

Contributors

AKEEL BILGRAMI is the Sidney Morgenbesser Professor of Philosophy at Columbia University.

MATTHEW J. BROWN is the Jo Ann and Donald N. Boydston Professor of American Philosophy at Southern Illinois University.

ANJAN CHAKRAVARTTY is the Appignani Foundation Professor of Philosophy at the University of Miami.

ALEKSANDRA HERNANDEZ is Assistant Professor of Philosophy at the University of Miami.

IAN JAMES KIDD is Associate Professor of Philosophy at the University of Nottingham.

JANET A. KOURANY is Professor of Philosophy and of Gender Studies at the University of Notre Dame.

DAVID LUDWIG is Associate Professor of Philosophy at Wageningen University.

GURPREET RATTAN is Professor of Philosophy at the University of Toronto.

ALAN RICHARDSON is Professor of Philosophy at the University of British Columbia.

STÉPHANIE RUPHY is Professor of Philosophy of Science at the École normale supérieure – Université Paris Sciences et Lettres.

THOMAS UEBEL is Emeritus Professor of Philosophy at the University of Manchester.

CATHERINE WILSON is Emeritus Professor of Philosophy at the University of York.

Introduction

Anjan Chakravartty

Traditions of humanistic thought and activism, and traditions of scientific investigation and practice, admit of such variation across time and place that it can be a dizzying task to grapple with one or the other let alone both together. That said, understanding at least some of the many, deep connections between these richly varied traditions seems crucial if we are to think well – perceptively and wisely – about how we may hope to connect them today and in the future. It is not hyperbolic, I think, to suggest that humanistic aspirations for the dignity and well-being of persons and the very integrity of the world in which we live have never been more intimately connected to the nature and consequences, whether intended or unintended, of our science and technology. This book is an attempt to illuminate a number of striking relationships between humanistic thought and science – and to help us consider the promise and challenges inherent in the foremost ambition of contemporary humanism *for* science: that it may serve to improve the human condition, broadly conceived.

Described this way, the project of this volume could easily have been a massive undertaking, encompassing a detailed intellectual history spanning millennia concerning the development of humanism, science, and relationships between the two. Our goal, though, has not been to write the encyclopedia this suggests but, more modestly, to motivate something of a philosophical agenda. It is fair to say that while issues *relating to* science and humanism pervade the history of ideas, the topic itself has been for the most part neglected in recent years and in contemporary discussions of science and society, even while the inhabitants of our planet are beset by mounting challenges on a global scale. Given the importance this suggests of a return to thinking about connections between humanist aspirations and scientific practice, and before describing more concretely some of the seeds we are attempting to sow here, let me make two further, preparatory remarks – one about reading the book, and another about the partiality of what we have been able to tackle.

First, then, who are these essays for? We have tried to make them accessible to a wide audience. While we hope and expect that philosophers, scientists, historians, and other scholars will be interested – after all, the discussions here range over the philosophy of science (and values), the history of philosophy and science, ethical, social, and political matters, and the history of ideas – we also hope to reach people beyond the academy. Granted, the ease with which these chapters may be read will, of course, vary somewhat with one's background interests and willingness to look up the occasional, unfamiliar term, but we have tried to avoid excessive jargon and to explain more technical ideas where they are helpful. It is worth noting the reason for this, I think: To the extent that we as a species share a collective interest in making this world a better one, it is incumbent on us, in as many ways as we can, to bring experts and nonexperts together to think about how we are going to proceed, collectively, and nowhere is this more urgent than in the domain of science and humanism.

Even with the best of intentions, though, we have only scratched the surface of the many topics that could easily fall within the remit of our title. As already noted, the subject matter is spatiotemporally immense, and we have done our best to sample some key parts of it. Furthermore, since no one is an expert in everything, some chapters focus more centrally on humanism, some more centrally on science, and some more centrally on issues crucial to understanding relations between the two. In other words, not all of these chapters engage in equal measure with all of these things; and yet, together, they are all pieces of the larger puzzle of science and humanism. With this in mind, I hope this book will be read as I envisioned it, optimistically, as something intended to be greater than the sum of its parts: While readers should certainly feel free to zero in on bits and pieces of it, the volume as a whole asks us, I hope – readers and authors alike – to think about how the larger picture fits together, or *should* fit together. No one chapter can do this, but each one is poised to help, and to this end I believe we have put a good deal on the table, and have wedged open some doors to further questions, discussions, and answers.

Turning now to the content, let me start with the obvious fact that many different but connectible views have been associated with the term "humanism" historically. Similarly, the term "science" has been associated with an amazing diversity of practices of inquiry. The common thread running through this book is a persistent strand of humanism conceived as a broad approach to thinking about and acting in the world (as well as various, more specific views of humanism, some in conflict with this strand), which has coevolved with the development of the sciences: the idea that in order to understand the natural and social worlds in which we live, and ourselves, and to act in pursuit of the common good, special prominence should be given to science. Whether in the form of knowledge and technologies to help feed and shelter people, to treat disease and injury, to address harms done to our planet, or to promote tolerance, equality, and freedom, the application of the fruits of scientific investigation by reason is crucial to shaping a better, collective future.

This is a powerful idea, but on reflection, it is immediately apparent that there is nothing intrinsic to the sciences that suggests they are or must be causes or facilitators of social or moral progress. Indeed, in some (and some notorious) cases, scientific work has been a cause or a facilitator of substantial harm. This reality calls for careful scrutiny and vigilance; our ideals for the interweaving of the sciences and humanistic thought call for articulation in every age. In hopes of promoting this sort of contemplation presently, let me proceed now to describe how the contents of the book are organized – into three parts – marking some natural divisions between spheres of thinking about science and humanism. Part I is concerned with certain relations between scientific and other forms of theorizing and practice, such as philosophical and moral reflection, which are intimately connected to humanism. Part II explores links between science and humanism in some venerable traditions of philosophy this past century, as possible sources of inspiration today. Part III engages with contexts of scientific practice, examining roles that philosophers, scientists, and others have played and *could* play in addressing (for example) the welfare of women, minorities, marginalized groups, and the environment, as well as our ethical and social responsibilities.

Appreciating that there are different ways the book *might* have been organized, and given that the thematic unity of each of the parts just described is inevitably somewhat abstract, let me elaborate next on some rationales for structuring the volume this way, and on the more specific issues discussed in each chapter.

Part I, "Interrelations: Scientific and Other Forms of Knowledge." The essays in Part I are all, though in different ways, concerned with relations between scientific and other forms of belief and knowledge (philosophical, moral, etc.), the latter of which are central to conceptions of humanism. Chapters 1 and 2 are natural companions, drawing on histories of humanism to discuss various affinities and antipathies of this tradition to science. In Chapter 1, "What Is Science For? Modern Intersections of Science and

Humanism," Anjan Chakravartty argues for a return to a vital aspect of a long arc of intellectual history connecting a broad, humanistic worldview and scientific inquiry, which seems to have gone missing in the present: a conception of the aim of science in terms of human and planetary welfare. Shifting from the notion of a general worldview, however, to more specific views that have been associated with humanism in philosophy, the extent to which the latter are congenial to the sciences has varied significantly. In Chapter 2, "Varieties of Philosophical Humanism and Conceptions of Science," Ian James Kidd explores this range of positions, focusing on those that are or have been critical of science and scientific modernity.

Completing Part I, Chapters 3 and 4 are also in a sense companions, each considering a different aspect of the notion of *scientism*, commonly associated with an overly strong confidence in the certainty or scope of scientific knowledge. While it is fair to say that humanists have not, necessarily or as a matter of course, subscribed to scientism historically, it is not unusual to hear scientistic-sounding claims in the public sphere of humanism today. The upshot of Chapters 3 and 4 is that a humanistic understanding of science should not be scientistic: The first of these chapters targets a version of scientism according to which the sciences are the ultimate adjudicators of objectivity; the second targets a version according to which the sciences are capable, in principle, of describing the natural world completely. In Chapter 3, "Scientism and the Limits of Objective Thinking," Gurpreet Rattan argues that while scientific thinking may *exemplify* the limits of objective thinking about the world, it does not itself *determine* what these limits are. They are determined, he contends, by norms that govern critical reflection more generally. In Chapter 4, "Scientism: Reflections on Nature, Value, and Agency," Akeel Bilgrami maintains that the presence of value in nature cannot be understood by means of scientific inquiry and description alone, and that appreciating this is crucial to understanding how we act in the world.

Part II, "Inspirations: Philosophies of Science and Its Social Role." Part II turns to influential movements in the recent history of philosophy to bring a number of accounts of science, values, and implications for humanistic understanding and action to the foreground. The spotlight here alternates between the two most important, explicitly humanist philosophical movements of the earlier twentieth century: logical empiricism and American pragmatism – the first two chapters exploring the formulation and development of certain humanistic dimensions of each of these movements, respectively. Chapter 5, "Scientific Humanisms: Sarton, Reichenbach, and the Crisis of Western Science after World War

I," by Alan Richardson, considers various, overlapping reflections on science in the soul-searching aftermath of the Great War, including the logical empiricism of Rudolf Carnap and Hans Reichenbach, and the New Humanism of historian of science George Sarton. In Chapter 6, "John Dewey, Humanism, and the Value of Science," Aleksandra Hernandez traces the concurrent development of Dewey's progressive, pragmatist humanism, which championed the sciences in the service of human flourishing, responding in part to hostilities toward science expressed by conservative literary humanists of the time.

Chapters 7 and 8 likewise examine logical empiricism and American pragmatism in turn, but with the goal of linking some of their philosophical contentions, in a more targeted way, to specific matters of present concern. The past several decades have witnessed a groundswell of feminist (and other) literature taking particular interest in forms of bias and inequity in relation to science, and in Chapter 7, "Sociopolitical Engagement and Scientific Value Freedom: The View from the Left Vienna Circle," Thomas Uebel charts thematic connections and differences between an advocacy for social, economic, and political reforms by prominent logical empiricists this past century and theorizing about similar ambitions more recently. Chapter 8, "The Pragmatic and the Religious Functions of Science," by Matthew J. Brown, draws another sort of connection between thinking about the sciences and an issue of perennial humanist interest – in this case, the question of religion. Brown contends that a pragmatist emphasis on science as a problem-solving activity has the potential, especially in Dewey's hands, for elaborating how the sciences may fulfill social and personal roles that are often (and more typically) associated with religion, concerning questions of meaning and our place in the world.

Part III, "Interventions: Scientific Knowledge and Social Imperatives." After the more conceptual discussions of Part I, and the reflections on some especially significant, theoretical views from our recent past in Part II, Part III is devoted to thinking about relations of science and humanism in a more concrete vein, wrestling with particular issues and cases of science in society. In Chapter 9, "The Present Plight of Science, and Our Plight," Janet A. Kourany poses several questions about the capacities of the sciences to help facilitate desirable outcomes given the many serious challenges we face, some of which threaten the integrity of science itself, as a prelude to thinking about how these capacities may be enhanced. Then, focusing on these and related questions in part through an examination of the case of global agriculture, Chapter 10, "Science and Justice: Beyond the New Orthodoxy of Value-Laden Science," by David Ludwig, urges us to go beyond thinking about the roles values play in composite systems of scientific inquiry and science-and-technology-based applications, to think about what it would mean for these systems to be *just*, not least in relation to those directly impacted by them.

These themes of learning from cases, and bringing the sciences to bear in society (and in the world as a whole) to help combat marginalization and promote more widely distributed well-being, continue through to the end. In Chapter 11, "The Human Sciences and the 'Theory of Women'," Catherine Wilson considers a history of "scientific sexism" implicated in nineteenth- and twentieth-century subordinations of women, and the genesis and development of more recent science that, in contrast, is straightforwardly supportive of moral and political values of equality, serving to oppose these earlier forms of injustice and oppression. Perhaps one day, a natural humanist inclination may lead us to establish more representative sciences that, as a consequence of their own principles of operation, investigate our natural and social worlds in ways that help us to address such issues of broader flourishing more systematically. In Chapter 12, "Toward More Inclusive Science: New Challenges and Responsibilities for Scientists, Philosophers, and Citizens," Stéphanie Ruphy examines an increasing demand (in many liberal democracies, for instance) for greater participation by citizens in scientific inquiry in hopes of better aligning research with the needs of society, and the new responsibilities for each of us this would entail.

In closing, let me return to my earlier remarks about the motivation for a book like this one, at this time. Arguably, though not taking center stage as a clear or pivotal problematic in its own right for over half a century, a return to relationships between science and humanism has been prefigured in other discussions more recently. A number of the chapters in this volume could be read in isolation as fitting contributions to these allied literatures – say, regarding the nature of scientific knowledge and its relations to other, putatively distinct forms of knowledge, or regarding the roles of science in society, or regarding connections between values and the sciences. Viewing these issues through the lens of humanism, however, encourages us to think in a more synthetic way about what science *is*, and what it could be, as a now-dominant component of our many cultures – to stand back from those allied literatures and see the sciences more transparently in these more synthetic terms. It is my hope that bringing these essays together with this explicit framing in mind may induce us to think about the contributions that such studies may make, beyond the insights they surely afford within the confines of otherwise separate discussions, to a spirited consideration of something greater.

Thus, by design and however partially, the collection of essays to follow covers extensive ground. As noted at the start of this Introduction, over a long stretch of intellectual history, a great many facets of the nexus of science and humanism have emerged. In the face of persistent and growing challenges in our own times, it seems crucial that we consider these many facets anew, in the present, for our own good and for that of the planet. In compiling a varied sample of the space of issues constituting this nexus of science and humanism, I hope we have succeeded not only in conveying its breadth but also in prompting further questions and deliberations on these and related matters. I hope that readers will engage with these chapters and ponder the aspects of science and humanism to which they are most relevant, reflecting and building upon them where they agree, and contesting and improving upon them where they do not, all in the spirit of a critical yet collaborative project, and with the ultimate goal of working toward an articulation and fulfillment of our best and most inspiring humanistic ambitions for science in the twenty-first century.

PART I

Interrelations

Scientific and Other Forms of Knowledge

CHAPTER 1

What Is Science For?

Modern Intersections of Science and Humanism

Anjan Chakravartty

> The great scientific revolution is still to come. It will ensue when men collectively and cooperatively organize their knowledge for application to achieve and make secure social values.
>
> John Dewey, "Science and Society"

Intersections (?) of Science (?) and Humanism (?)

It is uncontroversial that humanistic thought and scientific inquiry have been entangled throughout a very long arc of intellectual history. Beyond this, however, significant challenges await anyone hoping to understand let alone articulate the nature of these entanglements. Since "science" and "humanism" are labels that are commonly applied to traditions of theorizing and practice that predate the eighteenth- and nineteenth-century introduction and use of these terms in their modern senses, respectively, and since both of these traditions have evolved and speciated a great deal from antiquity to the present, any attempt to untangle the many complex relationships between them amounts to a formidable task.

Thankfully, and while endeavoring not to shy away from any of these complexities en route, my focus in this chapter is much narrower. My interest here is in what the history of these relations between science and humanism reveals about the (arguably) peculiar way in which their connection is typically viewed today, as being entirely asymmetrical. If, on a first pass common to dictionaries and encyclopedias, we take humanism to be a worldview emphasizing the interests, capacities, and welfare of humanity, as well as our potential for learning about the world as a means to confronting the challenges we face and promoting human flourishing (e.g., Lacey 1995: 375–376), the importance of the sciences to humanism is abundantly clear and, indeed, this is commonly, explicitly asserted. But what of the complementary relation of the importance of humanism to science? It is striking that while, for most of the more recent history of

Western societies, this latter relation was often acknowledged as something worthy of praise or criticism, in our times it may seem a strange thing to hold that humanist values and ambitions are at all relevant let alone important to what we think of as science.

Taking the past of both science and humanism as a prelude to a consideration of their connection in the present, my current aim is twofold. First, of historical interest and less controversially, I review certain developments in the intellectual history of the West since the Renaissance that were pivotal to establishing a widespread (though hardly universal) commitment to the idea that the sciences are among our most potent means for enhancing human and planetary flourishing. Also of historical interest but more controversially, I endeavor to illuminate just how strange it is, in historical perspective, that we have now drifted away from a complementary commitment to the idea that humanist ideals could or should be pivotal to our conception of the sciences. Not least given the serious, in some cases existential, crises we have brought upon ourselves and our planet in the relatively short duration of our existence as a species, I argue for a return to these ideals as a plausible basis for a normative conception of the aims of science today. This furnishes a partial answer to one of the three questions tagged obliquely in the heading of this section – the question of how we might best understand the *connection* between science and humanism.

The other two questions concern how, to this end, we should understand the extensions of the terms "science" and "humanism," in light of the historical evolution and speciation of these traditions of theory and practice mentioned earlier. Regarding the sciences, I am somewhat prescriptive. In Middle English the term "science" simply meant something like knowledge, derived from the Old French term, itself derived from the Latin word "*scientia.*" It did not take on something resembling its modern sense until the eighteenth century; subsequently, William Whewell coined the term "scientist" in the nineteenth century. For present purposes, however, I use the term "science" in the anachronistic way it is commonly used when we speak of ancient or medieval science, or when we apply it more specifically to traditions of natural philosophy, the precursors to what we now recognize as modern science. This is to elide modes of inquiry that have been transformed in numerous ways and very significantly over time, as well as substantially different forms of investigation across the highly specialized subdisciplines of the sciences. For my purposes, it suffices to recognize as "science" all that is commonly of interest to scholars of the sciences, past and present, in this looser and less pedantic though anachronistic way.

Regarding the question of what humanism is, here I attempt to be more descriptive, which occupies the following several sections. I begin by clarifying what "humanism" has come to mean in our contemporary setting, first and foremost in the eyes of the most influential humanist societies and organizations in the public sphere, where the very idea of humanism is intimately tied to the sciences. Next, in the manner of a film that begins, tantalizingly, with an enigmatic glimpse of the last scene, before going back in time to tell the story of how we got here from there, I rewind the clock to consider the history of this contemporary affiliation of science and humanism, and their coevolution, in terms of formative developments in the Renaissance and growing connections during the Enlightenment. Finally, I turn from this synoptic history of ideas to what I take to be a weighty question for today, which should be assessed, I contend, in the full light of the past: What is science *for*, exactly? I conclude with some thoughts about what this assessment entails for the future of both science and humanism.

Contemporary Humanist Invocations of Science

Earlier I described a first pass at humanism in terms of "a worldview emphasizing the interests, capacities, and welfare of humanity, as well as our potential for learning about the world as a means to confronting the challenges we face and promoting human flourishing." Sharpening up and drilling down to the core of the position, one might put a (still) highly abstract and compressed summary this way: Humanism is a worldview emphasizing reason and science as a basis for understanding the world and our place in it, and for making it a *better* place. In various ways, the rest of this chapter is an attempt to elaborate this summary and to make it more concrete, in order to exhibit key relationships between humanism and science.

Against a backdrop of scholarly debates about these relationships – fueled by different philosophical views which, each in its own way, claims allegiance or opposition to one of a number of different characterizations of humanism (more on which later) – in the lay public domain there is, and has been for much of this past century, an impressive convergence on the matter of what humanism is. One easily accessible window into this convergence is provided by an extensive overlap in descriptions of basic principles offered by the largest national and international humanist organizations concerning the worldview they espouse. A number of common themes appear, expressed in terms of variations on central commitments to or respect for: secularism; critical thinking; science as a source of knowledge (often associated with a vaguely specified naturalism); ethical deliberation and action; freedom and democracy – all of which, in keeping with the first-pass gloss on humanism given in the previous section, are conceived to play a crucial role in the service of human well-being, broadly conceived.

I cite some of these humanist-society pronouncements as evidence momentarily, but first, let me offer a more detailed, philosophical synthesis of what I am calling here a broadly shared worldview, which I take this evidence to support. It is helpful, I think, to collect the various aspects of the view into two families of commitments, each made up of interwoven domains of philosophical interest:

(1) metaphysics and epistemology; and
(2) value theory (most prominently, moral, social, and political philosophy).

What makes the conjunction of what may appear disparate aspects of humanism so interesting, and what explains the fact that in the history of ideas, it has seemed natural to collect these many, seemingly separable commitments under one heading as a worldview, are the ways in which these two families of commitments were (and are) linked to one another to envision an agenda for improving the human condition. Of course, any such agenda must of necessity extend to considerations beyond humanity, since humans do not exist in isolation but are embedded in the world, which brings human relationships with other life and the planet into the picture. Furthermore, the abidingly aspirational nature of the agenda inevitably renders its completion something of an ideal, toward which one can only work. But with these caveats in hand, let me turn now to what I have identified as two families of commitments.

Metaphysics and Epistemology

It is tempting to address questions of metaphysics (concerning the fundamental nature of the world and what it comprises) and epistemology (concerning the nature of knowledge and how we acquire it) independently. These are, after all, distinct subdisciplines of philosophy. However, in the context of humanism (as in many others), it is difficult to separate them, because the epistemological and metaphysical dimensions of humanism are tightly connected. An emphasis on reason and the sciences as a basis for investigating and understanding the world and our place in it, and for making it a better place, is closely associated here with a naturalistic orientation, which manifests as an endorsement of human capacities for substantive inquiry at the expense of supernatural doctrines or revelation. In the limit, this orientation manifests as a skepticism about or a denial of the supernatural altogether. A privileging of human reason and inquiry, with a focus on what observation and interaction with the world can reveal about it, is thereby bound up with judgments about what we are justified in saying, with genuine warrant, about reality itself.

Value Theory

The humanistic worldview is also centrally preoccupied with moral questions and adopts an explicitly ethical stance, promoting goods such as individual and social freedoms, welfare, happiness, and fulfillment, as well as the pursuit of cultural, economic, and other developments that would facilitate the wider distribution of such goods. This emphasis on *improving* the extent to which these desiderata are satisfied in society naturally brings major issues of social and political philosophy to the fore: peace, democracy, civil liberties, decent standards of living, and activism targeting the implementation of such goals and the ethical priorities they embody. This mandate is linked in several ways to the naturalistic orientations in metaphysics and epistemology described earlier. In just the way that a humanistic epistemology has implications for an account of the natural world, it also has implications for an account of the value-theoretic world. Reasoned, rational discourse is regarded as key to setting ethical priorities, not the dictates of supernatural or nonsecular doctrines, and what we learn from scientific inquiry into both the natural and social worlds must inform how we fashion social and political institutions to realize these ends.

Though expressed in different ways and without the philosophical framing I have just given the core commitments of contemporary humanism, the largest humanist organizations today present the worldview to which they subscribe in exactly these terms. According to Humanists International (2023), for example, "Humanists base their understanding of the world on reason and science, rejecting supernatural or divine beliefs"; they "believe in respecting and protecting everyone's human rights," and that "we have a responsibility to respect and care for one another, and to protect the natural world." Similarly, the American Humanist Association (2023) states that "Humanism is a nontheistic worldview with ethical values informed by scientific knowledge and driven by a desire to meet the needs of people in the here and now. At the foundation of those values is an affirmation of the dignity of every human being." And in much the same spirit, Humanists UK (2023) holds that "the word humanist has come to mean someone who trusts to the scientific method when it comes to understanding how the universe works and rejects the idea of the supernatural"; a humanist "makes their ethical decisions based on reason, empathy, and a concern for human beings and other sentient animals."

What I have described in this section is, I believe, an accurate summary of the dominant, popular conception of humanism today. Indeed, to extend this claim further, this understanding of humanism in the lay public domain is very much in sync with how it is understood – as a worldview – in recent and contemporary professional philosophy. In saying this, however, it is important to note that there are, in fact, philosophical views that associate humanism with much more specific philosophical claims, and not all of these claims are congenial to the marriage of science and humanism described here (see Chapter 2).[1] I return to this in later sections of this chapter, where I argue that some of these views are confused about the nature of humanism, or about the compatibility of science and humanism, or both. With this promissory note, let me turn now to a crucial clarification of the contemporary humanist worldview just sketched.

Interlude: Science, Religion, and Epistemic Authority

In part because the humanistic worldview is associated with such wide-ranging (albeit interconnected) commitments, across metaphysics, epistemology, and value theory, it is unsurprising that individual humanists are often most interested in or identify most with a proper subset of them. Some are especially exercised by legal issues concerning human rights and social justice, some by political institution-building to establish and protect democracy, and so on. One particular fixation, however, is a source of substantial confusion about humanism and requires separate clarification. It is not uncommon this past century to find humanism labeled as "*secular* humanism," with the intention of giving special emphasis to distinctions between it and other worldviews associated with various religious traditions. This all by itself is unproblematic, but it is often misrepresented in ways that *are* problematic, by proponents and critics of humanism alike, as

[1] It is not uncommon to find antipathies to science also expressed in other humanities disciplines, such as literature, both historically (see Chapter 5) and in the present.

expressing a blanket opposition to or rejection of religion. As a characterization of humanism, however, this is incorrect both historically and in the present, as well as muddled in ways I will attempt to clarify here, briefly.

Historically, religious identification and even some religious beliefs have been viewed by many as compatible with humanism.[2] The fact that this may seem less plausible now owes in part to a growth in the prominence of naturalistic orientations with respect to metaphysics in the tradition as a whole, but even recently, in the North American context (for example), the growth of humanist organizations was substantially supported by liberal religious groups (Weldon 2020), and there are still those who identify as religious humanists today. The Unitarian Universalist Humanist Association publishes a journal entitled *Religious Humanism*, and the American Humanist Association's Center for Education offers a course with the same title. Given how broadly we now understand, in this era, what it means to be "religious," this should not be surprising. Many who identify with a religious tradition do so primarily for reasons of social, community, or cultural affiliation, or attachment to a heritage. And many who fall under these descriptions do not hold the theological beliefs associated with these traditions or otherwise – an explicit commonplace in many religions including (for instance) Hinduism, Judaism, and Buddhism.

The key to understanding how humanist and religious commitments are sometimes compatible is to take note of an underlying point that is often overlooked: a matter of *epistemic authority*. When there are conflicts here between different traditions of investigation and belief formation regarding the world, where does authority lie – with naturalistic modes of inquiry and knowledge, or with supernaturalistic ones? If our most up-to-date cosmology estimates that the age of the universe is at least 13.7 billion years, but a religious text suggests that it is more like 6,000 years, or if our most sophisticated evolutionary biology gives an account of causal mechanisms giving rise to adaptations, but creation narratives attribute this causation to a God or gods, or if naturalistic descriptions characterize the behaviors of various systems in the world in terms of certain principles or laws, but supernaturalistic descriptions include violations of them in the form of miracles, which way does one lean? What is crucial here, from the

[2] For details on the closely connected question of complex relationships between *science* and religion in the early modern period leading gradually to an epistemically ascendant position for science in modernity, see the extraordinary, four-volume series by Gaukroger (2006; 2010; 2016; 2020). See also Brooke 1991.

point of view of compatibility, is simply to note that humanism recognizes the epistemic authority of a naturalistic – and ultimately scientific – orientation in cases of conflict.

This, of course, allows for some but not all religious commitments. Peter Lipton (2007) articulated his own "religious atheism" in terms of a commitment to "using the [religious] text as a tool for thought," and more specifically, as a resource to help facilitate independent moral reflection. More broadly, this amounts to belief in the claims of our best science,[3] and "acceptance" regarding contrary religious claims. Acceptance is not belief, but it nevertheless involves a form of commitment in virtue of the instrumental value that something has in relation to an aim or a goal. While not believing the content of Judaic texts that conflicts with our best science, Lipton found it helpful nonetheless to reflect on them in thinking about ethical matters. He elaborated on this with a thought-provoking analogy, citing British astronomer Sir Arthur Eddington's famous example of his two tables: The first, his everyday table, has clearly apparent dimensions, color, and other familiar properties, but the second, his scientific table, is mostly empty space and made up of electric charges with a combined "bulk" of less than a billionth of the everyday table. One might believe in the scientific table, and simply accept the idea of the everyday table for everyday purposes (Lipton 2007: 32; cf. Eddington 1928: xi–xii).

This is just one way of preserving an affiliation with the religious in the context of a humanistic worldview. Other alternatives are familiar. One might reinterpret religious doctrines as needed in such a way as to view their content nonliterally (e.g., as metaphorical) rather than as literal assertions, thereby sidestepping conflict with the sciences. One might, following Stephen Jay Gould (1999), describe science and religion as "non-overlapping magisteria" having entirely different domains of interest – a domain of facts, and a domain of purposes, meanings, and values, respectively – which are thus never contradictory. Given the arguably transparent purchase of both the scientific and the religious in both domains, however, and clear examples of conflict (a few given just a moment ago), the plausibility of this recipe seems dubious unless it can be reworked to integrate further strategies for conflict dissolution, such as

[3] Speaking of belief in relation to "our best science" is commonly associated with varieties of "scientific realism," which typically assert a more fulsome range of warranted beliefs than varieties of "antirealism" (see Chakravartty 2017). Here, however, I speak of scientific beliefs in a way that is neutral concerning philosophical debates between realists and antirealists, since both are, in their own ways, champions of scientific knowledge (see Chakravartty and van Fraassen 2018).

those just noted. Both scientific and religious worldviews are interpretable as furnishing descriptions of human beings and our embeddings in the world in terms of both facts and values (see Chapter 8).

The upshot of the clarification offered in this section is that while there are strong, natural affinities between humanism and positions that are deeply skeptical of supernatural commitments (e.g., atheism, agnosticism), and strong associations of humanism with secularism, understood as incorporating a rejection of any such commitments wielding untrammeled authority in our epistemic lives or otherwise, there are surely ways of thinking about religion that render it compatible with humanism, thus doing justice to the outlook of those who, historically and in the present, have identified themselves as religious humanists.

Renaissance Rediscovery and Facilitations of Science

I promised at the outset to take a scenic route to raising a question about the aim of science today, backlit by a historical past of connections between science and humanistic thought. Having sketched a contemporary portrait of humanism, which grants significant epistemic authority to the sciences, my aim now is to follow a strand through an evolving rope of humanism over time, during which the importance of science grew steadily. This is intrinsically historically interesting, but also and more importantly for present purposes, it showcases a long-standing tradition of understanding the nature and mission of science itself through the lens of humanism. Let me begin in the Renaissance, associated with the fourteenth century (sometimes earlier) through the early seventeenth century, a period of remarkable intellectual and cultural development leading from the Middle Ages to what we now regard as the early modern period and setting the stage for modernity more generally. While traditions of humanist thought can be identified not only in Europe but in China and India going back to antiquity, and in the medieval Islamicate world, for more proximate influences on the present coinciding with parallel developments in the sciences, the Renaissance is a helpful place to start.

In the Middle Ages, Latin scholars studied (among other things) earlier Arabic and Greek science and mathematics, but Renaissance intellectual culture was largely focused on the humanities, at least initially. This may make the latter seem an unpromising marker from which to begin an exploration of connections between science and humanism. Indeed, the term "humanism" was not yet in use, and the Italian term "*umanista*" was applied specifically to scholars who studied the languages, texts, cultures, and thinking of classical antiquity, much of which had been lost or ignored previously, in terms of disciplines we now associate with the humanities. This broad scholarship was facilitated by a rediscovery of Latin texts, with interested parties seeking out and hunting through the libraries of Europe to find them, after which came an influx of ancient Greek texts brought by scholars to Italy after the conquest of Constantinople by the Ottoman Turks. Together this facilitated a fusion of interest in Greek philosophy and Roman *humanitas*: roughly, an esteem for (the nature of) humanity, serving as an ideal in the education of a virtuous person. An education thus conceived took the form of *Studia Humanitatis*, comprising grammar, rhetoric, poetry, history, and moral philosophy.

Conspicuously, this did not include science per se, but these developments did of course have profound influence beyond the curriculum (e.g., in art and architecture), sowing the seeds of an entanglement of science and humanism. One might think of this in two ways, first in relation to the growth of humanism in its metaphysical and epistemological dimensions, yielding fertile conditions for the development of science; and second, in relation to its value-theoretic dimension. Regarding the first, the rediscovery of and engagement with ancient texts showcased values that scholars found expressed there, perhaps most inspiringly an ardent respect for human dignity, exemplified in capacities for self-expression, and for inquiry, fueled by the application of reason. This increasingly placed humanity, not supernatural forces or God or revelation, at the center of an understanding of how we learn about and interact with the world. As Protagoras had asserted in antiquity, "man is the measure of all things," a view which naturally erodes a conception of reality on which humans are epistemologically marginal, and opens the door to a more naturalistically oriented metaphysics.

In addition to being conducive to the growth of science generally, these epistemological and metaphysical developments were instrumental to more specific consequences. The rediscovery of ancient texts included scientific and mathematical works, which, as Pamela Long (2016: 496–498) observes, contributed to transformations in natural philosophy and "changes in the most basic assumptions of cosmology, physics, astronomy, biology, and almost every other branch of the study of nature" (Long 2016: 486; cf. Grafton 1990: 103–105). Anthony Grafton (1990: 103) notes that "humanists discovered and printed the passages in Cicero and Plutarch that showed that distinguished ancient thinkers had been willing to contemplate a heliocentric rather than a geocentric cosmos" – proposing that the sun, not the earth, is located at the center of the universe, and that the planets revolve around the former, not the latter – which inspired Nicolaus Copernicus' heliocentrism, bolstered by his own empirical findings (Kwa 2011: 53). This in turn inspired the groundbreaking astronomy of Johannes Kepler (Grafton 1990: 109). These influences were not only theoretical but also practical. The study of geometry in antiquity by Renaissance humanist mathematicians led to major advances in military engineering, including the design of canons, bastions, and fortifications (Kwa 2011: 54).

What is most telling for the moral of this chapter, though, is a striking feature of how science in this period was entangled with the *value-theoretic* dimension of humanism, in ways that go beyond inspirations and affordances for naturalistic orientations in metaphysics and epistemology. As Alan Lacey (1995: 375) suggests, it was "by introducing social, political, and moral questions" that, in the fifth century BCE (and here quoting Cicero), the Sophists and Socrates "called philosophy down from heaven to earth." It is thus hardly surprising that a Renaissance humanist attention to all-too-human concerns should pervade at least some conceptions of the sciences, which were then in the process of substantial development. This took two closely related but importantly distinct forms: an understanding that rational inquiry in the mold of science, given its epistemic authority, may serve as a means to enhancing human welfare; but in addition to this, that it *should do so*. This dual humanist understanding of science is expressed in the idea that "science can and *must* contribute to the community that nourishes it"; Renaissance humanists "had a substantial hand in the development of the notion, widely held by the seventeenth century, that science has profound social impact and *responsibility*" (Grafton 1990: 109, 117, emphasis added).

Sir Francis Bacon's *Novum Organum* (2000/1620) is widely appreciated for its articulation of a method for inquiry in natural philosophy based on observation, experiment, and induction in exactly this period, toward the end of the Renaissance. It is less widely cited for the fuller conception of science that accompanied this, according to which the fruits of such inquiry would benefit humanity in myriad ways, from improved health and longevity, to the development of forms of transportation, to better social relations, to more effective interventions in and control of our environment (see Chapter 9). The potentially negative connotations of "control" in this context – of humans exercising power over nature – are also important to consider, and I return to this in the following section. Independently of how we may think about this today, however, let me conclude this section by noting, once again, here in Bacon's conception of science, a further and explicitly *normative* contention that is irreducible to mere power or control. The "true and legitimate goal of the sciences," said Bacon, is not knowledge for its own sake, or profit or recognition, but rather 'to endow human life with new discoveries and resources' so as to improve the human condition (Bacon 2000/1620: 66, Aphorism LXXXI).

Enlightenment Ideals and Deepening Connections

Let us move forward now to relationships between science and humanism in the Enlightenment, the "Age of Reason," associated with the (later) seventeenth and eighteenth centuries or the long eighteenth century (extended at both ends) – a time during which "[t]he humanist mode of thinking deepened and widened" (Kolenda 1995: 341). This is an apt description of humanist conceptions of science in this period more specifically, which deepened in terms of yet more explicit advocacy for naturalistic orientations in metaphysics and epistemology, and widened in the value-theoretic domain, with more fulsome articulations of the relevant values and thus, by implication, the nature and mission of science as seen through a humanistic lens. Regarding the former, many draw tight connections, for example, between methodological prescriptions for inquiry championed by natural philosophers such as Copernicus, Galileo, Bacon, and Newton during the Scientific Revolution (in which the sciences made significant advances toward what we now call modern science), and appeals to the use of reason.[4] But with a normative moral concerning the aim of science potentially in view, let me focus here on the question of values.

The Enlightenment is often presented in terms of an exploration of and a commitment to certain values or ideals, including: human dignity, equality, and rights; freedom and democracy; cosmopolitanism and tolerance; social and political reform in the service of these values and, concomitantly, a rejection of traditional forms of authority including religious authority and an embrace of secularism. An astonishing number of works

[4] See Nola 2018: 47. The Scientific Revolution is typically associated with the late Renaissance and early Enlightenment. More precise dates are sometimes proposed, and the term "revolution" is sometimes contested given the gradual nature of these changes, but I do not consider these issues here. See Cassirer 1951/1932 (especially chapters 1 and 2) on evolving, humanist conceptions of reason and rationality influenced by developments in scientific inquiry and methodology during the Enlightenment, and Bronowski 1968 (36–38; also Bronowski 1956) on the influence of evolving conceptions of humanism on changing conceptions of nature, from something to be dominated and exploited (in the Renaissance) to something of which we are a product and a part (in the Enlightenment).

discussing these ideals arose in this period, many of which would later become hugely influential in social and political theory, and some representing the origins of sociology and economics. Enabling, epistemic values in the background of this took the form of commitments to reason, evidence, and critical thinking, and even the goal of educating the whole of society – ideals which, combined with greater freedom to question previous doctrine, are commonly cited as the fulcrum of relationships between Enlightenment humanism and science. David Cooper (1999: 7–8) notes that humanism at this time is often identified with "rational subjectivity," the idea that humans have the potential to be autonomous, rational "adjudicators of truth and value," and that "on this characterization, the scientific image is the paradigmatic expression of humanism."

What is perhaps most fascinating about the link between science and humanism during and after the Enlightenment, however, is not related to epistemic values so much as social and political ones. It is important to acknowledge here that in response to the positive, value-theoretic aspirations concerning the promise and proper function of science sketched earlier, some critics of humanism have strongly contested *any* such portrayal as misleading or Pollyannaish. Indeed, it is sometimes held that Enlightenment values were (and perhaps still are) responsible for attitudes, policies, actions, and science that, as it happens, brought about the *degradation* of human dignity and cultures, as well as barbaric relationships with other forms of life and the devastation of our planet. These contentions represent a pressing, prima facie challenge to the reasonableness of any humanist narrative according to which, over a long sweep of intellectual history, the sciences were (and are) regarded as powerful means by which to seek exactly the opposite. This is a crucial issue to which we must now turn.

I suspect some may be tempted simply to dismiss the contentions I have just mentioned as ill-formed. If the humanist conception of science is merely hopeful or aspirational, and critiques of humanism, the sciences, and their applications do not concern hopes and aspirations but rather actual, grievous, historical outcomes, is there a failure here to connect? Granted, aspirations and outcomes are different sorts of things, but this observation alone is unhelpful at best, prevaricating at worst: Articulating the senses in which humanism and these critiques are, in fact, connected is instructive about what is at stake. As a first step in this articulation, let us consider more precisely the relevant concerns.

Earlier I described humanism, conceived as a worldview, as having an explicitly ethical agenda, but various critics have argued that, informed by

Enlightenment values, humanism has been responsible for a number of clearly unethical consequences. These concerns may be collected, thematically, into three (overlapping) categories of ostensible harms:

(1) *harm to people*, caused by appeals to or implementations of particular ideals of reason (or rationality) that result in human suffering by means of prejudice, discrimination, colonialism, or imperialism;
(2) *harm to other life*, caused by preoccupations with human reason (or rationality) that result in the promotion of only human welfare and flourishing and, concomitantly, a disregard for or cruelty toward other life; and
(3) *harm to the environment*, caused by preoccupations with only human welfare and flourishing and, concomitantly, a disregard for or destructive exploitation of the environment and the planet more generally.

Though the details of specific charges levied under these headings vary substantially, it is fair to say that these categories of harms comprise a fairly exhaustive summary of concerns about humanism, and in cases where the sciences are charged with complicity in these harms, they are subject to these same concerns.

The worries indicated here are serious, but some of the critiques expressing them are not. Some attack views that are not endorsed by humanists nor plausibly described as humanistic. Here, one may justifiably adopt what Cooper (1999: 3) suggests as a constraint that "must be respected for the characterisation to be one of humanism . . . [:] the views criticised must have been described as humanist ones by people who have actually held them." In other words, serious criticism should target views identified as humanistic by self-described *humanists*, not merely by critics of something passed off as humanism for purposes of criticism. Consider, for instance, the polemics of Douglas Ehrenfeld (1981/1978: 5), whose sweeping critique is premised on the notion that humanism is committed to "an unquestioning faith in the power of reason" and an "irrational faith in our own limitless power." It is difficult to imagine how one could even begin to square such proclamations of human infallibility and omnipotence with naturalistic orientations in metaphysics and epistemology, which plainly suggest otherwise.[5] Reasoned discourse and scientific inquiry

[5] Cf. Law (2011: 4) on the error of equating humanism and utopianism. Accounts (such as Ehrenfeld's) that go on to associate humanism with laundry lists of failure and narratives of social

are inherently critical pursuits, in which beliefs and methods must be perpetually open to scrutiny.

Some critiques of humanism are more charitable, genuinely engaging with claims advocated by humanists in various contexts historically. This is susceptible, however, to spurious conflations in which the positions cited, properly identified with very specific issues or parochial theses, are then mispresented as humanism *simpliciter*. This runs together the more specific and parochial with the broader conception of humanism as a *worldview* described earlier. Since the more specific and parochial are not equivalent to the broader worldview, and since the former are often marginal or rejected as outmoded in the latter, it is specious to cite worries about these specifics as insuperable for the worldview more generally. For example, some twentieth- and twenty-first-century environmental ethics targets views concerning the "essence" of humanity, which aim to explain features of human thought, action, and morality. This "essentialist humanism" is charged with a worrying anthropocentrism leading to harms to other life and the environment (see Snaza 2017: 16–17). It is a mistake, though, to conflate *this* with a humanist worldview. Many have argued instead that human nature is complex, that it has no particular essence, and that the value-theoretic dimensions of humanism must extend beyond humanity narrowly construed, to the teeming world of dependencies in which humans are embedded.

Other critiques are premised on dubious claims of cause and effect between humanist attitudes and dreadful states of affairs. Enlightenment humanism in particular is sometimes blamed for hordes of dysfunctions: epidemics of self-absorption and excessive individualism; brutalities of colonialism and imperialism; exploitations of other humans and nonhumans; catastrophic pollution; anthropogenic climate change; and accompanying all of this, general moral decay.[6] Now, there are aspects of this that must be taken seriously, not dismissed out of hand (more on which shortly); but it is also important to note just how strained some such claims can be. Max Horkheimer and Theodor Adorno (1994/1947), for instance, argued that the Enlightenment was responsible for the rise of Nazism – an extraordinary assertion of causality between a misleadingly selective (and arguably confused) account of the prominence of certain

decline are often dubious (cf. Noonan 2022: 17–18). For the opposite extreme, associating humanism with a laundry list of successes and a narrative of social progress, see Pinker 2018.

[6] See, e.g., MacIntyre 1988 and Gray 1995. For a skeptical commentary on these and other, related critiques, stemming from a broad range of perspectives (conservative, libertarian, liberal, Marxist, postmodernist, etc.), see Badger 2010.

values and totalitarian mass murder two centuries later. Even if one takes seriously the claim that some of these values were, and perhaps still are, vehicles for dominating people and the planet, the astonishingly reductive leap from humanist values to Nazi terror is difficult to make sense of as anything other than a desperate attempt to grapple with the magnitude of such evil (cf. Nola 2018: 60–64).

All of this, however, leaves at least one class of criticisms of humanism intact – criticisms that engage with actual exemplifications (not uncharitable reconstructions) of widely shared humanist principles (not parochial theses) that have played plausible, concrete (not farfetched or ineffably abstract) roles in causing harm. These criticisms share a common ground: Humanist values are *themselves* abstract; in order to act on them, they must be spelled out and operationalized. Translating even well-intentioned but abstract principles into precise policies and concrete actions is inevitably shaped by the particular, spatiotemporal, cultural contexts in which these translations occur. This yields value-driven attitudes and practices that from later or other perspectives may seem terribly confused or even appalling: notions of equality, rights, or freedoms that exclude Indigenous people, otherwise racialized people, women, or some sexual orientations and identities; notions of toleration that exclude people belonging to certain linguistic, social, sectarian, or cultural groups; and so on. The very notions of reason and rationality, at various times and places, have been conceived in ways that have fueled discrimination, exclusion, incarceration, conquest, and slavery.[7]

Where does this leave a fair assessment of Enlightenment humanism? I submit that any such assessment must involve serious engagement with apparently conflicting perspectives. Failing this, humanists risk a blindness to historical and present wrongs done in the name of their own values, and critics of humanism risk seeing nothing else. Both extremes of partial perspective are undermined by a failure to do justice to a crucial aspect of the humanist worldview, noted earlier (initially) in connection with Renaissance humanism: a critical attitude toward received claims, doctrine, and dogma. Immanuel Kant (1996/1784) famously described enlightenment in terms of an emergence or a liberation from an immature state in which one is unable to think for oneself; in line with this, many apparently

7 Furthermore, this is hardly exclusive to the distant past. See Chapter 11 for a more recent history of "scientific racism," "scientific sexism," and, pivotally, redemptive contributions by later scientists, often women, that "awaited the political and social changes that brought women, who asked new questions and noticed new phenomena, into the natural and social sciences."

conflicting perspectives on humanism are reconcilable upon reflection. A charitable and defensible conception of humanism must incorporate an assiduous understanding of its value-theoretic dimension: one on which humanist values have a dual nature. At a certain level of abstraction, they are goods to be sought, but their contextual operationalizations must be subject to sustained vigilance, critique, and reformation.[8]

Perhaps some will find this combination of resolute aspiration and amelioration intolerable. After all, in some cases, words or concepts become so infused with harmful connotations that the best way forward is simply to discard them. (Consider now discarded terms once used to describe mental illnesses, sexual orientations, or racialized groups.) Many values, however, conceived in ways that transcend particular historical manifestations, are not sensible candidates for disposal. Conceived more abstractly, their positive senses are too deeply entrenched; this *makes* them goods to be sought and *drives* criticism and reconstruction of their concrete manifestations. In this spirit, postcolonial theorist Edward Said (2004: 9–10) rejected "dismissive attitudes" toward ideals such as justice, equality, and liberty – powerful inspirations for liberation movements this past century – found in postmodernist criticism, and rebutted attributions to humanism of a strict, "totalizing and essentializing" emphasis on individual thought and reason (e.g., by Marx, Nietzsche, Freud, Saussure, Lévi-Strauss, and Foucault), in contrast to embeddings of individuals in "systems" (e.g., Marx's "capital," Freud's "unconscious") that exert controlling influences on them. Surely, the capacities of both individuals and systems must be part of any compelling study of reason and rationality.

Let us take stock. Having considered how Renaissance intellectual culture facilitated the development of science, humanism, and relationships between the two, this section has given substantially more attention to the humanist side of the equation. This is not to downplay connections of humanism and science during the Enlightenment – which, as I mentioned, deepened with respect to naturalist orientations in metaphysics and epistemology, not least in light of articulations of methods of reasoning and inquiry furnished by natural philosophers during and after the Scientific Revolution. Also, as noted earlier in passing, concerns about harms done in the name of Enlightenment humanism to

[8] Views advocating criticism as a means to reformulating humanism in practice include various angles of approach, much like the unsparing critiques of humanism mentioned earlier. For recent examples, see Simpson 2001 on engaging postmodernist criticism, Pierce 2020 for a discussion of Black humanism, and McAleer and Rosenthal-Pubul 2023 for a defense of conservative humanism.

people, other life, and the environment are not obviously or uncontroversially separable from concerns about the sciences, which were in some ways integral to enacting many of these harms. Looking forward now, the emphasis on certain values and, in particular, on what I have called their dual nature, is essential to understanding why they persist, and in what forms, in ways relevant to science in the present.

In the third and most recent iteration of the "Humanist Manifesto" (originally published in 1933 and updated in 1973), the American Humanist Association (2003) extends a concern for human welfare "to the global ecosystem and beyond," asserting "a planetary duty to protect nature's integrity, diversity, and beauty in a secure, sustainable manner." Likewise, in the third and most recent statement (earlier ones appearing in 1952 and 2002) of "fundamental principles of modern Humanism," Humanists International (2022) asserts a "duty of care" that extends beyond humanity to "all sentient beings" and a responsibility "for the impact we have on the rest of the natural world," and seeks – perhaps implicitly reflecting on past wrongs – not "to impose our view on all humanity," but "to cooperate with people of different beliefs who share our values, all in the cause of building a better world." Most telling for present purposes, after advocating for "the application of science" to these ends, there is a qualification: "remembering that while science provides the means, human values must define the ends." Let us turn now, from Enlightenment values in historical context, to their extension in the relationship of science and humanism leading up to today.

Modern Intersections of Science and Humanism

I began this chapter by reflecting on contemporary humanism and its invocations of science before proceeding to sketch a synoptic history of the evolving entanglements of these two traditions, all with the ultimate goal of motivating a question about the *aim* of science in the present. What is science for, fundamentally – what is its *telos*, or end? Throughout the chapter I have been concerned to highlight not only the idea that the humanist worldview, in its various incarnations over time, has incorporated an appeal to science as a means by which to realize humanist aspirations, but also the idea that it is part of the very nature of science that it should play this role. Though the former idea is contestable, it seems uncontroversial that many subscribe to it, even with the addition of qualifications borne of healthy caution and an attentiveness to the maturity and rigor of any given domain of scientific inquiry and practice. My focus in this final section, however, is the latter idea, about the aim of science, which I suspect many people today may find strange or even unsettling.

As we have seen, this was not always the case. Indeed, tracing the history of science and humanism now closer to the present, in the nineteenth century and during the first half of the twentieth century, it was not at all strange among scientists and philosophers to think that the function of the sciences is to aid in making the world a better place (see Chapters 4–6). By way of illustration, let me mention two major philosophical traditions in this period, both of which counted philosophers as well as scientists among their proponents and discussants. The first is logical empiricism, which crystallized with the birth of the Vienna Circle, a highly interdisciplinary, scholarly collective that took shape in Austria in the 1920s and 30s, whose thought (together with that of allied colleagues in Germany, the United Kingdom, and elsewhere) came to represent the founding movement of the philosophy of science as a self-aware discipline.[9] Many influential members of the Circle and colleagues abroad were staunchly dedicated to social, political, and economic reforms, in line with what they later described as "scientific humanism" (Carnap 1963a: 83).

In their manifesto, "The Scientific Conception of the World," Otto Neurath and other founding members of the Circle noted that while "questions of life" were not in the forefront of their scholarly discussions, there was nonetheless substantial agreement on such questions borne of their shared worldview (Neurath et al. 1973/1929: 304–305): "endeavours toward a new organization of economic and social relations, toward the unification of mankind, toward a reform of school and education, all show an inner link with the scientific world-conception; it appears that these endeavours are welcomed and regarded with sympathy by the members of the Circle, some of whom indeed actively further them." The paramount objective was "unified science" (Neurath et al. 1973/1929: 306): "to link and harmonise the achievements of individual investigators in their various fields of science. From this aim follows the emphasis on *collective efforts*." The idea, in essence, was that in order to leverage the sciences to address "questions of life," there must be effective collaboration between different areas of inquiry and expertise, and by developing the means to this end, "The scientific world-conception serves life" (Neurath et al. 1973/1929: 318).

[9] See Stadler (2015: 7, 31, 47, 281) on the cultural context of Austria in the background of humanist commitments in the Vienna Circle. Sadly, I cannot engage here with many important figures in the wider background, such as nineteenth-century French sociologist Auguste Comte (to mention just one).

Herbert Feigl (1949: 136) identified this overarching preoccupation with value-laden aims of science with a confluence of philosophical approaches during this period, including not only empiricism but also pragmatism and others, converging in "a broad movement that one may well be tempted to regard as the twentieth-century sequel to the enlightenment of the eighteenth century." It involved "a synthesis of the scientific attitude with an active interest in the whole scale of human values" (Feigl 1949: 137) – a goal championed by the foremost figure of American pragmatism at this time, John Dewey. (For connections between logical empiricism, pragmatism, and discussions of values, see also Frank 2021.) Given the growth of the sciences into tools of previously unimaginable power, Dewey (1985/1931: 201) posed a question: "Here is the instrumentality, the most powerful, for good and evil, the world has ever known. What are we going to do with it?" His answer, not least given that science itself "has created a new social environment," is that science must "face the issue of its social responsibilities" (Dewey 1985/1931: 202; cf. Morgan 2016): systems of insurance to spread risks; preventative medicine; public hygiene; reduction of superstition (e.g., supernatural causes of plagues, famine, disease), and so on.

It is natural to wonder here whether this view of science, as crucially working toward humanistic ends, is relevant only to applied science as opposed to "pure" or "basic" science. Certainly, more applied science was and is a principal focus for some, but it would be a mistake to think that basic science is thereby excluded. Leaving aside the tenuous status of the distinction to begin with (scientific practice classified as "basic" or "applied" typically incorporates a great deal of both), on a common rendering of it – basic science targets "knowledge for its own sake" and applied science targets knowledge intended to facilitate previously envisioned applications – basic science is entirely consonant with humanism. For creatures like us, with an impressive capacity for and (often if not always) an ardent desire for knowledge of the natural and social worlds in which we live, knowledge for its own sake is already, all by itself, something that can be profoundly fulfilling and constitutive of well-being. It helps us to understand and to appreciate the world, ourselves, and relations between the two. "Knowledge for its own sake" is not, after all, an expression that can be taken too literally. Knowledge does not have sakes, but people (and other agents) do.

Another possible concern about prospects for a humanist understanding of science stems from the charge that an important form of humanism is actually *antithetical* to science. Cooper (1999) argues that there is an opposition between the "scientific worldview" and the dominant, contemporary form of "philosophical humanism," which he calls *existential* humanism. On this view, not only our descriptions of the world but, indeed, the world itself are products of human thinking and agency (Cooper 1999: 10). Many versions of what I earlier (note 3) called scientific antirealism do in fact suggest this, rejecting the common realist idea that the sciences describe a world that is independent of our conceptions of it. As an exemplar, Cooper cites William James, who identified humanism with pragmatism, contending that truth and reality are, for us, inextricably interwoven with experience; truths reflect how we "make" the world by carving it up so as to facilitate our purposes (see "Pragmatism and Humanism" in James 1995/1907). But while the *nature* of truth is disputed among philosophers, this is independent of the question of whether science generates *truths* – antirealism is not anti-*science*. Typically, realists and antirealists agree on scientific descriptions of the world that serve as a basis for action, even if they disagree about how best to analyze the concept of truth.[10] The alleged opposition of existential humanism and science is thus a non sequitur.

Given that normative accounts of the aim of science generally and humanist accounts more specifically were widespread for hundreds of years leading up to the recent past, our present situation seems highly irregular. In the present, and despite recent interest in the roles values may play in several aspects of scientific inquiry, the notion that we should take seriously the thought that the aim of science is ultimately normative has effectively disappeared from view. Where did it go? Partial answers to this question have been given, especially relevant to academia in the United States (home to American pragmatists and many leading logical empiricists, who had earlier fled fascism in Europe): shifting political winds and McCarthyism in the 1950s, which were hostile to any advocacy of social or political reforms branded as progressive, left wing, or socialist (Reisch 2005); changing priorities for research funding after the successful launch of Sputnik 1, the first artificial Earth satellite, by the Soviet Union in 1957 (Howard 2003); the strict separation of discussions of values from discussions of the cognitive content of science, on the part of some logical empiricists (Vaesen and Katzav 2019; cf. Dewulf 2021).

[10] For a discussion of the variety of philosophical understandings of truth associated with different approaches to thinking about scientific knowledge, none of which are opposed to science or incompatible with humanism, see Chakravartty 2018.

All of this said, and acknowledging the often heavy weight of historical inertia, answers to the question of why humanist conceptions of the aim of science effectively disappeared do not themselves appear to answer the further question of why, after all this time, they have not regenerated. Today, in the philosophy of science, discussions of the aim of science give the impression of being premised on an implicit assumption to the effect that questions about aims are appropriately – and exclusively, it seems – to be answered descriptively, by reflecting narrowly on the immediate, proximal functions of scientific theories and models. This is to suggest that if we study these things carefully enough, we will *see*, or at least be well equipped to theorize about and contest, the proper end or ends of science. Hence a contemporary focus on what seem purely epistemic features of successful theories and models: prediction, explanation, understanding, empirical adequacy, truth, knowledge, etc.[11] Advocates of these views argue about which are correct, or which is primary, or whether contextualism or pluralism regarding these views is tenable in application to different parts of science. None of this, however, is well suited to giving an account of the ultimate aim of science, or so I will now suggest.

Prediction, explanation, understanding, and so on are *instrumental* features of science. Prediction, for instance, is always the prediction of something *to* some end, an end we care about; lacking this motivation, we would have no use for it. It is what we do with predictions, guided by our reasons for making them, that illuminates our more distal aims. When we bring theorizing and modeling to bear in making predictions about global mean surface temperatures, or the effects of synthetic compounds on human physiology, we do so with intentions – for example, to facilitate planetary health or human health. Furthermore, as mentioned earlier, these intentions pertain not only to existential threats and everyday challenges but also to explanations and understandings we may hope to possess for the sake of nothing more nor less than a profound sense of longing. There are uncountable numbers of truths we *might* seek, but only some we do seek, and in some cases this is simply a matter of aspiring to forms of awe and contentment that can only be experienced in terms of a better understanding of ourselves and the world, as revealed by science. All of this, from the more practical to the more transcendent, may be part of a humanist conception of the ultimate aim of science.

[11] I am unable here to explore the full range and content of these views, but an impressive number of them are considered in discussions of scientific progress; see Niiniluoto 2024. See also de Regt 2017 and Potochnik 2017.

The distinction between more immediate, proximal aims of science and more distal, ultimate aims makes room for a humanist account of the latter, but it also invites us to consider the urgency of reviving such an account. Scientific inquiry, like any human endeavor, is driven by hopes and desires for certain outcomes. The question here is not whether there *are* such ends, but rather what they should be. The humanist worldview furnishes an answer to this question, but others are clearly possible. For instance, setting humanism to one side, we might instead establish an increasingly free market of science in which, to a large extent, the most powerful (often private) interests and corporations set the agendas of inquiry. As it turns out, this has already happened, significantly improving the welfare of certain members or strata of society and certain peoples or parts of the world, but with often alarming consequences for others.[12] The choice to make the sciences de facto servants of unfettered commercialization (and supporting ideologies) as opposed to humanist values has, it seems, been made, or largely made, in the present. But this choice could be unmade. Rather than forgetting a rich heritage of humanistic aspirations for science, unfolding over hundreds of years up to the recent past, we might consider resurrecting these ideals instead.

Going this route, however, will require significant courage of conviction. As argued in the previous section, humanist ideals are by their nature abstract, admitting of more concrete conceptions and operationalizations that inevitably reflect the historical and cultural contexts in which they arise. If the actions they guide are to yield something better than what has come before, they must be subject to stalwart questioning and rethinking. In this spirit, briefly and in conclusion, let me describe three of what I take to be the most pressing desiderata for remaking a humanist conception of science for the twenty-first century. The first concerns how we think about the bounds or scope of science itself; the second, relatedly, concerns relationships between Western science narrowly construed and other forms of systematic inquiry; the third concerns a constellation of practical issues raised by the ambition to implement a humanist agenda for the sciences.

One major obstacle to renewing a humanist conception of science is a growing suspicion of science in the public sphere (Kennedy and Tyson 2023). Some of this may stem from a perception that the fruits of scientific labor benefit only some and not others (more on which momentarily), but

[12] For just a few recent studies, see Krimsky 2003, Brown 2008, Oreskes and Conway 2010, and Wylie 2022. Cf. Sarewitz (2004: 400): "it is only after values are clarified and some goals agreed upon that appropriate decisions about science priorities can emerge."

some is surely attributable to a growing view that the sciences have no special epistemic authority. In some quarters, science is even described as something akin to a religion, equally well characterized in terms of dogma and faith. Of course, this is facile; it fails to reckon with the epistemic potency of empirical evidence and reflexively critical investigation. That said, such views are often concomitant with and conflated with a rejection, not of science per se, but of *scientism*: an especially strong endorsement of the epistemic authority and jurisdiction of the sciences. Scientism is quite reasonably taken to be a bad thing when the strength of its endorsement is excessive, amounting to a kind of hubris regarding the certainty and scope of scientific knowledge. Scientism is thus a much more plausible target of suspicion than science, and it is correspondingly crucial that humanists understand this distinction and take it seriously.

Certainly, lacking an awareness of the stage of an inquiry, the strength of the evidence, and the confidence of scientists in their own results, mindless deference or unthinking assertions of the truth or finality of scientific claims fails to grasp the nature of most scientific work as work-in-progress – even if it is, generally, our best bet epistemically and for acting in the world. Moreover, the notion that all questions are *in principle* answerable by the sciences alone is justifiably controversial (see Chapters 3 and 4). For example, though the human sciences (psychology, anthropology, sociology, etc.) investigate and contribute to our understanding of the nature of value, morality, and meaning, it is at best a promissory note that they will, one day, be capable of doing so comprehensively or exclusively, and it is unclear why taking a stand on this should be important to the epistemological dimension of humanism, which prizes both reason *and* science. Both are key and, presumably, not all reasoning is scientific reasoning. Or more neutrally still: There is nothing about humanism that entails that it is or should be.

This openness to uses of reason that transcend the sciences is critical to the possibility of renewing a humanist conception *of* science. Consider, for instance, domains in which scientific expertise overlaps with traditional or Indigenous expertise (see Chapter 10 for the case of global agriculture), where a failure to bring reason to bear in connecting different sources of expertise productively threatens epistemic injustice, and even the oppression of those with genuinely systematic knowledge falling outside the narrower remits of Western science. To complicate matters further, the broader remit of reason includes more expansive contexts in which science and technology, as well as social, economic, and political relations, are inextricably mixed. In these inevitably complex settings, competing goods and values are the norm, and we must think about how to prioritize (cf. Holman and Wilholt 2022). Vannevar Bush (1945: 10–12), the director of the US Office of Scientific Research and Development during World War II, argued that if science were supported and scientists given complete freedom to do as they pleased going forward, huge benefits to society would result. This view, however, enormously influential in its day, has few if any adherents today. A laissez-faire attitude is *compatible* with societal benefits, but it is also compatible with massive inequities.

What, then, is the alternative? Is it what Dewey (1985/1931: 203) described as a "Baconian ideal": "the systematic organization of all knowledge, the planned control of discovery and invention, for the relief and advancement of the human estate"? From a humanist perspective, there is no alternative but to face up to the task of marshaling our collective reason to grapple with the practical challenges of implementing a fairer and more just administration of the sciences, not to mention their complex embeddings in technology, industry, commerce, and culture. This includes not only a transparent, conscious, resolute focus on placing science in the service of the good but also on directing it away from the service of harm (cf. Kitcher 2001: chapter 8; Kourany 2016; see also Chapter 12). It requires that we engage our highest capacities for reason, critique, discovery, and invention to make good on a humanist conception of science.

In his defense of humanism, in response to critics who questioned his advocacy of it given his scholarship on the cruelties of colonialism and postcolonialism, Said (2004: 28) observed that

> there can be no true humanism whose scope is limited to extolling patriotically the virtues of our culture, our language, our monuments. . . . humanism is not a way of consolidating and affirming what "we" have always known and felt, but rather a means of questioning, upsetting, and reformulating so much of what is presented to us as commodified, packaged, uncontroversial, and uncritically codified certainties.

These same observations apply in equal measure to the sciences, for there can be no true science whose scope is limited to extolling dogmatically the virtues of our current methods for gaining knowledge and the outcomes of inquiry. Science too is about questioning, upsetting, and reformulating, and it would be a mistake to think that unlike all other human practices, the sciences are somehow insulated from being shaped by and having consequences for the social, cultural, economic, and political dimensions of the societies in which they are practiced. The *modi operandi* of an evolving humanism and an evolving science are complementary, in ways that do justice to their long association, and which hold out hope for a conception of science on which, not merely by accident but by design, it is an engine for positive change in the world.

CHAPTER 2

Varieties of Philosophical Humanism and Conceptions of Science

Ian James Kidd

Introduction

There are many varieties of humanism. Some are strongly allied to the sciences, others are antipathetic to them, while others offer subtler positions. A survey of the philosophical scene, historical and contemporary, reveals many varieties of humanism with distinct conceptions of the nature of science and its significance to human life. As one commentator puts it, a complex concept, such as "humanism," can be "stretched like a pair of socks to fit any sized feet" (Kurtz 2001: 144). Fortunately, there are limits to which doctrines could be reasonably counted as humanist: These constraints are supplied by the humanism characteristic of that vibrant intellectual and cultural period during fourteenth- to seventeenth-century European history called the Renaissance. It established the themes or sensibilities that shaped the subsequent history of philosophical humanisms.

By emphasizing the variety of humanisms, I am opposing two claims commonly made on behalf of humanism. The first concerns science and humanism, specifically that "an alliance between the two has been a central strand in the humanist tradition" (Norman 2004: 31). Certainly, this is true of certain tendencies in that tradition, most obviously in the dominant, entrenched form today – naturalistic secular humanism: The world, described by the sciences, does not contain "supernatural," nonnatural entities, realms, or processes, which undermines the truth-claims of the religious institutions and traditions that once provided people with moral guidance and a sense of meaning; fortunately, science, reason, and secularism are more than capable of furnishing us with the values and guidance we need to collectively flourish. That is a sketch of a certain *kind* of humanism, which goes by different names. The philosopher Charles Taylor, who calls it "Enlightenment humanism," traces it to the European Enlightenment whose legacy "survives in naturalism" (Taylor 1989: 384).

If we take a broader and more pluralist view, we find varieties of humanism with distinct attitudes toward science, some dithyrambic, some ambivalent, and others sternly critical.

I am also opposing the claim that humanism must be seen as a good thing, that to be a humanist is something we should approve or applaud. One will think that if one sees it as the stance set against dogmatism, irrationalism, and other moral and epistemic sins – the stance promoted by, for instance, the British Humanist Association, for whom humanism is a commitment to "logic, reason, and evidence" and treating folk with "warmth, understanding, and respect." Of course, until those epistemological and moral commitments are cashed out, they cannot be unambiguously credited to humanism; and the possible existence of other varieties of humanism does not preclude the possibility that other forms of them might be central to religious and philosophical traditions, too. Still, many insist that humanism is a good thing. "Everyone likes to be a humanist," remarked a distinguished scholar of the Renaissance, "or to appear as one" (Kristeller 1990: 3). But much depends on the definition of humanism, which in certain quarters is no term of praise: A French critic could once condemn National Socialism by saying "Nazism is a humanism" (Lacoue-Labarthe 1990: 95). Many other critics of modernity – many environmentalists, for instance – agree with the sentiment expressed in the curt title of David Ehrenfeld's book *The Arrogance of Humanism* (Ehrenfeld 1981). So, in fact, not everyone does want to be or be seen as a humanist and not everyone regards it as an unambiguously positive doctrine. Indeed, there are many varieties of anti-humanism, invoking many concerns – moral and existential, epistemological and metaphysical (see Cooper 2002: chapters 9 and 10).

If one set of complexities comes from the variegated nature of humanism, another set comes from science itself. Almost a century of historically and sociologically informed studies have emphasized that science is better understood as disunified and pluralistic: What we tend to refer to with the singular term "science" is actually a complex and changing assemblage of theories, methods, practices, and projects of enquiry with various accompanying metaphysical commitments (Galison and Stump 1996; Kellert, Longino, and Waters 2006). Fortunately, what humanism is usually trading in are not accounts of specific sciences but certain *conceptions of science*: broad schematic accounts of scientific knowledge, practices, and ambitions and how they are related to the wider structures and concerns of human life. Some examples are the Scientific World-Conception developed by the Vienna Circle (see Chapters 5 and 7), the natural theological tradition that saw scientific enquiry as furthering appreciation of God, or the critical-rationalist vision of science developed by Karl Popper.

The concern of this chapter is with varieties of philosophical humanism and their own conceptions of the nature and significance of science. I describe three main varieties that are evident in twentieth-century European philosophy – humanism as essentialism, humanism as rational subjectivity, and existential humanism, an ordering inspired by the work of David E. Cooper (1999; 2002). I deliberately omit the dominant contemporary variety that one might call naturalistic humanism which figures in many discussions of modern humanism: Our understanding of the origins and essential needs of human beings and our conceptions of the good or flourishing life should be informed by the sciences and consistent with a naturalistic metaphysics. The metaphysical stipulation does a lot of work for humanists of this sort. It rules out conceptions of the human condition and the human good rooted in other, alternative metaphysical pictures, most obviously supernaturalistic and theological ones. Such naturalistic humanism was carefully articulated by Herbert Feigl, writing in 1949, for whom "remnants of and regressions to . . . prescientific thought patterns" can be weeded out, meaning we inherit a "mature humanism" where human nature and history are "progressively understood in the light of advancing science" (Feigl 1949: 148).

My aim is to show that there are philosophical alternatives to that variety of humanism; anyway, I have no aspirations to be comprehensive, nor to argue for or against any specific variety of humanism. I want to chart some of the varieties of philosophical humanism and describe the different stories they tell about the relationship of humanism to science.

Renaissance Humanism

Though contemporary humanists sometimes trace their ancestry back to the ancient period, philosophical humanism in the West first took substantive form in the Renaissance. True, the thought of the *umanisti* was not "the sum total of Renaissance thought and learning, but only a well-defined sector of it" (Kristeller 1990: 114). However, it was sufficiently influential that it radiated outwards, geographically and historically, through to contemporary varieties of humanism. I therefore agree with Cooper's judgment that contemporary humanisms must be "intelligibly descended from a tradition of humanist thought in the West" which began with the Renaissance, the later varieties being "plausibly construable as developments, perhaps culminations, of earlier humanist tendencies of thought" (Cooper 1999: 3). Renaissance humanists were intellectually vigorous and had diverse convictions, interests, and concerns, as one should expect for a period whose name means "rebirth."

Before looking closer at Renaissance humanist philosophy, it is worth considering the fact of its neglect relative to, for instance, the ancient and early modern periods. Doubtlessly, there are several reasons, but consider three that seem especially relevant to understanding the relationship of humanism to science. First, a sense of the Renaissance being populated by a motley crew of "sententious moralisers and *littérateurs*, by philologists and compilers [and] wild-eyed magicians" (Hankins 2007: 339). Certainly, most humanists were interested in the study and translation of ancient texts and engaged enthusiastically in moral enquiry, and many also pursued interests in magic and astrology (Copenhaver 2007; Kraye 1996). But the superior response to what seems exotic or absurd is not derogation but understanding informed by an appreciation of the contexts and concerns of specific figures.

A second reason for neglect of the Renaissance is the understanding of it as essentially an artistic movement devoted to literature and the visual and plastic arts, but without serious philosophical interests or aspirations. This is uncompelling. Enthusiasm for aesthetic pursuits often reflects and shapes philosophical developments, such as the new Renaissance moral ideals of creative self-expression (Mann 1996: 1ff.). Such creativity could be exercised through an imaginative appropriation of ancient art and architecture or the production of artworks or an enrichment of cultures of aesthetic appreciation (Hope and McGrath 1996: 161ff.). A final and related reason are remarks that Renaissance humanism was "neither a philosophy nor an ideology," but "a cultural movement centered on rhetoric, literature, and history" (Monfasani 1998: 533). This, too, is uncompelling. In addition to the fact of significant philosophizing during the period, most obviously concerning reflections on the conduct and aims of a moral life, cultural movements can inspire and in turn be shaped by philosophical developments.

I highlight these reasons since they are relevant to science. A contemporary humanist who sees the Renaissance as populated by "moralisers and *littérateurs*" concerned with art or magic and who lack philosophical interests is unlikely to see them as precursors of their own outlook. Indeed, some humanists ignore the Renaissance, instead tracing their roots to the more scientifically toned Enlightenment. It is notable that some high-profile contemporary humanists, such as Steven Pinker, often voice crassly philistine attitudes toward aesthetic endeavors: Consider his characterization of pleasure in music as "auditory cheesecake" (Pinker 1999: 534). I will not dwell on this. Instead, there are some deeper features of Renaissance humanist philosophy worthy of our consideration.

In what follows I describe three main themes prominent in and characteristic of at least the majority of Renaissance humanist philosophy. The themes are rich and can each be articulated and interrelated in different ways, but for convenience they can be understood as emphasizing human dignity, our independence from the divine, and our frailty. I survey some of them and argue in later sections that each recurs, albeit in modulated forms, in the contemporary, twentieth-century varieties of humanism described earlier. I also suggest that the frailty theme has a special role in shaping the character of humanist conceptions of science.

The dignity theme was a response to various challenges to the medieval tendency of articulating our status and worth in relation to God. We are made in God's image, our minds illuminated by His, uniquely capable, among all beings, of achieving salvation and *beatitude*: These are some versions of theological accounts of human dignity. By contrast, the humanists offered alternatives: We are dignified – indeed, *interesting* – in our own right creatively self-expressive creatures, whether in the new social roles of artists, inventors, or men of *virtù*. We participate in civic and political life animated by increasingly impenitent desires for "the ancient prizes of fame and glory" (Hankins 2007: 125).

Perhaps the most famous statement of this theme is Pico della Mirandolla's Oration on the Dignity of Man of 1486, an amazingly syncretic text combining material from all sorts of ancient philosophies and religions, which has an angel declaiming to Adam

> [T]he nature of all other beings is constrained . . . ; But you, constrained by no limits, may determine your nature for yourself, according to your own free will [. . .] We have made you neither of heaven nor of earth . . . so that you may, as the free and extraordinary shaper of yourself, fashion yourself in whatever form you prefer. (Mirandolla 2012: §§ 19, 20, 22)

Granted, Pico defines our distinctness and dignity in a divinely bestowed capacity to "fashion" ourselves, while others, such as the philosopher Pietro Pomponazzi, defined them in terms of the uniqueness of our status as "the mediator between the material and spiritual worlds" (Blum 2007: 221). We are not yet at full-blooded assertions of our independence from the divine. Still, such proclamations were to open the way for metaphysical and moral doctrines of human independence that enabled later forms of secular and naturalistic humanism. Four hundred years later, Nietzsche judged human beings – true, authentic ones, anyway – as “new, unique, incomparable” because they “create themselves,” forging their own “table of values” (Nietzsche 2001: §335).

Assertions of human independence from the divine developed slowly and depended, naturally, on cultural and philosophical developments during the later early modern period, albeit accelerating during the Enlightenment. The Renaissance humanists opened the way by making possible new emphasis on distinctively human goals: New kinds of moral and cultural significance could now be attached to lives devoted to artistic self-expression, political accomplishment, rhetorical eloquence, technical endeavors, and civic vocationalism. All this was inherited by modern humanists who explain their goals in terms of, for instance, enhancement of human “life, health, happiness, freedom, knowledge, love [and] richness of experience” (Pinker 2018: 410). Contemporary humanist organizations offer similar statements with the proviso, often left implicit, that the range of those goods is constrained by the stipulations of a scientific naturalist worldview.

The dignity and independence themes converged in a culture of “self-assertion,” which was Hans Blumenberg’s useful term for an “existential program” animated by a self-conscious sense of human beings as emplaced within a historical situation affording new possibilities of life whose realization depends on human agency (Blumenberg 1983: 205). Self-assertion was encouraged by the intellectual and imaginative revitalization sparked by the retrieval of Hellenistic moral philosophies and the relaxation of the moral and metaphysical strictures of medieval Christendom. Critics protest that Blumenberg downplays other important historical, political, and social conditions that also contributed to these changes (Pippin 1998: 275). Still, modern humanists should find in this much with which they sympathize: a new historical sense that our future is uncertain, with undetermined outcomes human beings can influence through judicious exercises of reason, imagination, and will; a new outlook that sees the natural world narrowly in terms of human interests and needs – for energy, fuel, food, and so on – coupled to imperatives to control or modify the world for the sake of human convenience and preference; and novel interest in exploring and exercising the creatively expressive capacities of the human mind and body (cf. Cooper 2002: 36ff. and 43ff.).

It is crude to present these Renaissance humanist developments as the opening stages of a long crusade against the dogmatism of religious institutions. After all, many humanists of the period were devoted to improvement of the study and practice of Christianity, the most obvious being

Desiderius Erasmus. We should also be skeptical about historical narratives that draw straight lines connecting science, humanism, and secularism, since they are all too often guilty of historiographical sins such as triumphalism – a way of writing history from the position of the "victors" which leads one to distort the actual complexities of the historical process (Numbers 2009). Still, contemporary humanists, seeking to ally humanism and science, could welcome Renaissance humanist emphases on our dignity and independence from the divine. After that, however, they need to reckon with the third theme – human frailty.

From Dignity to Frailty

The frailty of human beings can have physical, epistemic, moral, or existential dimensions and my focus is our epistemic frailty – a conviction that our epistemic capacities are weak or infirm, incapable of attaining certain truths or sustaining ambitious epistemic goals. The medieval period had offered many sources of epistemic strength and confidence, such as the conviction that God vouchsafes both the integrity of our rational capacities and their fit with the rational intelligibility of the world, "an ordered structure ... oriented to man" (Blumenberg 1983: 139). Theologians debated the epistemological and metaphysical details and disputed the obstacles to our epistemic confidence, such as the implications of the Fall of Man. Conversely, for early modern natural philosophers, the deep worry was that human beings are, epistemically as well as spiritually, "damaged goods," corrupted by original sin, and the urgent question was whether or not we have "retained a capacity to discern intelligibility in the natural and moral orders" (Harrison 2007: 44).

Such inherited structures of epistemic confidence had been called into question in the fifteenth century thanks to various theological, social, and cultural developments. The sudden availability of ancient Greek skepticism thanks to the rediscovery of texts, the realization that serious rivals to Aristotelian Christianity existed that had been lost and never reimagined, and a new, disturbing sense of the contingency of our opinions and beliefs thanks to acquaintance with earlier and distant cultures were just some of them. As a great historian of skepticism puts it, little wonder that "early modern philosophy developed out of a skeptical crisis" (Popkin 2003: viii). I think that a humanist sense of our epistemic frailty emerged from this turbulent context of uncertainty and crisis and – more importantly – would shape later humanist conceptions of science up to the twentieth century.

For those sensitive to concerns about epistemic frailty, two related tasks were urgent: undertaking an appraisal of the nature, scope, and strength of our epistemic capacities and, more practically, working out how to act on the results of that appraisal. "To the humanist," explains one scholar of the Renaissance, "truth seemed particular, conditioned, and subject to many limitations" (Nauert 1995: 20). For them, there were many ways to respond. Some catalogued our personal limitations and foibles. Some emphasized the various contingencies shaping our practices and outlooks. Some scrutinized translations and revised historical and philological practice. Others questioned or rejected aspirations to universal, objective, or final knowledge and truth and, instead, saw judgments as reflections of the unique circumstances of culturally and historically situated enquirers. At its most radical, a sense of epistemic frailty took the dramatic form of denying that human beings are able to "elaborate a comprehensive picture of reality," which led to dramatic epistemological conclusions: "Most of the clearest-headed and most influential humanists regarded human intellectual activity as instrumental and showed little interest in metaphysics. The human intellect, they believed, is suited only to making response to specific problems – generally, problems of moral choice – that arise in the ongoing process of living" (Nauert 1995: 204).

If articulation of the nature of reality is beyond our reach, a next best option is working for a perspicuous understanding of the human condition. This requires knowledge of social practices and cultural history and of the artistic and literary works in which people explore and express their sense of themselves and their world. Granted, some humanists, such as Pico, still worked to develop ambitious metaphysical pictures, but this still involved "an eclectic survey of past philosophies and religions, a picture which no one rational mind could ever have generated" (Cooper 2002: 48).

I think the epistemic frailty theme helped to shape the relationship between the later varieties of philosophical humanism. Crudely put, the difference was between those who took either *quietist* or *activist* stances on our epistemic frailty: The quietists acquiesced in our frailty and rejected epistemically ambitious goals; the activists sought to ameliorate our frailties and accentuate our epistemic powers. Equally crudely put, the quietists rejected the epistemically ambitious conceptions of science which were endorsed by the activists (a tension we will see most vividly in the later discussion of existential humanism).

A quietist accepts our epistemic frailty and accommodates to it by cultivating attitudes and styles of conduct of a more modest and diffident sort. The French skeptic and humanist Michel de Montaigne advises us to abandon disputatiousness and febrile pursuit of certainty and instead cultivate diffidence, "a fear of making judgments," and always strive to be "teachable, zealous" (Montaigne 1991: 570, 564). Confronted with a variety of convictions, the wise person is exploratory, enquiring: They delight in the diversity of customs and opinions, restrain the impulses to utter surety, and submit "undogmatically to the customs and intuitions of society" (Hartle 2005: 195). For the French Christian humanist Blaise Pascal, too, human beings are "wretched," epistemically and spiritually, "equally incapable of knowing and of not desiring to know," albeit able to grasp their wretchedness and thereby attain a form of "greatness" (Pascal 1980/1670: §§ 75, 114). Here are two statements of a form of quietism rooted in a humanist sense of epistemic frailty.

The activist response to epistemic frailty involves attempts to mitigate or overcome it through a combination of self-transformative disciplines and the creation of supporting social systems of enquiry. Consider early modern English natural philosophy, the systematic project of understanding the operations of the natural world by identifying their principles by careful methodical enquiry. Its practitioners shared an acute sense of the "epistemic infirmities of the intellect" and sought purgative or curative "disciplines" and "regimens." Drawing on classical philosophies, the natural philosophers used their therapeutic conception of philosophy to help epistemically corrupted and enfeebled human beings "to conduct the mind in the right way toward the double acquisition of truth and of virtuous dispositions" (Corneanu 2011: 9). The *idola mente* ("idols of the mind") described by the English philosopher and early champion of science Francis Bacon, consist of inherent and acquired epistemic frailties – "weaknesses," "deformities" – which all "do violence to the understanding and confuse everything" (Bacon 2000/1620: §44). Natural philosophy helps us overcome them through the methodological disciplining of individual minds and the centralization of enquiry – a conception of science whose rich classical, Christian, and humanist influences should prompt us to rethink the tenacious myth of Bacon's anti-humanism (Vickers 2000).

I distinguished the quietist and activist responses to epistemic frailty for convenience, though obviously they exist in a dialectical relationship that changes with cultural, intellectual, and historical context. It is tempting to see the subsequent history of science as evidence that the activist responses won out, but that is too quick (Grafton 1996: 204ff.). Granted, maybe no modern philosophers of science would articulate epistemic frailty in terms of our corrupt, postlapsarian state. Other options are available, though, from transcendental or perspectival constraints on human knowledge or a sense of the historical contingency of what came to be our scientific inheritance (Kidd 2020).

I return to epistemic frailty, humanism, and conceptions of science later. What matters for now is appreciating that Renaissance humanism introduced new concerns about the scope and strength of our epistemic capacities. At the most extreme, there is the denial that we do or could ever produce a "comprehensive picture" of reality. This precludes the strong epistemic ambitions central to scientific realism – roughly, a conviction that our best theories and models describe, or are well on the way to describing, the world (Chakravartty 2017). Less extreme options include aspirations to provide more provisional, particular, pragmatic kinds of knowledge and understanding of the world, sufficient for certain modest cognitive and practical purposes. Between these lie a whole range of epistemological and metaphysical positions which is reflected in the varieties of humanism, including the three surveyed in what follows.

Essentialist Humanism

Martin Heidegger proposed that every form of humanism "presupposed [a] universal essence of man" (Heidegger 1993: 226). This sloganizes a variety of humanism which aims to identify "the essentially, universally human" (Davies 2006: 22). Some essentialist humanists articulate conceptions of our essence or nature, while others occupy themselves with the search for it, even if all agree that an account of our essence must be *ennobling*. The "anti-humanism" that Bernard Williams perceived in Lutheranism was explained by reference to its vision of human nature as "twisted," fundamentally corrupted by original sin (Williams 2008: 147).

When it comes to accounts of our distinctive essence, possibilities abound. A short list includes our capacity for autonomous agency, moral self-consciousness, spiritual relationship with God, or accounts of our being existentially concerned creatures who are "condemned to meaning" (Merleau-Ponty 1962: xix). We necessarily experience our lives and the world under the categories of meaning, significance, purpose, and value. For the Renaissance humanists, humans are essentially creative, self-assertive beings, able, as Juan de Luis Vives put it, to "bring forth extraordinary things" (Vives 1948: 392). Pico denied we had an essence in the sense of something that fixes our position in the cosmic hierarchy but still accepted our unique creative ability to "fashion ourselves." Such examples show that essentialist humanism can take many forms – scientific, theological, philosophical. Moreover, claims about essence are not constitutive of humanism, and some self-described humanists reject any talk of a human essence. For Jean-Paul Sartre, the "fundamental meaning" of humanism should be that "existence precedes essence" (Sartre 1966: 28). Our "essence" does not fix in advance the kind of person we will become or the kind of life we will lead: Our distinctive capacities of reflection and choice enable us to choose our own kind of "existence."

An interesting critic of essentialist humanism is the French Marxist philosopher Louis Althusser, who declares the belief in "a definite pre-existing essence" as a "philosophical myth of man" which should be "reduced to ashes" (Althusser 1998: 275). By emphasizing an alleged common essence, those "myths" downplay the importance of social and historical structures and obscure the differences between people under different material conditions – neither of these being acceptable to a good Marxist. Ameliorative projects demand diligent attention to the structural and material conditions of human life, not distracting attempts to discern some underlying essence allegedly common to bourgeois capitalists in Los Angeles and oppressed workers in Laos. In other writings, Althusser clarifies his target as "liberal-rational" humanism, a doctrine which exaggerates the power of individuals to use their rational powers to change the conditions and direction of their life: "the human subject . . . is not the 'centre' of history" (Althusser 1977: 201).

Althusser criticizes essentialist humanism because it obstructs or undermines a perspicuous social and structural understanding of human life and also propagates a stifling conviction that there is an "essential or best form" that life should take (Lewis 2018: §3.5). An essentialist doctrine can be criticized on the grounds of content, coherence, and consistency with our everyday understanding of human beings. Certainly, some philosophers who talk of our essence advance inconsistent claims (see Cooper 2002: 86ff.). During the twentieth century, most debates about essentialist humanism involved the biological sciences. The sociobiologist E. O. Wilson begins and closes his Pulitzer Prize-winning 1978 book, *On Human Nature*, by declaring that "human nature can be laid open as an object of fully empirical research," meaning that at last our "self-conception" can be "enormously and truthfully enriched," as we finally progress toward a "scientific humanism" (E. O. Wilson 2004: 2, 206).

A heated debate ensued among naturalists enamored by the idea of some definitive statement of human nature and their constructionist, postmodernist, and other rivals who all denied any nature or essence at all. What followed was a truculent clash of extreme doctrines, alleviated by the soberer account of human nature offered by Mary Midgley in her 1978 book, *Beast and Man*. She rejected both overconfident claims about the fixity of our nature and radical claims about our being utterly plastic creatures, emphasizing that both scientific study and everyday experience and practice show that we have "highly particular, sharply limited needs and possibilities," which delimit the "schemes of life" into which we can fit and flourish (Midgley 1994: 22–24ff.). We should reject polarizing caricatures, crass dualisms of "nature vs. nurture," and an empirically, conceptually, and methodologically myopic fixation on biology – at which point we can get on with the multidisciplinary project of developing an appropriately complicated account of our dappled natures (McElwain 2019: chapter 2).

A clear theme of these critiques is that accounts of the human essence or nature must be properly pluralistic if they are to capture our complexities. If science is to play a role, then it should not dominate the stage, and if it does we risk narrowing our understanding of those vital or essential features of human beings. John Dupré – a philosopher who criticizes misuses of and misconceptions about the biological sciences – calls our attention to the seemingly inexorable cycle of new research programs that promise to reveal all about human nature. Some recent culprits include sociobiology, evolutionary psychology, evolutionary biology, and genetics (Dupré 2001; Dupré and Barnes 2008).

Dupré argues that human nature consists of "the developmental cycles that currently constitute human life," which have biological and cultural dimensions requiring both scientific and humanistic illumination (Dupré 2001: 95). Since social practices and cultural history count among the essential determinants of our nature, seeing evolution as "the route to deep insight into human nature" is "profoundly mistaken" (Dupré 2003: 4–7ff.). "Imperialist scientism" would narrow our understanding of our complexity and distinctness and assign to the sciences work that ought to be shared across a range of disciplines (cf. Midgley 2002: 215). Similar criticisms are offered by Raymond Tallis – himself a self-described naturalist and humanist – who rejects the specific forms of scientism he dubs "Darwinitis" and "neuromania." A true humanist should demand from the sciences and the humanities "an image of humanity that is richer and truer to our distinctive nature than that of an exceptionally gifted chimp" (Tallis 2011: 10).

I hear clear echoes of the Renaissance dignity and independence themes in these twentieth-century varieties of essentialist humanism. Dupré, Midgley, and Tallis share a conviction that any satisfying account of what is essential to us requires a "radical epistemological pluralism," encompassing the natural and social sciences and the humanities, tempered by a principled reticence about the prospects for "any grand unifying theory of human nature" (Dupré 2002: S292–S293). Our essence is a complicated and evolving product shaped by biological, historical, and cultural factors – an attitude the Renaissance humanists would applaud.

Notice, though, that the biological debates are essentially epistemological – what can science contribute to our knowledge and understanding of human nature? This is interesting, no doubt, but there are two essentialist humanist accounts of science worth considering in this connection.

The first account can be called a *vital conception of science*. It sees the sciences as the primary engines for the realization or expression of our essence or nature, a means of drawing out our essential capacities. Science realizes our nature, rather than just describes it. A good example is Karl Marx's early writings in the economic and philosophical manuscripts which describe our *Gattungswesen* ("species-essence") in very humanist terms: We are essentially creative, embodied creatures for whom using our practical epistemic capacities to transform the world brings material satisfaction and, more important, overcomes our painful sense of estrangement (Cohen 2000: 379). Creative activities shape the world in our image, transforming it from something "independent and alien" into a realm increasingly intelligible by virtue of bearing the marks of human purposes and activity (Marx 2009: 139ff.). Science must be liberated from its "subordinated" condition of "serving material production" for the sake of the bourgeoisie (Marx 1986: 318). Once that is achieved, it can be recognized as our preeminent vehicle for exercising our epistemically and practically creative powers and realizing "man's *real* nature," our "true anthropological nature" (Marx 1994: 110).

I see the early Marx as offering an account of our "species-essence" as creative beings which is clearly related to Renaissance humanist themes and also includes a rich conception of science as an existentially transformative enterprise. Science should be valued, not simply as a source of biological knowledge, but as a vital enterprise that "receives its purpose" from its ability to further "the evolution of all human powers as such" (quoted in Adams 1991: 267). Few contemporary philosophers of science would endorse this existentially charged account of science even if some appreciate Marx's other contributions to more mainstream issues in philosophy of science (Farr 1991; Kidd 2021). Still, it offers a further essentialist humanist conception of science.

A second account of essentialism, humanism, and science has a different character: It denies any substantive, central role for science in the effort to understand or cultivate our nature or essence. Some contemporary moral philosophers argue that what is essentially definitive of us is our *ergon* – a term that is too narrow if translated as "function" – coupled to some conception of *eudaimonia*, an account of what it means for us to flourish as the distinctive kind of beings we are (Roughley 2021: §5). Humans beings have certain set needs and dispositions, reflecting the sorts of creatures we are: Insofar as our needs are satisfactorily met and those dispositions are given meaningful expression, we *flourish*. Martha Nussbaum, for one, has argued there are some "functions" of humans "so important, so central, that their absence will mean the absence of human being" (Nussbaum 1995: 94). Such functions are primarily identified through diligently conducted moral reflection illuminated by empathetic humanistic understanding, rather than the natural sciences. Indeed, in later writings, Nussbaum argued that the already modest role of science is even further reduced by educational developments which downplay the humane aspects of science – imaginative, creative, critical aspects (Nussbaum 2016: 2).

It should be clear, hopefully, that the nineteenth and twentieth centuries included several varieties of essentialist humanist doctrines which provoked lively debates about their epistemological, political, and existential dimensions. I have only sketched out some examples; more could be offered. The point here is simply that there is a variety of essentialist humanisms, each telling a distinct story about the nature and significance of science.

Rational Subjectivity

Humanism can also be understood as a doctrine that identifies our most important and ennobling characteristic as rational subjectivity. To be a human being is to be a subject, which means recognizing oneself as a rational creature, one able to self-reflectively understand its own existence. For rational subjectivists, our central, defining, and most ennobling feature is our capacity for reason, and "the sovereignty of rational consciousness" is a "pillar" of humanism (Davies 2006: 60). An appreciation of our rationality, for Pinker, marks out an "Enlightenment humanism," one that promotes "fairness, autonomy, and rationality" because the enhancement of our powers of reasoning enhances our moral capacities: The rationally sophisticated person is better able to "detach [themselves] from a parochial vantage point" (Pinker 2011: 639, 656). Other scholars agree that rational subjectivity characterizes humanism. Charles Taylor maintains that rational subjectivity is central to a "modern humanism" which encourages images of human beings as uniquely "capable of . . . courageous disengagement" from social context and emotional needs, whose self-conscious understanding of their status as rational agents elicits "admiration and awe" (Taylor 1989: 94). Many champions of humanist doctrines of rational subjectivity adopt, as a motto, Immanuel Kant's injunction to aspiring enlightened people to "use your own understanding" (Kant 1991: 54).

A celebration of rationality as the defining feature of human subjectivity is familiar thanks to its prominence within mainstream contemporary humanisms and its enduring place in the recent history of European philosophy. Cooper notes "Cartesian and Kantian images of the lone spectator surveying and adjudicating, from a withdrawn and superior vantage point, the totality of beliefs, practices and norms that constitute the milieu of everyday life" (Cooper 1999: 8). Many today reject these images, but they could only do that because they were once widely entertained.

Many humanists present doctrines of rational subjectivity as a positive thing, not least for its obvious connections to the sciences, at least if they are understood in certain ways. Since this position is familiar, I consider here an alternative – more critical – account of humanism as rational subjectivity from the writings of Michel Foucault.

Actually, humanism is understood in two senses in his writings, the first – which is not my concern – takes *humanism* to refer to "modern thought about man, our concern for him," specifically the idea that human beings are unique: We are both located within the empirical order of the world, like an object, but ones capable of knowledge and subjectivity – what Foucault describes as a "strange empirico-transcendental doublet" (Foucault 1970: 318ff.). This is something whose full implications were drawn out by Kant, whose doctrines of transcendental idealism are an effort to understand and reconcile the "apparent duality of our nature" (Kant 1991: 207). In what follows, though, I focus on Foucault's second sense of humanism – as the name of a doctrine characterizing human beings in terms of rational, autonomous subjectivity.

"The theory of the subject is at the heart of humanism" (Foucault 1977: 222) and those who reject humanism thereby reject "the theory of the knowing subject" (Foucault 1970: xiv). To be a subject means, among other things, being self-consciously capable of understanding and representing ourselves, one another, and the world. Exercising our epistemic abilities, on this account, enables us to make sense of the world, which, in turn, better places us to guide our lives in rational ways. Illuminated by truth, a rational subjectivity can be autonomous – "self-regulating" – and thereby aspire to freedom. Foucault, of course, urges us to oppose this "*metaphysical* illusion of a self-empowering ego," which he thinks has become entrenched across "modern, occidental culture" (Ingram 1994: 217). A human being who falls for those enticing and self-aggrandizing illusions condemns themselves to try and live within a narrow pattern of development and conduct. Ideals of rationality, after all, get codified in standards of rationality or reasonableness and Foucault sees two problematic consequences. First, the ideal of conformity to the standards of rationality comes at the awful cost of surveillance, anxious self-monitoring, and subjection to increasingly intrusive disciplinary regimes. Reason imposes itself in codified standards which get expressed through institutionalized social practices, the classic examples being the prisons and hospitals described by Foucault.

The establishment of specific images of what a rational subjectivity should be creates a second unfortunate consequence: Those who resist conformity are classified as deviants – as wild, irrational, untamable – and thereby oppressed or destroyed. Anyone who does not fit the strictures of rational subjectivity, as defined, becomes a dangerous deviant. Following many critics, Foucault presents the Enlightenment as the aggressive imposition of an ideal of "a single *rational* trajectory along which humanity fulfills its essential nature," one with grim implications for women, "the mad," and other "deviants" who fall outside its strictures (Ingram 1994: 218). Worse still, their oppression was disguised by a rational subjectivist "grounding of reason, history, and truth in the figure of the transcendentally free and creative subject" (Owen 1994: 221). Epistemologically, politically, and culturally, humanist doctrines of rational subjectivity are oppressive: hence Foucault's often-quoted anticipation of the happy day when "Man" – the rationally autonomous image of him – will be "erased, like a face drawn in the sand at the edge of the sea" (Foucault 1970: 387).

Foucault is obviously critical of humanist doctrines of rational subjectivity and there is much to say about them. I confine myself to exploring their relationships to science to show, hopefully, that there are lots of ways of thinking about the connections of science, humanism, and rationality, including ones that emphasize potential tensions between them. For some self-described humanists, after all, science, humanism, and rationality are mutually reinforcing. A good example is Pinker's book *Enlightenment Now*, which insists on natural harmony between the four themes in its subtitle, *Reason, Science, Humanism, and Progress.* I question this by considering what I call the *ratiocentrism* and *dehumanization* criticisms.

The ratiocentrism criticism challenges the tenability of characterizations of the nature and conduct of human beings and their existence in terms of rational subjectivity. The very general concern is that our rational capacities are too complex and interconnected with our social, affective, imaginative, and practical capacities for it to be plausible to try and privilege *rationality* as the locus of our subjectivity. I suspect few philosophers today think of us as essentially isolated rationalities floating about in an asocial void, sometimes stirred by affects or imagination. Such conceptions of rationality are truncated and too abstract, deriding constitutive aspects of human beings such as our affective capacities as superficial contingencies.

Rational subjectivity is also, for critics, a poor fit with the everyday experience and conduct of life. As a distinguished philosopher of emotion puts it, "the rational ideal of careful deliberation" really "seems utterly remote" from how we actually conduct our lives, especially during times of "mental turmoil" (Goldie 2012: 146). A vision of humans as autonomous rational beings also incorporates an epistemic individualism rendered increasingly untenable once we appreciate just how hugely dependent our knowledge-practices are on other people: We depend on other people for criticism, ideas, information, and other epistemic goods; many epistemic projects are too large or complex to be performed by a single person; moreover, a lot of our everyday epistemic activities rely on social practices and institutions. Such criticisms are common to many philosophical communities – feminists, pragmatists, existential phenomenologists, Wittgensteinians, and others, all skeptical, in their own ways, of attempts to define subjectivity in terms of cool rational capacities hived off from our embodied, social, affective engagement with the world. For John Cottingham, only those who are in the grip of a "ratiocentric bias" could find such truncated conceptions of rationality and their thin visions of human subjectivity attractive or compelling (Cottingham 2009: 250).

That is a general statement of the ratiocentric criticism, the specific variations of which get articulated in many ways, depending on the sensibilities, concerns, and commitments of different critics. It is worth noting, though, that very few philosophers of science would really endorse rational subjectivity. Granted, some still interpret science as a sleek engine of reason, as "the one realm of accomplishment of which we can unashamedly boast before any tribunal of minds" (Pinker 2018: 385). For these ratiocentric humanists, the scientific enterprise is "the achievement *par excellence* of detached rational investigation" (Cooper 1999: 8).

The problem with this is that images of science as the institutional systematization of detached rational enquiry are no longer tenable thanks the investigations of sociologists and historians of science, feminist epistemologists of science, and postpositivist philosophies of science (see Chapters 9–11). If science is a search for truth, we should be truthful about science, including what were once called embarrassingly "extra-rational" dimensions.

Granted, some do reject science and reason, either sincerely or provocatively, and the "Science Wars" of the 1990s saw heated attacks on, and defense of, science (one episode saw a physicist write a spoof exposé of the "intellectual impostures" of certain criticisms of the sciences – Sokal and Bricmont 1998). Care is needed, though, to clarify what is actually being rejected and why. Paul Feyerabend's book *Farewell to Reason*, despite its provocative title, was specifically criticizing "faulty" accounts of scientific rationality made untenable by studies of scientific practice which showed that "scientists do not proceed 'rationally' in the sense stipulated by abstract models" (Feyerabend 1987: 1). Rejecting faulty conceptions of science and rationality is not the same as rejecting science and rationality *tout court*. Unfortunately, this obvious point was occluded by the noisy and ideologically charged "Science Wars" of the 1990s, though hopefully we are now past the bad old days when "what philosophy of science was offering as an account of scientific rationality was of surprisingly little relevance to actual science" (Kourany 2010: 107). It is a nice irony that careful studies of science helped challenge the varieties of rational subjectivity to which some humanists cling.

I turn now to the second, "dehumanization" criticism of humanist doctrines of rational subjectivity, which presents them as offering diminished and dehumanizing conceptions of humanity. Foucault's influential critiques of *épistémès*, "discursive regimes," and the "modern era of 'bio-power'" exemplify the dehumanization criticism. Doctrines of rational subjectivity, first, construct human beings in narrowly rational terms, rendering us conveniently susceptible to monitoring and control. For Foucault, "proper" behavior is modeled and predicted by medical and psychiatric sciences that are themselves part of a "political technology of the body," directed at "the subjugation of bodies and the control of populations" (Foucault 1978: 140). Second, sciences are presented as rational enterprises immune to the prejudices and sentiments that sully the rest of the social world. Conveniently, they appear as the only means of achieving the objective truths that ought to be the basis of rational social practice and political policy. Moreover, this image of science conveniently conceals those "all-too-human" factors animating science, most obviously the "drives" for power, emphasized by Nietzsche, that inspired Foucault's methods of "archaeology" and "genealogy" (cf. Flynn 2005: 30–38ff.). What matters is appreciating that "knowing subjects and truths known are the product of relations of power and knowledge," a crucial insight that exposes "an aspiration to power" that, unchecked, ends in "the suppression of all conflicting voices and lives" (Rouse 2005: 107). Hence the "sense of listening" to the many silenced voices who offer *exposés* of the conceits of doctrines of rational subjectivity and also challenge official narratives of socially progressive rational enquiry (Foucault 1981: 8). Think, for instance, of calls for greater inclusion of persons with disabilities who point to more complex conceptions of subjectivity, or historical studies showing the role of sexist biases in scientific enquiry, past and present.

This dehumanization critique of rational subjectivist humanism is complex and only an example of wider discourses premised on insidious connections between certain conceptions of science, rationality, and humanism. We should, though, see rational subjectivist humanism as a continuation of the Renaissance project of articulating our distinctive, ennobling aspects. The nomination of rationality was intelligible and, up to a point, sensible, even if it later took truncated ratiocentric forms that obscured other aspects of humanity – affects, imagination, sentiment, intuition, spiritual impulses, moral sensibility. Celebration of rationality might also sometimes play valuable strategic roles: the sociologist of science Steven Shapin criticizes a past hagiographical tradition fixated on images of the "genius" scientist, a rational superhero stripped of sentiment and subjective partiality. An appeal to rationality played a role, though, during earlier hostile times when science needed the "protection and celebration" that could come by stressing its "essential rationality and . . . unique status among other forms of human endeavor" (Shapin 2010: 11–12ff.). The trick here is doing this while avoiding entrenching distorting conceptions of science and rationality. According to one account, the value-free ideal, which explained the superlative rationality of science partly in its alleged immunity to social and political values, helped protect American philosophy of science from the ideological strife of the 1950s, then unfortunately became entrenched in ways that delayed appreciation of the essential role of values in scientific practice (Reisch 2005; for studies of values in science, see Part III).

It is ironic that, if these commentators are right, humanism as rational subjectivity can promote both distorting conceptions of science and dehumanizing visions of humanity. I take no stand on whether one could amend those doctrines to avoid these risks. What matters for my purposes is simply to note that this is a further variety of humanism with its own accounts of science.

Existential Humanism

I want to consider one final variety of humanism – *existential humanism* – which David E. Cooper has constructed from figures in existential phenomenology, pragmatism, neo-Kantianism, and several other nineteenth- and twentieth-century traditions, as well as Nietzsche and several others (Cooper 2002: chapter 5). The core claim is that an ineradicable role is played by human perspectives, life, and practice in shaping not only our understanding of the world but also – more radically – the world itself. For Sartre, our distinctiveness lies in the fact that we are beings "by whom it happens that *there is* a world" (Sartre 1957: 552). Existential humanism understands the *world* in terms of the articulated and intelligible world of our experience and engagement: The *human* world is therefore "inseparable from subjectivity and inter-subjectivity" (Merleau-Ponty 1962: xx).

Here I only sketch some general features of existential humanism, directing those who want further details to Cooper's own elaborations and defenses (Cooper 2002: chapters 8–10). Some of the main ones are that the concepts we apply to the world necessarily reflect human values and interests. Consequently, those concepts cannot be extricated from human traditions and forms of life and are only intelligible in relation to our purposive practices and ambitions – no sense can be made of what it is for something to exist, therefore, except in relation to those purposes, practices, and perspectives. As Nietzsche put it, "we have only drawn the concept 'real, truly existing' from the 'concerning us'" (quoted in Poellner 1995: 89). A creature that lacked interests and concerns could not have a world at all; nothing would be "lit up" for them as distinct or salient – a possible object of experience, evaluation, and interaction. Hence the existential phenomenological characterization of our ways of "being-in-the-world" in terms of embodied "operative intentionality" (Merleau-Ponty 1962: xviii). We inhabit a world of practical possibilities, revealed through embodied engagement – climbing, carrying, exploring, walking – which gain significance through their contributions to our life-projects. For the existential phenomenologists, "being-in-the-world" is immersed, engaged, active, and only becomes detached and spectatorial for special purposes. For this reason, existential humanists reject the primacy assigned to rationality by rational subjectivists and their conceptions of rationality.

A lightning sketch omits many details, but two clarificatory points are worth marking, which concern an obvious criticism and a connection to science. First, the existential humanist is not claiming that the world is somehow a product or construction of human beings. One does find such promethean rhetoric out there. Some pragmatists and, at times, Nietzsche indulge in it, as does Umberto Eco when speaking of our "gradually constructing ourselves a World" (Eco 1999: 20; cf. Cooper 2002: 103ff.). Our relationship to the world is much more *intimate*: The world is "always, already" there, as Heidegger puts it; our engagement with it is usually unreflective, smooth, and supple – not at all one of our going about "imposing" order on some formless mass. Indeed, our being-in-the-world is experienced as comportment within a world already, as it were, up and running – something existential phenomenologists convey by characterizing the human world as a theater of possibilities, "referential totality," or "cradle of meanings" (Heidegger 1962: §17; Merleau-Ponty 1962: 499).

Second, existential humanism is clearly a late descendant of Renaissance humanism. What it aspires to is an account of our distinctiveness and dignity – our unique manner of being-in-the-world enables us to inhabit and share a world which is experienced as a dynamic space of significant possibilities for personal and collective agency. This world is given its color and animation through its being "lit up" in virtue of our purposive practices which are, themselves, components of the existential life-projects embedded within a rich intersubjective world – an unquestionably *human* world. Its inhabitants are, indeed, embodied human beings and not a transcendental I, some abstract "constitutor" in whom "nothing human is to be found" (Husserl 1970: 183). Existential humanist attitudes to the divine vary considerably, though pronounced theological commitments are only visible among the "religious existentialists," such as Karl Jaspers and Gabriel Marcel. Finally, the frailty theme remains in the epistemic and existential forms of, for instance, the phenomenological claim that the structures of meaning that "light up" the world for us are not anything that we created, and which can, at times, collapse in the horrible experiences of emptiness – of the sudden collapse of the sense of things *mattering* – that Heidegger called *Unheimlichkeit* (cf. Cooper 2002: 249ff.).

I finish here with existential humanist conceptions of science. Central to scientific enquiry is a disengaged spectatorial stance on the world, one dependent on, and therefore derivative of, our everyday ways of experiencing and engaging in the world: "cognition in the . . . spectator sense . . . presupposes existence" (Heidegger 1982: 276). Scientific enquiry therefore takes for granted the prior richness of our experiential world, whose structure and intelligibility owe to our concerns and interests. As Husserl puts it, science is rooted in the "meaning-fundament" of the "life-world," the *Lebenswelt*, the shared structure of enthusiasms, interests, needs, and presuppositions constitutive of a certain form of life (Husserl 1970: 121).

Existential humanists draw two conclusions about science. First, scientific descriptions and explanations of the world, while valuable for certain purposes, presuppose a background way of experiencing the world, one they cannot account for in their own terms. We can "stop and stare . . . in the scientist's manner," but this entails abstracting things from that "relational totality" of interconnected structures of possibility even to the point that it is "dimmed it down to [a] uniformity" (Heidegger 1962: 114, 178). The objects and processes studied by sciences, such as animals and the weather, initially appear for us through more basic ways of being-in-the-world. Science is therefore derivative, albeit useful for certain cognitive and practical purposes; problems only arise when this dependence gets forgotten (Ratcliffe 2013). This has two implications: Phenomenology plays the vital role of describing the tacit, background structures of meaning and experience that the sciences presuppose, hence Heidegger's remark that the sciences are "utterly incapable of gaining access . . . to their [own] essence" (Heidegger 1977: 177). The "essence" of science, then, is revealed by phenomenology, making it the most fundamental method of enquiry (Cooper 2002: 193–200ff.).

A second conclusion existential humanists draw about science is the need for what we might call an *existential critique* of science. Many existential humanists celebrate the richness of our being-in-the-world: We experience a world suffused with meaning and significance that resonates with our moral and emotional experiences. By contrast, the world described by the sciences seems to them flat, thin, cold, and devoid of those features, such as meaningfulness, that are vital to *living*. The rhetoric used by existential humanists will seem overwrought to those who do not experience the scientific worldview as alienating and empty – Heidegger speaks of the "distress" of our age of science and Husserl declares a "crisis of European sciences" that precipitates a "barbarian hatred of spirit," the erosion of the deep values which stir in us nobler sentiments and feelings.

Such critiques have two aspects: The distinctive character of human existence is itself threatened by internalization and privileging of scientific descriptions of the world, including we human beings. Heidegger, indeed, judges that what is "messing up" modern thought and culture is "the dominance and primacy of the *theoretical*" (Heidegger 1987: 87). By ignoring or forgetting the richness of our "primordial" being-in-the-world, we thereby fail to appreciate a fundamental truth about our existential situation, about the unique sorts of creatures we are. Such concerns are, again, expressed in different ways by different existential humanists, as in Husserl's warning that falling victim to a dualistic picture of "nature . . . alien to spirit" destroys our sense of intimacy with the world, replacing it with a sense of estrangement, one so radical it can precipitate a slide into cultural "barbarism" (Husserl 1970: 390, 121ff.). Our dualistic and disenchanted existence, for these critics, encourages a calculating, exploitative stance on the world and feeds painful feelings of alienation, hence Heidegger's characterization of the contemporary human condition being one of "distress."

A second aspect of the existential critique of science concerns the tendencies of these scientistic attitudes to gradually occlude other ways of experiencing and making sense of the world. The world, recall, is a rich theater of possibilities lit up by virtue of the myriad practices and projects of human life – the pursuit of religious conviction, say, or appreciation of beauty. When scientific ways of thinking prevail, warn existential humanists, this experiential richness gets dimmed down and treated as superficial, "primitive," or "confused" (Cooper 2002: 337–345). As the later Heidegger famously put it, "ways of revealing" the world central to scientific enquiry start to "drive out" other ways, even to the point that the fact that it is *a* way of revealing a particular stance – one among others, suited only for certain purposes – gets forgotten, hence its "monstrous" character (Heidegger 1987: 26ff.). Nature, for instance, gets narrowly "revealed" in relation to human concerns: "the earth . . . reveals itself as a coal-mining district," the Rhine as "water-power supplier" (Heidegger 1977: 14ff.). The possibility of alternative ways of experiencing places, creatures, and things is therefore gradually driven out, dimmed down, until one sees cattle as meat-on-legs, chickens as "egg units," and human beings are talked about and treated as "human resources" (cf. Zimmerman 1990). Crucially, this expresses a corrupted variety of humanism which "explains . . . whatever is, in its entirety . . . in relation to man" (Heidegger 1977: 133). In Paul Feyerabend's evocative term, it is a "conquest of abundance," a process of experiential impoverishment made possible by the entrenchment of inflated conceptions of the nature and significance of science (Feyerabend 1999).

I hope that even this sketch confirms that existential humanism is a complex doctrine with distinctive conceptions of the epistemological, cultural, and existential status of science. It is also a genuine form of humanism: Our distinctiveness lies in our unique ways of being-in-the-world as existentially concerned creatures, "condemned to meaning," who experience the world as a theater of possibilities. In most of its forms, there is no theological dimension, even though some responses to its intrinsic sense of epistemic and existential frailty are expressed in religious terms. The French Catholic and existentialist Gabriel Marcel, for one, urges us to cultivate an "ontological humility" (Marcel 1949: 132).

Whatever one makes of existential humanism, we must recognize it as a contemporary variety of philosophical humanism which offers distinctive critical accounts of science. Indeed, it is one of the latest in an ongoing history of diverse forms of humanism, which offer different ways of understanding and evaluating scientific knowledge and ambitions. Ironically, what we find in the history of philosophical humanisms is what we ought to expect of creative, self-expressive creatures: endless variety.

CHAPTER 3

Scientism and the Limits of Objective Thinking

Gurpreet Rattan

On one side, *humanism* opposes religion, or, more specifically, the idea that religion is the ultimate and universal source of cognitive and practical significance. This opposition to religion makes an alliance between humanism and *science* quite natural.[1] So does the use of a suite of tools, concepts, and ideals in science that are very much congenial for the humanist: observation, reason, evidence, argument, experimentation, progress, objectivity, universality. Indeed, this convergence between humanistic values and science might encourage not only their alliance but even an identification of sorts, at least for cognitive significance: Science is the ultimate source of cognitive significance, or as I term it later, *objective thinking*. Call this view of the role of science, *scientism*, and the interpretation of humanism, *humanism as scientism.*

On the other side, though, humanism is concerned with the subjects (and outlook) of the *humanities*: with language, and languages; literature, poetry, theater, music, dance, and art; ideas, ideologies, history, and society. And these subjects, taken at face value, are very much unlike those of science. The suite of tools, concepts, and ideals at use in the humanities are diverse and varied, including different forms of analysis and critique, creativity and new forms of expression, the articulation of new frameworks for thinking and acting, and much more. But at least one thick strand in the methods of the humanities has the humanities focused on *interpretation* and *criticism* – of words and texts, of ideas, practices and movements, of societies and their history.

This tension in humanism is the background to this chapter, the topic of which is a critique of scientism. One result of my critique is that the idea of humanism as scientism is a bad one. A more positive result is to make room for an understanding of humanism as fundamentally involving

[1] Cf. Ladyman (2018: 109): "Scientism can be seen as a struggle for science's self-determination in seeking to liberate territory from the forces of superstition and the supernatural."

interpretation and criticism, while at the same time locating this dimension of humanism in science itself.[2]

* * * * *

Scientism is philosophically interesting not only because it is an "ism,"[3] and so purporting to be a general framework for thinking in some domain, but also because of the particular domain for which it purports to be a general framework, namely for the domain of objective thinking itself. On this view, science *determines* the limits of objective thinking by *constituting* what it is to think objectively at all. Thinking is *validated* as objective by the fact, if it is one, that the thinking is scientific. Otherwise, the thinking is *invalidated*, at least with respect to its status as objective.[4] In a slogan, objective thinking starts and ends, constitutively, with scientific thinking.

"Objectivity" is understood in many ways. My own use emphasizes connections to impartiality, rationality, universality, and knowledge, to draw from Janack's (2002) analysis. I say more about *objective thinking* later but, roughly, objective thinking is thinking that uses concepts and methods appropriate for acquiring knowledge. This understanding of objective thinking is not meant to rule out, for example, knowledge that there is beer in the fridge, on the grounds that such mundane thinking is not scientific thinking; nor, however, it is much concerned to vindicate such thinking. The interesting contrast is not between scientific and mundane thinking (see de Ridder 2018 on "high-grade" and "low-grade" knowledge in connection with scientism), but rather, between the objective thinking and knowledge in science and the thinking that goes on in other areas of our intellectual lives: in thinking about how to live and live together in ethics and politics, about beauty and what is beautiful in aesthetics, about the ultimate nature of reality in metaphysics.

This chapter argues that although there is a connection between science and the limits of object thinking, the connection is *not* that science determines those limits. There *are* limits to objective thinking, but as I suggest near the close of this chapter, the account of these limits is

[2] For some discussion of the rivalry between scientism and humanism, see Stenmark 2018.

[3] I compare my own way of understanding "scientism" to some other ways later in the chapter. For a sense of the diversity here, see one or more of the recent edited anthologies on scientism: by Maarten Boudry and Massimo Pigliucci (2017), Jeroen de Ridder, Rik Peels, and René van Woudenberg (2018), and Moti Mizrahi (2022).

[4] My use of "validate" (and related) derives not from the validity of arguments in logic, but from Kant's "objective validity" (1929/1781/1787: A239–242/B298–300).

considerably different from the kind of account scientism gives, according to which the limits of objective thinking are determined by conformity to some particular kind of thinking, namely, scientific thinking. I argue instead that the connection between science and the limits of objective thinking is that some scientific thinking makes use of thinking *at* the limits of objective thinking, and this makes it the case that science *participates in* thinking at the limits of objective thinking. But it does so without *setting* those limits. The key to my argument is to identify a mismatch between, on the one hand, the conception of science that underpins scientism and, on the other hand, a conception of science that is informed by an analysis of the full extent of thinking at work in scientific thinking.

Using Thomas Kuhn's classic terminology, I argue that the conception of science that underpins scientism is a conception of science as *normal science*. Riffing on Kuhn (I discuss Kuhn in detail later), normal science is the kind of science we all know and love, with exemplary achievements and a community of practitioners employing widely accepted methods and engaged in mutually comprehensible practices resulting in a steady stream of results and progress. I argue that an analysis of the kind of thinking, both individually and collectively, that goes on in normal science shows normal science to be a plausible basis for putting forth science as an "ism" – scientism.[5] Roughly, we are impressed by science when it functions like this and tie progress in science to it possessing a distinctive claim on the truth (cf. Putnam 1975).

But science in times of crisis or revolution becomes *extraordinary science*. Extraordinary science is science interpreting and critically reflecting on its own methods and practices. In Kuhn's presentation, this is typically in response to the accumulation of anomalous results, confusion about the ameliorative possibilities for dealing with these anomalous results in the existing disciplinary matrix or paradigm, and defection by practitioners to competing paradigms. I argue that an analysis of the kind of thinking, both individually and collectively, that goes on in extraordinary science shows extraordinary science *not* to be the plausible basis for a plausible "ism."

Notably, though, crises abate and are sometimes overcome through conceptual and methodological innovations produced by the interpretative and critical reflective perspective extraordinary science takes on the disciplinary matrix or paradigm. The advances extraordinary science makes through its use of the interpretive and critical reflective perspective show

[5] I talk about something (mainly science) being "the basis of an 'ism'" in this chapter, but throughout this means that something does (or does not, as the case may be) form the basis for a *plausible* "ism."

that perspective to be of cognitive value. This means that any account of science that leaves out the kind of thinking that goes on in extraordinary science does so on pain of invalidating the intellectual achievements of extraordinary science.

So here is where we are: If science is the basis of an "ism," then the thinking in extraordinary science is invalidated, and if the thinking in extraordinary science is validated, then science is not the basis of an "ism." Since the thinking in extraordinary science is valid, any account of the limits of objective thinking should validate it, making it the case that science is not the basis of an "ism." This is my main critical argument against scientism.

Extraordinary science does, though, I argue, *participate in* thinking at the limits of objective thinking. The thinking in extraordinary science *encounters* the limits of objective thinking. So what determines the limits of objective thinking? I close by explaining how the argument of the chapter can be generalized to show that limits of objective thinking cannot be determined by a requirement that a thinker accept or adhere to any positively specified methods. I outline briefly what another approach to the limits of objective thinking might look like.

What Is Scientism?

My understanding of scientism analyzes things in a different way from most current literature on scientism. My idea is that scientism is a view about the determination of the limits of objective thinking, with science and its concepts and methods constituting what it is to think objectively at all. My understanding of scientism has its ancestry in logical empiricism, which used empiricism as a critical tool to distinguish cognitive significance from different kinds of meaning or content, including practical and expressive significance (see later in the chapter).

My general idea about scientism comes apart from some recent ideas, but I believe in a way that increases its interest and plausibility. At the same time, though, I believe that my idea is roughly consonant with mainstream ideas already in the literature, even if it emphasizes different connections. De Ridder et al. (2018), for example, open their volume on scientism with the rough formulation of scientism as "the view that only science can provide us with knowledge or rational belief" and, after recounting some common conceptions of scientism, write that "[s]cientism can thus stand for a number of exclusivity claims about science." My idea comports with their rough formulation, and falls under their general description about exclusivity, assigning science an exclusive claim to the domain of objective thinking.

My understanding of scientism derives from the role of science in discussions about the limits of cognitive significance and intelligibility in the first half of the twentieth century, in the philosophy of logical empiricism. Logical empiricism was elaborated in different ways by different practitioners; I follow a strand that is particularly influential, due to Rudolf Carnap. Logical empiricism was a critical philosophy, and it found much philosophy, notably ethics and metaphysics, to be outside the realm of cognitive intelligibility.[6] These domains were contrasted with science, but the contrast was not that science is empirical and the others not.

What distinguishes the cognitively significant discourse of science from ethics and metaphysics is not so much the use of an *empirical* method but the use of an *intersubjectively authoritative method*, whatever its character, whether a priori or empirical.[7] An intersubjectively authoritative method is a method that has widespread implicit and even explicit endorsement in that community, and which governs the cognitive use of language (for expressing knowledge). Methods of empirically testing and verifying hypotheses, for example using redshift and luminosity information about galaxies to determine their relative velocity and distance, are intersubjectively authoritative methods, but so are, for example, the numerical methods used to solve the Navier–Stokes equations for incompressible flows in fluid dynamics.

The existence of intersubjectively authoritative methods figure in the account of why the statements of science have an intersubjectively constant meaning (in terms of a "criterion of cognitive significance") and of what those meanings are ("meaning is method of verification").[8] Experience and empirical method play a role here, but experience and empirical method are secondary to the general idea of intersubjective authority, whether that authority concerns empirical or a priori methods. Intersubjectively authoritative methods can also explain the meaningfulness of statements of mathematics, since there are a priori methods that can be set out that

[6] For some classic texts, see Carnap 1932, Ayer 1936. Semantics and aspects of epistemology too are outside the realm of cognitive intelligibility, in so far as questions of semantic analysis and epistemic method are *external* questions (see Carnap 1950a) and so a matter of practical decision, not cognitive attitude. Once we practically decide on some language and epistemic method, internal questions can follow.

[7] For discussion of "intersubjective accountability" in logical empiricism, see Uebel 2020.

[8] See Hempel 1951; Schlick 1936.

explain the meanings of mathematical languages.[9] This is reflected in logical empiricism in the fact that the criterion of cognitive significance includes criteria for mathematical claims. It is true that on their view, the criteria are wholly analytic or definitional, making no tie between the cognitive use of language and experience and empirical methods; and it is also true that logical empiricists often accepted the pragmatic or conventional nature of the analytic.[10] But the latter idea is an additional, non-compulsory step to take in a view about the analytic; moreover, what matters here is not that these criteria are not empirical and instead analytic but that there are indeed such criteria, whether empirical or analytic, underpinning mathematical meaning and content. What matters is the presence of intersubjectively authoritative method.

The distinction between science, on one hand, and ethics and metaphysics, on the other, does not imply that the latter have *no* content; the conclusion rather is that they fail to have cognitive or, as I call it, objective content. Objective contents are adequate to be the contents of belief and knowledge, and appropriate for use in acquiring knowledge of the world. Sentences (or many sentences) using language such as "witches," "curses," "horoscopes," and "phlogiston" are inappropriate for acquiring knowledge. The logical empiricist idea is that ethical ("One ought not to lie") and metaphysical ("Ordinary objects are physical objects and not ideas") sentences are also inappropriate for acquiring knowledge. In all such cases, no objective contents are expressed. These sentences may have a kind of content, but, the logical empiricists maintained, because these discourses lack any significant intersubjectively authoritative method, they cannot have the kind of content appropriate to inquiry, belief, and knowledge – they cannot have objective content.[11]

I hope I have said enough to indicate the intellectual background to the idea of scientism that I am working with. However, the conception of scientism according to which science constitutes and thereby sets the limits of objective thinking does conflict with some ideas in the recent literature.

[9] Carnap (1950a) explicitly includes languages of number and of formal semantical notions as meaningful languages, despite their apparent commitment to abstract objects, because these languages are governed by intersubjectively authoritative methods for their application.

[10] Again, see Carnap (1950a), who deflates the nature of the commitment involved to abstract objects by making acceptance of the language a practical rather than theoretical matter.

[11] "The thesis that the sentences of metaphysics are meaningless, is thus to be understood in the sense that they have no cognitive meaning, no assertive content. The obvious psychological fact that they have expressive meaning is thereby not denied" (Carnap 1932: 81).

One conflict is with philosophers who take mathematics to be a problem for scientism. For example, Alex Rosenberg (2018: 84) writes: "Scientism as a philosophy faces two great challenges: First, how to accommodate mathematics. If numbers are abstract objects with which we can have no causal relations, it is difficult to see how we acquired any mathematical knowledge."[12] The view seems to put, what should be from scientism's point of view, the metaphysical and epistemological cart (qualms about abstract objects, commitment to a causal theory of knowledge) before the methodological horse (in the form of the highly mathematized practice of science).[13] The conception of scientism that I have described which prioritizes intersubjectively authoritative method includes mathematics because, though not empirical, it is subject to intersubjectively authoritative method.

Other philosophers take an *expansionist* ambition to be central to scientism, according to which the concepts and methods of traditionally nonscientific areas of discourse or inquiry are replaced by scientific terms and apply scientific methods. Note, for example, the expansionism in Rik Peels' definition of scientism (which itself aims to unify three other definitions from Peacocke 1993, Radnitzky 1978/Churchland 2011, and Stenmark 2001 – hence "Scientism$_4$"): "Scientism$_4$: The view that the boundaries of the natural sciences should be expanded to include academic disciplines or realms of life that are widely considered *not* to be the domain of science" (Peels 2018: 47). The conception of scientism according to which science sets the limits of objective thinking is consistent with this idea, and even coheres with it in some sense, but it doesn't entail it. It is consistent and coheres with the idea because we might think that the explanation for why expansionism is in order is that such expansionism will extend the domain of objective thinking and knowledge, and that this is a good thing. It doesn't entail this kind of expansionism because certain domains of discourse may possess some other value for our lives different from cognitive value (e.g., for a van Fraassen-style [2002: 171] "spiritual journey of discovery"), and so may appropriately possess a different kind of content from objective content (e.g., Carnap's *expressive meaning*).

12 The second challenge concerns eliminativism about the mental (see Paul Churchland 1981).

13 Assuming numbers are abstract objects. But numbers could be properties to which we have empirical access (see Yi 1999); or bare determiner semantic values more friendly to a rationalist epistemology of arithmetic, and to which numbers as objects are related via cognitive type-shifting coercion (see Hofweber 2005; 2016: chapter 6; for related work see also Sher 2013).

Why Scient*ism*?

Why might science be the basis of an "ism"? Why scient*ism*?

First let's say some things about "isms" and the way that scientism is an "ism." As I am thinking about it here, "isms" are not propositions but instead constitute conceptual and methodological frameworks or stances for thinking about some domain (cf. Carnap 1950a; van Fraassen 2002). What is special about scientism is that it is an "ism" not for this or that area of thought but for the domain of objective thinking itself. This way of being special is not the same as being the most general kind of "ism," governing everything, like *materialism*, *idealism*, and *solipsism*. However, if all of reality is thinkable, as an "ism" scientism may nevertheless, as it were, reach *through to* all of reality via the idea that objective thought represents reality. Scientism is more general than "isms" such as *anarchism* and *capitalism*, which govern approaches to particular domains (government and economic organization), and is at the same level of generality with an "ism" such as *skepticism* in concerning the qualities and quality of representation and cognition.

So, scientism understands science as the framework for objective thinking and this connection itself is explained through an account of objective content that ties objective content to intersubjectively authoritative methods. Once we see this, we can trace at least in outline the attraction of scientism to the presence of intersubjectively authoritative methods in science. The general idea, which I elaborate later, is that science is the basis for an "ism" because the use of intersubjectively authoritative methods gives science desirable epistemological and sociological features. We are so impressed with science that we think that its status should not be that of just another discipline but that of a framework, and not just any framework but for objective thinking itself.

I try to get a handle on just what the intersubjectively authoritative methods in science consist of epistemologically, and how they contribute to an attractive sociology, by critically reviewing Thomas Kuhn's discussion of *normal science*, as well as the distinction between it and *revolutionary* or *extraordinary science*, in his classic, *The Structure of Scientific Revolutions*.

Before beginning, though, I would like to say a little bit about my use of Kuhn. It is a surprising feature of recent work on scientism that it contains almost no discussion of Kuhn.[14] This is a surprising hole in the literature

[14] The three edited anthologies mentioned in note 3 contain a mere half dozen or so references to Kuhn. From the other direction, one book (Wray 2021a) and two collections of papers (Wray 2021b; Melogno et al. 2023) contain exactly zero references to scientism.

given Kuhn's influence on the philosophy of science, but also because one might think that scientism should welcome a broadly scientific – the biological, psychological, sociological, historic, and economic factors that influence the institution of science and as a result condition scientific practice – view of science itself. I review Kuhn rather extensively. The argument of this chapter in effect charges scientism, despite being post-Kuhnian, with being, philosophically, pre-Kuhnian. The hole in the literature is filled with a spade with a critical edge.

At the same time, the main argument of the chapter, though colorfully filled out by Kuhn, relies only on some basic Kuhnian ideas about different phases of scientific practice and an idea about the kinds of thinking they involve. Kuhn, at least initially, connects these with phases of normal and extraordinary science. Philosophers have argued that the structure of science may not conform exactly to Kuhn's structure for scientific revolutions, and that normal science makes its own form of revolutionary advance (see Toulmin 1970). Kuhn later acknowledges this, emphasizing that the key to his view "is a certain sort of reconstruction of group commitments," and not necessarily revolution (Kuhn 2012: §1). This is consistent with the argument of this chapter, and its key idea that there is a substantial distinction between working within a set of basic commitments and working in a way that evaluates and challenges those basic commitments, and that these differences make a sociological difference, even if that difference is not to start a revolution.

Again, many philosophers reject Kuhn's ideas about "incommensurability" and the attending specter of relativism, but one does not need to accept those ideas to accept the idea that comes into play later, namely that communication becomes more difficult in periods of transition, and that this makes it appropriate to think that different languages are in play. However, these languages may permit intertranslation, albeit with difficulty, *pace* incommensurability. In any event, a commitment to incommensurability or relativism plays no role in my argument (and indeed is antithetical to my use of Kuhn).

With those initial clarifications and disclaimers in mind, let's return to Kuhn. Following Kuhn, I highlight two epistemological elements of intersubjectively authoritative method. The first is the presence and role of exemplars of scientific achievement. The second, which for Kuhn is epistemologically posterior to exemplars but crucial for understanding their significance, is a commitment to systems of rules – at conceptual, theoretical, experimental, and instrumentational levels – for conducting exemplar-grounded scientific inquiry.

The attractive sociological elements include a conception of science involving a community of inquirers, whose members are capable of mutual understanding, and further, consensus and agreement, and whose work cumulatively builds and progresses. These ideas about the sociology of science are commonplace in everyday parlance – "the scientific community," "scientists have discovered," "according to scientists."

I now work out these epistemological and sociological ideas in more detail, with explicit reference to Kuhn's discussion.

"Normal science," Kuhn writes, "means research firmly based upon one or more past scientific achievements, achievements that some particular scientific community acknowledges for a time as supplying the foundation for its further practice" (II: "The Route to Normal Science").[15] The past scientific achievements, or some distinguished subset of them, constitute the *exemplars* for the scientific community – "a set of recurrent and quasi-standard illustrations of various theories in their conceptual, observational, and instrumental applications" (II: "The Route to Normal Science"). Exemplars are the gateway to new scientific forms of life with novel, unanticipated results pregnant with new problems to solve. Kuhn gives the transition from Ptolemaic to Copernican astronomy and the transition from classical to relativistic and quantum physics (see also Kuhn 1978) as exemplar-driven transitions to new *disciplinary matrices* (Kuhn 2012) or *paradigms* for the practice of science.

Exemplar achievements "are sufficiently unprecedented to attract an enduring group of adherents away from competing modes of scientific activity," but also "sufficiently open-ended to leave all sorts of problems for the redefined group of practitioners to resolve" (II: "The Route to Normal Science"). Here we see epistemological and sociological elements of exemplars and exemplar-driven science.

Epistemologically, the role of exemplars is to generate scientific problems as *puzzles*. According to Kuhn, science as normal science should be construed as a kind of "puzzle-solving." Puzzles are "that special category of problems that can serve to test ingenuity or skill in solution" (IV: "Normal Science as Puzzle-Solving"). Although arriving at a solution may require skill and ingenuity, less up for grabs is whether there are solutions. Kuhn argues that "one of the things a scientific community acquires with a paradigm is a criterion for choosing problems that, while the paradigm is taken for granted, can be assumed to have solutions" (IV:

[15] I give references to *Structure* in this way, to highlight the setting in the book for the quotations from Kuhn I give.

"Normal Science as Puzzle-Solving"). Puzzles are thus distinguished from some other problems in promising well-defined solutions.

Further, "to classify as a puzzle, a problem must be characterized by more than an assured solution. There must also be rules that limit both the nature of acceptable solutions and the steps by which they are to be obtained" (IV, "Normal Science as Puzzle-Solving"). According to Kuhn, rules are "abstracted" (V: "The Priority of Paradigms") from the exemplars.[16] Achievement for the scientific community is modeled on the exemplars and involves following conceptual, theoretical, experimental, and instrumentational rules implicit in those exemplars. These rules, together with the exemplars, constitute the new forms of scientific life – the disciplinary matrix (Kuhn 2012) or paradigm. For a current example, in research in artificial intelligence, deep learning, large language models, and big data constitute primary elements of the disciplinary matrix, building on exemplar results about restricted Boltzmann machines, backpropagation, and deep belief networks (Hinton, Osindero, and Teh 2006).

The net result is that normal science is an exemplar-driven activity in which problems that are relatively assured of solutions are chosen for investigations that require high levels of skill and ingenuity while at the same time using shared rules that are implicit in the exemplars of the normal science. The exemplars and shared rules set up a disciplinary matrix within which the activity of normal science proceeds.[17] This summarizes Kuhn's influential account of what intersubjectively authoritative method in science looks like in at least some detail.

I want now to indicate how these epistemological features of normal science intertwine with an attractive sociology.

Sociologically, disciplinary matrices or paradigms are instantiated materially in textbooks and lab and instrumentational manuals that recount exemplars and contain courses of study, problems, and instructions that form the basis for inculcation into the scientific community. As Kuhn puts it, "by studying [the exemplars] and by practicing with them, the members of the corresponding community learn their trade"

[16] As Ian Hacking in effect observes in his Introduction to the 50th anniversary edition of *The Structure of Scientific Revolutions*, the role of exemplars introduces a kind of rule-following problem (Wittgenstein 1953). Kuhn also connects his discussion to Wittgenstein, but to his idea of a conceptual practice being held together by family resemblances instead of sets of necessary and sufficient conditions (V: "The Priority of Paradigms").

[17] Rules can play this role consistently with acknowledging Kuhn's rejection of the commonplace view that science is a continuous and cumulative project that moves closer and closer to the truth by application of "the scientific method" (cf. Bird 2018: §2). In this way Kuhn finds a place for the commonplace view in his system, while at the same time fundamentally challenging it.

("Introduction"), and later, "[t]he study of [exemplars] ... is what mainly prepares the student for membership in the particular scientific community with which he will later practice" (II: "The Route to Normal Science").

Exemplars and their role in gaining membership into the scientific community also help explain the mutual understanding and, further, the high levels of agreement and consensus in science. As Kuhn puts it, scientists "who learned the bases of their field from the same concrete models ... seldom evoke overt disagreement over fundamentals ... [and] are committed to the same rules and standards for scientific practice" (II: "The Route to Normal Science"). Mutual understanding, in the form of agreement over fundamentals, as well as a more general agreement and consensus, are the result of the exemplar-based inculcation into the community and propel the research tradition forward.

Finally, commitment to the exemplars and rules that make up the disciplinary matrices or paradigms of normal science explain the progress that is seen in normal science. Kuhn suggests that it is almost definitional of normal science that it progresses – "we tend to see as science any field in which progress is marked." Why though? Why should the properties of normal science give rise to progress? The answer, Kuhn tells us, "depend[s] in part upon an inversion ... Does a field make progress because it is a science, or is it a science because it makes progress?" (XIII: "Progress through Revolutions"). The inversion is that scientific activity, including progress in science, is a result of the kind of community that forms around exemplars in normal science, and the role of those exemplars in providing a model for future achievement and success according to rules shared by the community. The community sets itself up for success not in the abstract, by having the correct methodology for success, but concretely, by having exemplars that it can follow to solve problems that the disciplinary matrix or paradigm has licensed as problems to solve, in part by limiting the problems to solve to those that can be solved by the rules implicit in paradigms. The epistemology and sociology of normal science underwrite this progress.[18]

We have been considering the prospects for scientism understood as a framework or stance for objective thinking itself. We took the inspiration for this idea to come from logical empiricism, which tied objective content

[18] Kuhn balked at the idea that it is progress toward the truth, but thought that we could still say, as Hacking puts it, that there is "progress *away from* less adequate conceptions of, and interactions with, the world" (Introduction: §13, "Progress Through Revolutions").

(or cognitive significance or meaning) to the presence of intersubjectively authoritative method. The idea was that science is the appropriate basis for an "ism" because the use of intersubjectively authoritative method possesses attractive epistemological and sociological features. We then used Kuhn's conception of normal science to understand what intersubjectively authoritative method in science consists in, and then, again following Kuhn, made explicit the intertwined sociology. The discussion of Kuhn confirmed the idea that normal science has attractive epistemological and sociological features in so far as concretely specified exemplars, rule-based inquiry, mutual understanding, agreement, community, and progress are epistemologically and sociologically attractive. The net result is that if the science in scientism is normal science, it is understandable why science is thought to be the basis of an "ism."

Why *not* Scient*ism*?

However, Kuhn did not think that normal science exhausts science. Kuhn thought that scientific activity also includes what he called *extraordinary science*. I discuss extraordinary science in more detail in this section, but for orientation, it is useful to note that Kuhn's distinction between normal science and extraordinary science is the fulcrum around which the argument of this chapter turns. I make this explicit in the next section.

Let us consider Kuhn's account of extraordinary science in relation to the epistemological and sociological features of normal science.

Extraordinary science happens

> when ... the profession can no longer evade anomalies that subvert the existing tradition of scientific practice ... then begin the extraordinary investigations that lead the profession at last to a new set of commitments, a new basis for the practice of science. The extraordinary episodes in which that shift of professional commitments occurs are the ones known in this essay as scientific revolutions. They are the tradition-shattering complements to the tradition-bound activity of normal science. ("Introduction")

On Kuhn's picture, as normal science proceeds, anomalies accrue and the inability of the disciplinary matrix or paradigm to explain or otherwise accommodate them becomes manifest. This "blurring of the paradigm" precipitates a crisis, from which the new paradigm emerges as a "reconstruction of the field from new fundamentals, a reconstruction that changes ... many of its paradigm methods and applications" (VIII, "The Response to Crisis"). The epistemology of extraordinary science does not include acceptance of and adherence to paradigms and their methods with an eye to trace out their applications, but instead reconstruction of fields from new paradigms with new methods and applications.

This difference in epistemology is one aspect of a more general difference between normal science and extraordinary science, namely a difference in the kind of thinking that goes on in each. Normal science of course includes proposing hypotheses gathering of evidence, performing experiments, etc., but will also include forms of reflective and metarepresenta tional thinking in the honing and revising of hypotheses and theories that is ubiquitous in the course of everyday scientific work. But in normal science, this is done against the background of a disciplinary matrix or paradigm. Extraordinary science, by contrast, involves significant reordering and reprioritization of commitments, opacity instead of transparency about proper methods, an absence of substantial common ground in disagreement over paradigms, and factionalizing instead of the fostering of scientific community. Choosing between paradigms is not "determined merely by the evaluative procedures characteristic of normal science," since "these depend in part upon a particular paradigm, and that paradigm is at issue" (IX, "The Nature and Necessity of Scientific Revolutions"). It is these kinds of more foundational and controversial but also more nebulous concerns that are addressed with interpretation and critical reflection in extraordinary science.

Further, Kuhn thinks that the thinking directed at evaluating the paradigms themselves approximates much more the kind of thinking that goes on in philosophy than the kind of thinking that goes on in normal science. In the transition to extraordinary science, "scientists [turn] to philosophical analysis as a device for unlocking the riddles of their field" and make "recourse to philosophy and to debate over fundamentals" (VIII, "The Response to Crisis"). Extraordinary science employs a form of thinking appropriate for when foundations are shifting and thinking must make progress without solid ground under its feet. It is thinking that does not proceed from first principles but examines first principles themselves.[19]

With extraordinary science comes a deterioration of intersubjective understanding and communication breakdown. In the transition to extraordinary science, scientists "whose discourse had previously proceeded with apparently full understanding may suddenly find themselves responding to the same

[19] There are familiar problems about how to understand the epistemology of first principles. Working these out, however, is not required for our purposes, which accepts that, but does not explain how, the thinking in extraordinary science is valid.

stimulus with incompatible descriptions and generalizations." These differences in response lead to problems that are "first . . . evident in communication" but which are "not merely linguistic, and [which] cannot be resolved simply by stipulating the definitions of troublesome terms" (Kuhn 2012: Postscript, 5, "Exemplars, Incommensurability, and Revolutions"). This kind of phenomenon can be seen in common examples, for example concerning the question of the connection between the availability of guns and crime, on which American conservative and liberal paradigms not only disagree (see Pew Research Center 2021) but, more deeply, on which those adopting them encounter significant problems of intersubjective comprehension.

However, when extraordinary science tries to move past the barriers to intersubjective comprehension, extraordinary science looks very different from normal science. Again, what is called for is something more like philosophy, in its interpretive, analytic, critically reflective, and even dialectical aspects. Kuhn tell us that members of different disciplinary matrices "recognize each other as members of different language communities and then become translators." Though translation will be difficult, Kuhn does not close off the possibility, and allows that translation "is a potent tool both for persuasion and for conversion" (2012: Postscript, 5, "Exemplars, Incommensurability, and Revolutions").

Unsurprisingly, the epistemology of extraordinary science is intertwined with its distinctive sociology, one conditioned by the loss of mutual understanding. The breakdown of a paradigm is accompanied by significant attention to and disagreement about foundational issues and replaces community not only with a "proliferation of competing articulations" and "different language communities" (VIII: The Response to Crisis) but with a loss of identity and alienation. Kuhn recounts Wolfgang Pauli's words of resignation in the face of the difficulties of accounting for atomic spectra that led to the development of Heisenberg's matrix mechanics: "At the moment physics is again terribly confused. In any case, it is too difficult for me, and I wish I had been a movie comedian or something of the sort and had never heard of physics" (quoted by Kuhn, VIII: "The Response to Crisis").

So, the erosion of a paradigm and rules implicit in it occasions a crisis that includes the splintering of the scientific community into different schools, among which communication is difficult enough to suspect that different languages and systems of meanings are at play, and this in turn has the potential to produce a kind of existential crisis and a loss of a sense of identity among scientists.

Making It Explicit

So: exemplars, disciplinary matrices or paradigms, and the rules they implicitly contain are not only the basis for programs of research but also of mutual understanding, significant agreement, and the progress of science. Extraordinary science, because it interprets and critically evaluates, rather than takes for granted, the paradigms and the rules themselves, foregoes these features of normal science. If Kuhn is right about this, extraordinary science lacks the epistemologically and sociologically attractive features of normal science. But possessing these features is what made it plausible that normal science is the basis for an "ism." So, if Kuhn is right, there is no reason to think that extraordinary science is the basis of an "ism," and no reason to think that science conceived in total, as including both normal and extraordinary phases, is the basis of an "ism."

We can write this up as an explicit argument for which the main lines of support for the premises can be indicated. After that, I outline some larger lessons for the limits of objectivity and how, should they exist, they can and, especially, cannot be determined. I conclude by discussing the significance for humanism.

Here is the argument.

P1 The *science* in *scientism* means either *normal science*, or *normal science and extraordinary science.*

P1 sets up the argument, which takes the form of a dilemma for scientism, one that resolves in a way that undermines scientism. The dilemma turns on the question of how to interpret the *science* in *scientism* and, in particular, on how to interpret the distinction between normal and extraordinary science. The distinction is justified by Kuhn's analysis of science. As I've indicated, my overall argument does not require that Kuhn's analysis of science is correct in all its details. But it does require a distinction about different phases in scientific practice, and that these phases highlight the use of different kinds of thinking in science.

P2 If the *science* in *scientism* means *normal science*, then science can be the basis of an "ism" and scientism becomes plausible, but only at the expense of making the thinking in extraordinary science invalid.

P2 recognizes that there is something attractive about normal science, and it is justified by the argument, based in Kuhn's account of normal science, that normal science possesses attractive epistemological (exemplar-driven, rule-following) and sociological (community, mutual understanding, agreement, progress) features. But this comes at a cost. As a reminder, to say that some form of thinking is invalid is to say that it fails to trade in objective contents, the kinds of contents appropriate for belief and knowledge. According to scientism, whether a kind of thinking counts as objective thinking is determined by whether the relevant thinking is scientific. Further, we are supposing that the *science* in *scientism* is normal science, and the thinking that goes on in normal science is fundamentally different in kind from the thinking that goes on in extraordinary science, which is more like philosophical thinking in involving interpretive, analytical, and dialectical aspects. P2 records the consequence of this understanding of a scientism that is based exclusively on normal science for the status any kind of thinking that does not conform to the kind of thinking that goes on in normal science.

P3 If the *science* in *scientism* means *normal and extraordinary science*, the thinking in extraordinary science is valid, but at the expense of undermining the idea that science is the basis of an "ism."

P3 recognizes that what is attractive about normal science, and what made plausible the idea that normal science could be the basis of an "ism," is missing in extraordinary science. This was justified by the argument, based in Kuhn's description of extraordinary science, that extraordinary science lacks the attractive epistemological and sociological features of normal science: It is not paradigm-driven and rule-following but challenges paradigms and rules, and involves the fragmentation of scientific communities, significant disagreement, the breakdown of communication, stalled progress, and even loss of identity and alienation.

C1 Either science is the basis of an "ism," or the thinking in extraordinary science is valid, not both (P1–P3).

C1 makes the dilemma for scientism explicit. To put the dilemma in different terms from those in C1, science can either be an "ism" but exclude a part, maybe the intellectually most profound part, of science itself, or science can include *all* of science, but then not be the basis of an "ism."

A dilemma is an exhaustive disjunction (throughout I have assumed that science should not be identified with extraordinary science alone for purposes of scientism) with unattractive disjuncts. Spelling out the disjunction a little, it says that either science is the basis of an "ism" and extraordinary science is invalid *or* extraordinary science is valid and science is not the basis of an "ism." For all that has been said so far it is open to the proponent of scientism to accept the first disjunct. But the next premise closes that option off.

P4 The thinking in extraordinary science is valid.

P4 needs more discussion. I briefly discuss two considerations here. My point is not to provide a comprehensive defense of P4 but to make it plausible and to indicate the costs of rejecting it.

The first is that unless extraordinary science were valid, the current practice of science would seem to have a significant *objectivity deficit*, since the current practice of science is the result of extraordinary science replacing former paradigms and rules by new (the current) ones: If extraordinary science is not valid, this acceptance of new paradigms and rules is based on invalid thinking (again, thinking that does not trade in objective contents and that is not appropriate for belief and knowledge of how things are). The idea that the thinking that led to a new paradigm is invalid, but the paradigm and its rules are nonetheless valid, seems difficult to accept.

Second, though extraordinary science is not the basis for an "ism," extraordinary science is, for Kuhn, cognitively valuable.[20] So Kuhn is on board with the idea expressed in the previous paragraph. However, the kind of thinking that extraordinary science involves is not proprietary to science. As Kuhn put it, in times of crisis and revolution, scientists "turn to philosophical analysis" and make "recourse to philosophy."

More specifically, we can say that the kind of thinking that occurs during periods of crisis involves a significant dose of a kind of topic-neutral, completely general, reflection of thought on itself, where thought takes its commitments, normative standards, and standing as subject. What this amounts to in science is reflection on accepted paradigms and rules that form the foundation for the practice of normal science. But, again, this is not a kind of thinking or cognition that is special to science. It is instead a kind of thinking or cognition that evaluates the commitments, standards, and standing of thought in any domain. We can summarize this by saying that extraordinary science makes use of the resources of *interpretation and critical reflection* as it transitions from the paradigms and rules of one tradition of normal science to another. This identification, or partial identification, of the thinking in extraordinary science with interpretation and critical reflection suggests that unless there is reason

[20] Many inspired by Kuhn saw him as highlighting arational and even irrational considerations. Kuhn himself did not see things this way – see for example XIII: "Progress Through Revolutions" and his interaction with Lakatos (1970), in Kuhn 1970: "Either we are both defenders of irrationality, which I join him in doubting, or else, as I suppose, we are both trying to change a current notion of what rationality is" (Kuhn 1970: 39).

to think that interpretation and critical reflection are in general without cognitive value,[21] or that they are widely applicable but just not in science, the kind of thinking in extraordinary science should be validated.

Finally:

C Science is not the basis for an "ism" for objective thinking.

From these premises, our conclusion about the prospects for scientism follows.

Science and the Limits of Objectivity

On the view according to which science is the basis of an "ism," there is a tight connection between science and the limits of objectivity according to which science, by constituting what it is to think objectively, sets or determines the limits of objectivity. But if science is not the basis of an "ism" and so does not set the limits of objectivity, is there *any* connection at all between science and those limits? And if science does not set or determine the limits of objectivity, what does so instead?

At this point we can make the following conditional claim: If interpretation and critical reflection have some substantial connection to the limits of objectivity, then science will have such a connection to the limits of objectivity. This is because science includes extraordinary science, and extraordinary science makes uses of interpretation and critical reflection. However, this connection will be looser than that envisioned by scientism. The connection would not be that science sets or determines the limits of objectivity, but that science, through extraordinary science and its use of interpretation and critical reflection, sometimes operates at and thereby participates in and encounters those limits.[22] To repeat, though, this depends on the idea that interpretation and critical reflection themselves have a tight connection to the limits of objectivity. Do they?

There is an objection to the view that critical reflection has or could have the kind of tight connection to the limits of objectivity that we have argued science lacks. The objection is that the argument against scientism arguably generalizes and licenses the conclusion that the limits of objectivity are not set or determined by *any* positively specified methodology. The reason is that whatever positively specified methodology one may accept or adhere to in hopes of setting or determining the limits of objectivity (the

[21] For views along these lines, see Hilary Kornblith 2012, Whiting 2019.

[22] Perhaps, more strongly, normal science itself trades in objective contents only because of its connections to extraordinary science and critical reflection. For related discussion, see Rattan 2016.

way that adherence to scientific methodology is supposed to set or determine the limits of objectivity), it will always be possible to take that positively specified methodology as itself an object of evaluation and critically reason about *it*.[23] When we are in engaged in interpretation and critical reflection, our thinking takes positively specified methodology as a subject of critical evaluation. This means, though, that the positive methodology in question does not occur in thinking in a mode of full acceptance (because it is under critical evaluation), and so by the lights of the "ism" for objective thinking that the positively specified methodology ostensibly defines, interpretation and critical reflection about it are invalidated. So, again, if interpretation and critical reflection are valid and possible for any positively specified methodology, then no positively specified methodology can be the basis of an "ism."

This might be thought to be an objection to discerning some limit to objective thinking in interpretation and critical reflection, but really there is no objection. The way that the limits of objective thinking come up in interpretation and critical reflection is not by interpretation and critical reflection serving, with some positively specified methodology, to *police* the limits of objectivity. Another general conclusion of the chapter is that the limits of objectivity are not what they are because of any kind of methodological policing presence ruling *this* kind of thinking in and *that* kind out. Scientism falls really to this more general idea.

I want to suggest though, in closing, that when it comes to interpretation and critical reflection, there is reason to think that they could bear a tight connection to the limits of objectivity without some positively specified methodology for interpretation and critical reflection playing some kind of policing role. Instead, the tight connection to the limits of objectivity comes from the epistemology of interpretation and critical reflection itself. The epistemology of interpretation and critical reflection is a topic for another time, but for now we can briefly think of that structure as having a *subjective aspect* according to which the subject critically reasons about her own beliefs for the purpose of achieving comprehension and reflective justification for them; an *intersubjective aspect* according to which a subject critically reasons about another's beliefs for the purpose of being able to understand and learn from or instruct another, even under conditions of disagreement about paradigms; and an *objective aspect* according to which evaluation of a disagreement with

[23] Cf. Tyler Burge (1986: 720) on use- or role-theoretic accounts of conceptual content. "Our conception of mind is responsive to intellectual norms which provide the permanent possibility of challenge to any actual practices of individuals or communities that we could envisage."

another does not build in preference for one's own belief (or paradigm) merely because it is one's own. And now we might imagine that the limits of objectivity arise not from any methodological policing but are instead based on principled difficulties in reconciling the subjective, intersubjective, and objective aspects of interpretation and critical reflection. Again, science will participate in such limits, but it does not set or determine those limits, which are, to repeat, set or determined by principled difficulties in reconciling the subjective, intersubjective, and objective aspects of interpretation and critical reflection. This is an approach to the limits of objectivity that does not arise through methodological policing, and so is consistent with the considerations adduced here against scientism.

* * * * *

Let me conclude by returning to the issue of humanism. I said that a direct consequence of my argument is that the view of humanism as scientism is a bad idea. But what else? Does anything else follow for humanism? Well, one thing is that if science does not constitute the limits of objective thinking, then there is room to think that the humanities, too, can participate in objective thinking. Against logical empiricism, the possibility is now open for ethics and metaphysics, too, to participate in objective thinking, even though their methods are not those of science. And against scientism, no discipline in general need follow science in its proprietary methods to count as being engaged in objective thinking. We might insist that a necessary condition for engaging in objective thinking is to engage in interpretation and critical reflection, but interpretation and critical reflection do not provide any specific positive methodology conformity to which sets the limits of objectivity. Disciplines that do engage in interpretation and critical reflection about their most basic commitments will still encounter the limits of objectivity as a result of trying simultaneously to meet subjective, intersubjective, and objective norms in interpretation and critical reflection.

This, I think, is a cheerful conclusion for any humanism that wants to recognize the validity of the humanities and their methods, since it emphasizes the role of interpretation and critical reflection in understanding what it is to think objectively and what the limits are that objective thinking can expect to encounter. Since science too makes use of the resources of interpretation and critical reflection, the result does not reflect negatively on the status of science with respect to objective thinking. Science is thus included within humanism without, as scientism would have it, science setting the limits of objective thinking.

CHAPTER 4

Scientism: Reflections on Nature, Value, and Agency

Akeel Bilgrami

I

That there is a distinction we make between "scientific" and "scientistic" is reflected in the fact that many who have no phobia of science, indeed admire it greatly, have declared that they find scientism distasteful. Since a chapter, even a book, gives one insufficient space to say what "science" is, let me take for granted that we have some instinctive grasp of what we intend by that term, and ask: What is "scientism"? Even a glance at the writings that recoil from scientism would suggest that it is perceived to be a kind of overreach in the name of science, taking it to a place beyond its proper dominion.

One form of overreaching has tended to take the form of making large claims on science's behalf, claims that are philosophical rather than scientific, yet, in doing so, relying – by a sleight of hand, a fallacious conflation – on the authority of science. In this chapter, I explore one such claim, the claim that *there is nothing (no property of or in nature, no fact of or in nature) that cannot be brought under the purview of natural science's inquiries.* Such a full coverage of nature on science's behalf is a claim that is philosophical, not scientific, since no science contains that claim. Yet many consider the assertion that nature contains properties that natural science does not countenance to be unscientific. That is the just the sort of sleight of hand that is said to be typical of scientism. But even suppose that no one declares it to be unscientific. The question I explore in this chapter is whether it is overzealous on behalf of science – in the way that scientism is – to say that any denial of the claim which I have italicized above is bad *philosophy.*

It is worth noting (one would not understand the real nature of the claim if one did not) that this claim, despite the sort of overreach I've just mentioned, is often accompanied by quite genuine expressions of humility which admit to having brought very little of what is in nature effectively within the purview of natural science's explanatory scope. The point of the claim is not to say that what can be done has been done. The point of the claim is not even perhaps best presented by saying that it can be done. Rather it is to say that it is the business of natural science to cover in its inquiry all that is there in nature. It is this last point that the opponents of scientism are resisting.

I want to explore a familiar (and contentious) ground of such a resistance, which repudiates scientism by asserting that nature (and the world we inhabit, more generally) contains value properties and facts, for instance moral facts, and these are not the proper subject of science.

What are these facts and properties? This very question should perhaps be seen as expressing a prejudice. For now, let us just say in response that such facts are what are specified by sentences such as, "The Malabar forest is valuable," "Gandhi's civil disobedience was courageous," "The genocide of Jews was cruel," "The treatment of Palestinians by the Israeli government is inhuman," when these sentences are true; and when they are true, the predicates "is valuable," "is courageous," "is cruel," "is inhuman" are satisfied by the relevant value properties. I have said "for now" let us characterize these value properties in this relatively uninformative sense because, as the chapter unfolds, more substantive characterizations will emerge that elaborate the nature of values by linking them *intrinsically* to practical agency. The more tentative and uninformative characterization just given is intended merely to show that there is no reason to think (that it would, in fact, be a prejudice to think) that there is anything more mysterious about value properties than there is about any other properties which we unselfconsciously speak of in the sentences we utter, and so there is no particular need to withhold notions of fact and of truth, when speaking of values in sentences that deploy the vocabulary of value than there is in speaking of anything else in any other vocabulary we unselfconsciously deploy.

Of course, none of this will satisfy those (many) who simply deny that the notion of truth (and the objectivity it is supposed to reflect) so much as applies to sentences or propositions that contain an (irreducible) vocabulary of value. What I aim to provide in this chapter are grounds for withdrawing this denial. But before I proceed to do that, I should expend a few words to register skepticism about a widely held view that *only* certain considerations (not present in the grounds I aim to provide) can ground the proper applicability of the notion of truth. Thus Bernard Williams, echoing many others,[1] has proposed that only the natural sciences are characterized by a certain *accumulation and convergence* in judgment that is *necessary* to ground a proper application of notions of

[1] See Williams 1985. A similar argument was made earlier by Richard Boyd and by Hilary Putnam.

truth, objectivity, and realism to their propositions and the objects and properties and facts they posit and describe. The argument goes roughly as follows. Science is the *only* cognitive enterprise in which the results of inquiry have built upon each other over the centuries and cumulatively arrived upon a convergence. And there can be no other explanation of this convergence but to say that the judgments of science are tracking the truth and that their theoretical posits are *real* objects or properties in the world. Where there is no such convergence in judgment, we lack a proper ground for the attribution of objectivity and realism that is carried by the concept of truth.

Let me express my skepticism about this argument by asking: When did we *start* converging in this cumulative way upon the truth? Did we start doing so from the very beginnings of human thought? Surely not. For centuries there were all sorts of false starts and false leads that were abandoned. In fact, Williams himself is quite explicit that it is only sometime in the seventeenth century that we began upon such a convergence. Why is that? Because it is only with science in the modern period that we were set on the right path, says Williams, and, once on that path, the results of scientific inquiry began to accumulate, building upon each other, toward a convergence. So now, let us further ask: What makes this path that opened up sometime only in the seventeenth century the *right* path? What is that word "right" doing there? If we become clear about what makes that path the right path, we would have established the objectivity and realism that William hankers for in science, and we do not need the rest of the accumulation and convergence as an argument for scientific objectivity and truth. For Williams' argument to be the argument it is, it must presuppose what he wants to argue for. He is presupposing *truth* in his assumption of a *right* path in the elaboration of the argument for truth via convergence. That notion of truth is not compulsorily tied to any considerations of convergence. Thus, the domain of value, even if it cannot be characterized by such a convergence, can nevertheless be host to the full prestige of *truth*.

Returning then to my anti-scientistic claim, which appeals to the presence of value properties, it is important for me to give a careful answer to the question why it is that value properties, if they exist in the world we inhabit, are not the proper subject of natural science. We can, of course, study values by methods of study that are disinterested and systematic in the way that science, at least when it is exemplary, aspires to be, but there is something fundamental about values that will necessarily be left out in such study. It is this fundamental fact about values that constitutes the real source of the resistance to scientism. Much of this chapter, therefore, seeks to elaborate this source of resistance more or less exactly in an argument.

The fundamental fact about value properties in the world (including nature) that makes the natural sciences beside the point is that when we perceive such properties in nature (in the world around us, generally), they prompt our *practical* agency, not our theoretical agency, not our agency that seeks to explain and predict in a detached and disengaged form of inquiry, but the agency that seeks to address the *normative* (in the sense of practical) demands those perceptible values make on us. Invoking Kant, we might say that value properties in nature and the perceptible world, generally, fall outside the scope of science because their scope really falls within practical reason or agency, the subject of Kant's *second* "Critique," quite outside the reach of physics and mathematics that are the explicit examples of the *theoretical* domain mentioned in the theme-setting "Preface" of his *first* "Critique." Kant himself did not put things as I have ("value properties in nature and the perceptible world prompt our practical, not theoretical, reason") because he thought nature and the properties of the perceptible world (what he called "phenomena") were entirely the domain of theoretical reason and, in particular, Newtonian science. That is why practical reason was relegated by him to what he called "noumena," for him a nonnatural realm. So, my appeal to Kant here is, in some strict sense, partial. But if we put aside the gratuitous metaphysics of the distinction between phenomena and noumena (a distinction forced on him only because of the sheer artifice of his equation of "nature" or "phenomena" with "that which is exhaustively the domain of Newtonian science"), the insight that there is *a disjointness* between practical and theoretical agency (needing two quite different critical philosophical inquiries or "critiques") is a natural starting point to explore the real source of the resistance to scientism.

Now, these considerations of agency that I am bringing to center stage may, with some right, be thought to offer a more radical and perhaps deeper path of resistance to scientism than I pursue in this chapter. As I said, what I am pursuing is a resistance to the scientistic claim that nature contains only what the natural sciences countenance by providing an argument for the counterclaim that nature contains value properties that natural sciences do not countenance, because unlike the properties that the natural sciences do permit they are intrinsically tied to our practical agency. The more radical path against scientism appeals to considerations of agency in a more pre-emptive manner than the resistance

I pursue by denying *the very articulability* of the scientistic claim we are discussing. It might go like this.

Natural scientific explanation, its laws and predictions and the ontology of objects, properties, and facts that these laws and explanations deliver, are an outcome of theoretical *agency*. Once these deliverances of this sort of agency is in place, we have a metaphysical picture of what the world contains, and then scientism appeals to just this metaphysical picture to say: "We do not find values in that picture, nor can we fit values in that picture." The radical path of resistance to such scientism says in response: You have things the wrong way round. You proceed as if the metaphysical picture is primary and comes fully formed, as if from nowhere. You, thus, present the ontology as what is given first, ignoring the fact that it is the deliverance of an exercise of (theoretical) agency. Science is first and foremost a *practice*. The practice, of course, has its outcome (it's deliverances, as I put it), but philosophy must make the practice the primary focus, not the outcome, since it is the practice that makes possible the outcome. And if, for that reason, it is made primary, we will find that in the practice, the practitioners are constantly and ineliminably speaking of how evidence *justifies* certain hypotheses, to generate its laws and predictions. That talk of justification is normative talk; it speaks to *values* of rationality. So, it is neither here nor there that we cannot find a place for values in the *metaphysical outcome* of this practice. The values are present in the very thing that makes the outcome possible, the exercise of theoretical *agency* in the practice of science. Philosophy must not make the metaphysics primary. It is the agency and practice that is primary.[2] If so, scientism cannot so much as get off the ground. I have considerable sympathy with this path of resistance to scientism because I have sympathy with its conception of philosophy, but since it will only carry conviction for those already possessed of such meta-philosophical commitments, I proceed more patiently without assuming those commitments at the outset.

Two more preliminaries – first an historically contextualizing point and then a ground-clearing one – before I lay out in detail the chapter's argument for the anti-scientistic conclusion that the perceptible world, including nature, contains value properties.

[2] Pragmatists have long stressed the primacy of practice, yet many have failed to see the more radical implications and succumbed to one or other version of the scientism we are discussing. The diagnosis for this deserves a careful accounting elsewhere.

John McDowell (1979) has, in recent years, attributed just such a conclusion to Aristotle.[3] But Aristotle is a very high philosophical location for it and high philosophy is only a narrow strand of intellectual history. It was a pervasive part of the worldview of a wide variety of folk and spiritual traditions (including popular Christianity) for centuries before and after Aristotle. These traditions, unlike Aristotle, mostly viewed the source of perceptible value properties in the world to be sacred and conceived nature as being shot through with value *because* it was shot through with the presence of the divine. It was only after the desacralization of nature in the modern period that such a view of nature began to be treated with a special hostility, not initially by those who proclaimed the "death of God" (that came somewhat later) but even before that by those who arranged for the "exile of God" ("Deus absconditus") to a distant place outside the universe of matter and nature in a strictly "providential" role.

This was not a purely intellectual hostility.[4] It was often motivated by political and material considerations. Those who continued to see nature as sacralized by God's presence were dismissed as "enthusiasts" both for seeking to make God democratically available to all who inhabited his earth rather than exclusively accessible only to the learned scriptural judgment of university-trained divines, as well as for placing metaphysical and theological obstacles in the way of prospects for taking from nature with impunity. I say "with impunity" for a reason. Human beings have, of course, been taking from nature ever since they came to inhabit it, but in every social world until this period, there were rituals enacted before and after cycles of planting (and even hunting) to show respect and reciprocation to nature for the gifts it presented. By contrast, with desacralization, taking "with impunity" seems a quite apt description of the human – at

[3] McDowell has developed the view he attributes to Aristotle along interesting and attractive lines, though for a radical disagreement between us on one central matter – the supervenience of value properties on the properties that natural science studies – see chapter 5 of Bilgrami 2006 and McDowell 2006. See also the exchanges in the symposium on my 2006 book in Baldwin 2010, Normore 2010, and Bilgrami 2010.

[4] There were intellectual issues at stake as well, and on one central such issue, that of *motion*, it is not obvious that the "enthusiasts'" view (the presence of God in all matter and nature) was any less warranted than the Newtonians' ("the exile of God"). Neither side of the dispute was getting prizes in this period for any kind of atheistic denial of God. Newton's laws were apparently compatible with the existence of God for all sides. The crucial point is that there is no reason to think that it was only God conceived as stationed at an external or Archimedean point, providing for motion as a clockwinder, that was compatible with these laws. The enthusiasts' quite different conception of God as present in nature and providing for motion as an *inner* source of dynamism, was quite as compatible.

any rate, European – outlook on nature. It was such an ushering out of God to this remote station, external to nature and matter, that made possible the scientistic claim that nature contained no properties that natural science (then known as "natural philosophy") did not study. It emerged in the seventeenth century, and grew into an entire outlook, a zeitgeist, as a result of worldly alliances formed first in England (spreading next to the Netherlands, and then to the rest of Europe) between the (high church) Anglican establishment, the institutions around science (such as the Royal Society founded in 1660 and somewhat later the Royal Institution), and commercial interests, determined to transform the very concept of nature into the concept of natural resources.[5]

This deracination of God from nature resulted over time in an illicit extension of the notion of desacralization to the more general notion of "disenchantment."[6] The exile of God, thus, led to evacuating nature of *value* properties as well, which was perhaps an unsurprising consequence in a time when values were pervasively assumed to have religious foundations. Thus, by the time of the eighteenth century, in high philosophy, Hume was presenting values as wholly derivable from our *states of mind* (our desires and moral sentiments, our capacities for sympathy, etc.), whereas the *world* we inhabit was a fully Newtonian world, bereft of all properties that fall outside the scope of explanation by Newton's laws. For all the vehement disagreements on the nature of values that contemporary Humeans and Kantians have registered in recent years, Hume and Kant were one on this particular issue – their radically different ethical and meta-ethical views *both* ruled out the possibility of even a *secular* enchantment of nature, that is to say, a conception of nature that contained value properties without any sacred source underlying them.

The second and ground-clearing point is this. It may seem that such a resistance to this form of scientism by appeal to values is pushing at open doors. Isn't the heyday of a no-holds-barred "naturalism," in which natural science claims this kind of exhaustive coverage, a philosophical outlook that has now passed? Has there not been a frequent acknowledgment that human subjects, because of their unique possession of "reason," language, self-consciousness, etc., are set apart in not being subsumable under the laws with which we aspire to explain the natural phenomena in the world

[5] For a fine account of these alliances, see Jacob 1981. See also Schaffer 1997 and Jacob 1978.

[6] It is a pity that there is no Latin expression such as "Deus deracinus," since "Deus absconditus" gives the misleading impression – at least to English speakers – of a fugitive fleeing, whereas it was a willful putting away of God to a remote outpost. But, in fact, "absconditus" does not mean what it sounds like to the English-speaking ear. It means, roughly, "put away for safeguarding."

they inhabit? This acknowledgment, which though it may have come late to philosophers in the English-speaking tradition (it was explicitly made much earlier in the "verstehen" and hermeneutical traditions in European philosophy), is now increasingly voiced by "analytic" philosophers.

But it is not these doors against which the anti-scientism in question is pushing. As I said, Kant's very claim, in his second work of "critical" philosophy, to a pure practical reason that was radically disjoint from theoretical (what Kant sometimes called "speculative") reason, the subject of his first such work, was already an acknowledgment of the limited coverage of Newtonian science. Rather, it is the very fact that Kant had to seek a distinct domain from the perceptible world of "phenomena" for practical reason that reflects the scientism being resisted. Hence, the acknowledgment by the hermeneutical tradition and more recently by analytic philosophers that human subjects are set apart from the rest of natural phenomena as objects of inquiry misses the point that what is being resisted is just the idea that values are a construction of human subjectivity, that is to say, of human states of mind (moral sentiments, to use the vocabulary of Hume and Adam Smith). The resistance is precisely claiming that human states of mind such as moral sentiments are themselves formed by the *perception* of values *in the world*, that is to say, the "phenomenal" and natural world that human subjects inhabit; in other words, it is claiming – to put it in Weberian terms – that it is shallow to think that it is merely *we* who are "enchanted" while we inhabit a *world*, including a natural habitat, that is disenchanted. So, the argument I try to lay out is for the conclusion that it is *only because* the world that human subjects inhabit *is* (to continue with this Weberian vocabulary) "enchanted" that human subjects are.

II

I have said that the fundamental path of resistance to scientism that I follow is the one that denies the claim that "there is nothing, no facts, no properties, in nature that fall outside the purview of the natural sciences" by asserting that nature (the perceptible world, generally) contains facts and properties described in irreducibly value terms, and these cannot be brought under the sort of detached inquiry that natural science undertakes in its explanations and predictions because our perception of value properties (or, more simply, values) in the world prompts our *practical* agency, not our disengaged and detached theoretical inquiry. *So, it is really by exploiting the conceptual tie between values and (practical) agency that one takes this path of resistance.*

One can first get a glimpse of the relevant considerations of agency if we consider an utterly familiar ambiguity we find in the following thought or proposition, which is so often on our minds and lips:

"I will do ... "

It could be interpreted in one of two ways:

(1) I intend that I will ...
(2) I predict that I will ...

These are radically different thoughts that can be expressed by the same words; different because (1) and (2) harbor entirely distinct points of view or perspectives on oneself. When one predicts that one will do something, one is taking a disengaged or detached point of view on oneself. One is viewing oneself to be an object rather than a subject. When one intends something, one has an engaged perspective on oneself, one takes oneself to be an agent. In (1), both occurrences of "I" are the I of agency. But in (2), only the first occurrence of "I" is the I of agency. In the second occurrence of "I" in (2), the personal pronoun denotes an object. This is because in (2) the subject in the first occurrence of "I," speaking or thinking these words or this thought, views himself or herself in a purely disengaged and detached way.

One can have both these perspectives (engaged and detached) on oneself, but not at the same time. In other words, one cannot at the same time both intend and predict that one will do something. The one perspective necessarily displaces the other. Moreover, and more important for the purposes of this chapter's argument, unless one had an engaged perspective on oneself, one would not be a practical agent. Or, to put it from the other side: (3) *If one only had or if all one had was a detached or disengaged perspective on oneself (as exemplified in [2]), one would cease to be a practical agent.* This is a point of real significance, which I exploit later.

Why do I use the term "practical" agent in making this significant point? Because, as I said, in (2) the first occurrence of "I" *is* the I of agency. But in (2) that agency is exercised in a purely theoretical way on oneself, explaining and predicting one's behavior. Predicting and explaining are, of course, agentive acts, so there is no denying that one is an agent when one is viewing oneself with detachment and predicting what one will do. But one's angle on oneself, being detached in this way, restricts one's agency in (2) to theoretical agency. It is only when one's angle on oneself is engaged, as in (1), that one is a practical agent. That is why, were we only to possess the perspective present in (2), we would not be practical agents, even if we were agents.

I have put this last qualification "even if we were agents" in this conditional and hesitant form because it is highly implausible that we could possess agency *at all*, agency of *any* kind, if we possessed no practical agency. The idea that we are agents who are only capable of detached observation and prediction and explanation but no practical agency whatever is, in the end, an incoherent idea, though I won't argue for that here.[7] The point I keep in focus till a little later, however, is the italicized point (3) – that we cannot really be agents in the practical sense at all if we only have a detached perspective on ourselves as in (2). Point (3) is a conditional and I am *pretending*, for the sake of argument, that the antecedent in that conditional is coherent, just so as to set up the conditional for the later use I want to put it to.

Now, so far, I have said that an elementary ambiguity in a certain very common thought or expression hides a deep philosophical distinction between two perspectives each one of us can have *on one's self*. But this perspectival distinction (detached and engaged) is a perfectly general one and need not be restricted to the idea of a perspective on oneself. Being general, it should extend and apply quite naturally to the perspective we have *on the world*. That is, we can have a detached perspective on the world or an engaged perspective on it.

In many of our ordinary observations we think of the world in a detached way quite informally ("There is a table in Akeel Bilgrami's study"), but when we do natural science that detached perspective takes its most regimented form and we predict and explain the objects, properties, and events in the world, bringing them under laws and generalizations, moving to a different vocabulary ("molecules").[8] This detached perspective, whether informal or systematic as in science, is simply an extension of (2) from a perspective on ourselves to a perspective on the world. We then have to ask, if that is what a detached perspective on the world is, what is it to have an *engaged* perspective *on the world*? Here one's agency cannot be purely theoretical as it is when one is viewing the world in a detached way as containing elements to be predicted and explained by being brought under laws and generalizations. It would have to be *practical* agency. If so, two questions arise. What is practical agency and what would the world have to contain (over and above the properties that are explained

[7] See the discussion of the superlatively disengaged subject, Oblomov, in Bilgrami 2006.

[8] This point about regimentation makes clear that science has no interest in these common-sense observations of properties or facts – about the furniture in the house, say. But these properties, despite science's lack of interest in them, do not pose a problem for scientism in the way that facts and properties described in value terms do.

by theoretical agency exercised in the detached perspective on it) if we are to have *that* form of engaged rather than detached perspective on it?

It is here that the link between (practical) agency and value comes to view. If we are to be agents, practically engaged with the world, the world must contain elements over and above the elements that natural science (with its detached perspective on the world) studies. It is those elements that we paradigmatically describe in the vocabulary of values (though see just a little later for what is – and what is not – the real significance underlying this). Examples can be multiplied. Someone living by the sea perceives a storm on the horizon. What he has perceived can be described in meteorological terms (condensation, H_2O, etc.), but it can also be described in value terms (as *a threat*, say). Or take an example from Gilbert Harman (1977: chapter 1). Someone is driving past an alley and sees some kids burning a cat. One can describe what she sees in detached terms (*Felis catus*, carnivorous mammal, combustion, etc.) or in value terms (as *cruelty*, say). Unless we *see* the world as described with value terms – that is, as containing such properties as threats and cruelty – we could not be engaged with it in the practical sense. Over and above the condensation (the approaching storm on the coastline) and the combustion (the burning cat) which are captured by the perspective in the extension of (2) onto the world, the world must contain value properties (perceptible threats and cruelties) to trigger the extension of the perspective of (1) onto the world. Thus, a fisherman who sees the horizon of the Bangladeshi coast in detached, meteorological terms will have only the extension of the perspective of (2) on the world, but if he sees it as a threat, he will have a quite different perspective on the world, an engaged or agentive one, perhaps prompting him to go to the local municipality to arrange for some form of protection. So also, someone may go to Calcutta and view another person's condition in detached terms of average daily caloric counts, but then may also perceive that that person is *malnourished*, or as *in need*. When he perceives the world from a perspective that describes it in value terms of this latter sort, he will be prompted to practical agency – perhaps to give money to Oxfam, say.

These are mere examples of how we may take the same distinction as is found in (1) and (2) and extend it onto the world. And I have used the vocabulary of science (caloric counts, condensation, combustion, etc.) and evaluative vocabulary (needs, threats, cruelty, etc.) in formulating the examples to make it clear that the latter describes properties *in the world* that natural science does not study. I should, however, say by way of caution that, though I have used such a contrasting vocabulary to make the distinction vivid, the distinction between the two perspectives is *not* a linguistic one but a *philosophical* one. Someone who has thoroughly internalized the link between a certain scientific description of the world (some average caloric measure that is counted by public health officials as a nutritional minimum, say) and a person's need, will, without any turn to the thought or vocabulary of needs in how he conceives the person's condition, be prompted to *practical* agency. The point is not that one keeps changing one's vocabulary or concepts as one moves from detachment to engagement or vice versa. The point is only that detachment and engagement are two distinct *perspectives* on the world (as well as on oneself) and the world must contain properties over and above what the natural sciences study in order for us to have the latter perspective on it. That we are paradigmatically using or thinking in the evaluative conceptual *vocabulary* when we perceive the value properties in the world is not what is essential. The distinction is not intended as a linguistic distinction but a philosophical one. Contrasting vocabularies are just an easy way to convey the philosophical distinction but should not be seen as essential to the distinction.

What this eventually points to is that the so-called fact–value distinction is really, at bottom, a distinction in perspectives: the detached and engaged perspectives. If values are properties in the world, the perception of those properties is an apprehension of facts. So, values *are* facts, and can't, therefore, stand in a distinction with them. The distinction, therefore, can be reformulated as a distinction between what makes the kinds of facts that values are distinct from the kinds of facts that natural science studies. And it is in elaborating this latter distinction that we have been invoking the perspectival contrast between detachment and practical agency. The reformulation has radical consequences. It puts into doubt the very intelligibility of what philosophers claim is the "supervenience" of values on the facts that natural science studies, broadly speaking the claim that where there is no difference in the facts or properties that natural science studies, there cannot exist a difference of values or value properties. This claim posits a dependency relation of values on facts. But if values are facts or properties intrinsically tied to a perspective of practical *engagement*, precisely the perspective that is missing in the *detachment* of our angle on the facts that natural science studies, the very idea of such dependency becomes incoherent. I say "incoherent" and mean it. The point about a perspectival disjointness is not to deny supervenience; it is rather to say that supervenience cannot be asserted or denied, no more than it can be coherently asserted or denied that duck facts are supervenient on rabbit facts. In fact, I would be inclined to say that if supervenience of this kind were a coherent notion, it would indeed be foolish and implausible to deny it. So, it is of real importance to register that supervenience is *not* being denied; rather, the deep and intrinsic link between value and the considerations of agency we have been stressing render the very idea of supervenience unintelligible. (For more on this issue, see the references in note 4.)

The crucial point, for the purposes of this chapter's argument, is that this deep and intrinsic link between agency and value should now have come fully into view. We have a perspective on the world that is an engaged or agentive rather than a detached one only to the extent that the world contains value properties over and above the properties that natural science studies and which trigger the engaged rather than a detached perspective on it.

To recapitulate the argument so far: Starting with a familiar ambiguity in a ubiquitous thought or proposition ("I will ... "), I've teased out of it, in small steps which introduced the notion of agency, how natural it is to think that values are visible properties in the world. But to show that something is a natural thing to think is not yet to give an argument for it. It is only to show that one may think it without strain. Can we do better and present an argument for the conclusion that the world is populated by values over and above the properties that natural science studies?

III

One way to come to an argument toward such a conclusion might be to raise a challenge for it and answer the challenge. I have relied on the link between practical agency and value to make my claims thus far. It is only as or qua practical agents, that is, as subjects capable of engagement rather than mere detachment in our perspective on the world, that we view the world as containing value properties. The challenge might, then, seek to disrupt this link, denying that in order to be agents of this sort we must see the *world* as suffused with values. Practical agency, it might be said, is a simpler phenomenon than I am presenting it to be. It is a matter merely of acting on our desires and other such states and mental dispositions (including the loftier form they take, our "moral sentiments"). I have the desire to help the poor, I give money to Oxfam. I feel fear and vulnerability, I appeal to the municipality. I feel a combination of sympathy and indignation, I get out of my car and stop the kids from burning the cat. And so on. Values, on this view, are derivable from these desires and other states of mind and our agency is merely acting on these desires and states of mind. There is no need to add the further complication I am adding, which is that the desires and other such states of mind such as moral sentiments must be *responses to value properties in the world*, in order for us to be agents.

So, what I am insisting on is that desires are *not self-standing* in the way that this challenge proposes. They are responses to something that prompts them; they are responses to value properties in the world. Desires *in us* are nothing if there are no desir*abilities* (and undesirabilities) or values (and disvalues) *in the world* as well. And our agency consists in the fact that these desirabilities or values in the world, when we perceive them, make normative demands on us which trigger the appropriate desires in us upon which we act, as practical agents. (It should go without saying, but in case it does not let me say it: It triggers them only if we are virtuous or rational. If, as I said, they are properly *normative* demands that the world and its properties make on us, then the prompting to agency by those demands is not a causal, at any rate not a *merely* causal, prompting.[9]) And the challenge to us, which views desires as self-standing, simply denies this, claiming instead that agency consists merely in acting on our desires and those desires do not answer to any *external* calling of desirabilities or values in the world.

Can our agency be adequately characterized in terms of desires viewed as self-standing in this way? This is a good challenge because the view of agency it offers as an alternative is simpler and, therefore, may seem to be more intuitive than the more complicated one I am insisting on. To respond to the challenge, let me introduce some more conceptual apparatus.

There is a curiosity that was first pointed out by Gareth Evans (1982) in an insightful passage in his book, *The Varieties of Reference*. When we are asked, "Is it raining?" we tend to look out of the window and respond. And (this is the curiosity) when we are asked, "Do you believe that it is raining?" we tend to do the same. We don't scan our interiors to see if it contains the belief that it is raining. We simply look outside and respond. In short, we tend to do the same thing whether we are being asked about the *world* or about a state of *mind*, such as a belief. Evans went on to draw

[9] In saying this, I am presenting another mark of what I have insisted on throughout – the irreducibility of value properties to nonnormative properties (purely causal ones in this example). I am taking such irreducibility for granted in this chapter, without argument. In Bilgrami 2006, I present an argument for it that combines considerations derived from Moore and Frege. See also the exchanges in Baldwin 2010, Normore 2010, and Bilgrami 2010.

very interesting conclusions from this curiosity about the nature of self-knowledge, but I want to exploit the curiosity for a different purpose on the specific theme of this chapter.

I think it is perfectly plausible to *extend* Evans' insight about beliefs to *desires* as well. If we are asked, "Do you desire *x*?" we don't, in the normal and routine case (allowing, as exceptions, other unusual contexts, such as for instance on a psychoanalyst's couch), scan our mental interiors to see if it contains the desire for *x*, we simply consider the desir*ability* of *x*.

Two quick points of clarification, before I proceed further with the argument. First, for the sake of simplicity and convenience, I work with just the term "desires" here, as philosophers so often do, to function as a sort of omnibus term that is capacious enough to take in a range of ("conative") states of mind, including "moral" desires or what have been called "moral sentiments." So also, I take "desirabilities" in the world to be the more general term that is capacious enough to include "values." Second, by "consider *x*" I mean *either observe x* if it is available in our vicinity and consider whether it is desirable *or*, if it is not available in perception, we may *imagine* its desirability – and I am assuming that imagination in these cases, as in all cases, depends on some background of previous perceptions of *x* or of other things and properties like or approximating *x*.

If I am right that Evans' point can be extended to desires in this way, the dependence of desires on desir*abilities* (or values) in the world that McDowell and a large number of other moral realists[10] have laid claim for, is, prima facie, established. But someone, determined to press the challenge further, may deny that what seems prima facie so, is so. This denial would stubbornly maintain that when we are asked about whether we desire something, unlike what Evans said about beliefs, we simply do not and cannot look to desirabilities since there *are no* such things or properties. On this view, *facts* (such as that it is raining) may rightly be viewed as "believabilities" (to coin a term that is the counterpart to what

[10] There is a vast amount of writing on moral realism presenting very different positions. To name just three: Some take moral properties to be real in what might be called "Platonist" terms, not intrinsically tied to motivation in practical human agency (Parfit), nor to their routine perceptibility by human subjects in the world around them (Moore, who thought they are the objects of a special moral "intuition"). Others take values to be properties in the world but seek to reduce them to physical properties or nonnormatively characterized causal-dispositional properties, or see them as standing in some dependency relation to them. The moral realist position I am arguing for is neither of these, but rather sees these properties as at once irreducible to nonnormative properties (even unassessable for supervenience relations with nonnormative properties) *and* tied intrinsically to human motivation and practical agency.

I have called "desirabilities"), and so Evans' point is right about beliefs, but since there is no equivalent to facts in the case of desires, my extension of Evans' point, which claims that there are desirabilities we look to, is unwarranted. Therefore, in responding to the Evansian question about whether we desire something, we must and do look inwards into our minds to see if it contains the desires being asked about.

I won't indulge the temptation I have here to say that this just begs the question and denies without argument what I have concluded from my extension of Evans' point, viz., that there is indeed a counterpart to facts in the case of desires, that is, desirabilities. This would only result in each side to the dispute claiming that the other is begging the question. I was supposed to give *an argument* that there are value properties or desirabilities, it will be said, and I have only given *an analogy* with beliefs in extending Evans' point; I have not given an argument. But something more specific can be said by way of argument to break this impasse.

Let's proceed, then, as if this challenge does not beg the question and ask, instead, what follows from its denial of my extension of Evans' point to desires. It would follow from the challenge's conclusion that we do always look into our minds in order to answer questions of the form, "Do you desire . . . ?" That would mean that we always step outside of ourselves and look at ourselves from the outside in, as it were, before we respond. We look at ourselves as objects to be scanned for whether or not we possess the relevant desire. In other words, we take the *detached* perspective *on ourselves*. And if there were no desirabilities, only desires in the self-standing sense, then our desires would *only and always* be available to us as such objects of detached self-scrutiny. Our entire relation to our desires could only be one in which they are given to us or available to us as desir*ed* by us. They would not be given to us or available to us via the desirabilities we perceive since there are no desirabilities to be perceived. We have no other way of being with our desires and experiencing them except by way of detached self-scrutiny of them. It is here that (3) strikes us with its relevance. That claim was: *if one only had or if all one had was a detached or disengaged perspective on oneself (as exemplified in [2]), one would cease to be a practical agent.*

The challenge, therefore, can only be successful by depriving us of our practical agency. But we manifestly do possess practical agency. So, the self-standing view of desires that the challenge assumes can only be true by denying something else that is manifestly true of us – that we are practical agents. That is why I had said that we could not so much as be agents, in the practical sense of agency, if desires were self-standing rather than responses to external callings, responses to the normative demands on us that come from desirabilities, or value properties in the world.

Though the argument, as I have presented it, is complicated, with the complications in place, the conclusion should be presentable in rather obvious terms.

Consider what is the philosophically significant difference between my thinking or saying as an observation of myself, "*x* is desired by me" and my thinking or saying "*x* is desirable."

In the former thought ("*x* is desired by me") the desire itself lacks motivational power for me *qua agent*. Even if the desire were to dispose me to act, that act will be something that *happens* to me. I will not *enact* it. It will be an "*act*" only by courtesy, as it were. It will not be an agentive intervention in the world. I will merely be the *carrier* of the intervention in the world. These are the effects on desire of the desire being only available to me as desir*ed*, as something that is the object of my detached gaze (rather than via my perceiving some desir*ability*). It is deadened or leadened in its *agentive* motivational power by being an object of a detached perspective on myself.

But now consider the latter thought ("*x* is desirable"). When I think that *x* is desirable, my desire, which is a response to that desirability of *x*, is not an object of my observation. Its being given to me, its availability to me, is indirect, it is not given to me as desir*ed*, but *via* my apprehension of a desir*ability*. This is what makes it possible that I have an agentive relationship with my desire, because this indirect way of being given to me allows the desire to have agentive motivational power. The crucial point, then, is that it is *only* when our desires are not directly given to us in our detached perspective on ourselves, but rather are *indirectly* given to us via our direct observation of desirabilities in the world, that our desires have agentive motivational power. And without that motivational power, desires cannot be the basis of our agency, as the challenge claims. So the simpler view, that desires are self-standing and not responses to desirabilities or values, simply cannot make its way to accounting for the agency we manifestly possess.

I admit that the argument I have elaborated in this section for the claim that there are value properties in the world ties value properties to the possession of practical agency, and it will not move a philosopher who is prepared to deny that we do possess such agency. In that sense the chapter's conclusion is modest. The argument has no efficacy against such a philosopher, and I have no argument against someone (an Alamo-style philosopher, prepared to bite all bullets) who denies that we are agents, except, I suppose, just to say, "Come off it." These just are the limitations of philosophy. No argument is efficacious against *all* comers: what analytic philosophers like to call "knock-down" arguments. At best, one can say: If you don't believe what my argument establishes (in this case, the meta-ethical position that the world, including nature, contains visible value properties that fall outside the purview of natural science), see how much else that seems true you have to give up believing (in this case, that we are subjects who possess practical agency).

IV

This chapter, despite the gestural note it strikes in its concluding paragraph, is not the occasion to explore the wider implications of the meta-ethical claim I have tried to establish – implications for politics, political economy, and the vexed subject of the environment, which I hinted at in my introductory remarks when briefly giving the early historical context of this chapter's themes.[11] Those implications are of the utmost significance and need patient working out, and yet the tradition of philosophy within whose idiom I have made the argument for the claim has shown little interest in relating meta-ethics to these wider subjects and issues. So I particularly regret not having the space to do so here. What I try to do instead, as I bring the chapter to a close, is to address some more immediate and much narrower philosophical implications of the claim, and respond to some sources of doubt about the claim.

Perhaps one immediate implication to be drawn is that ethics is, in one sense, primarily a *perceptual* discipline. I use the word "primarily" and mean it. When I say it is perceptual, I don't mean to suggest that *deliberative* and reflective elements are not important in ethics. They certainly are. But their role *nests within* a more basic perceptual understanding in which our moral agency responds to the normative demands of the value properties we perceive (or fail to perceive or misperceive) as we navigate the world we inhabit. It is when someone has different or conflicting perceptions of value that the role of deliberation (of ranking and weighing and assessing rational support or lack of support among values, etc.) comes into play. So also agents from different cultures or backgrounds may apprehend quite different normative demands in the very same perceptible situation, and when this happens, the relevance of

[11] I make an initial stab at drawing some of these implications in the section on "Enchantment" in Bilgrami 2014.

deliberation via cultural exchange similarly comes into play to resolve the conflicts.

What follows from putting perception in primary place on the subject of value, in this way, is that the relevant states of mind (which as we saw are not self-standing) are *at once* our conduits for *apprehending* the world *and* states that *motivate* our agentive responses to what is apprehended. That is to say, it is not as if one sort of state apprehends the value properties in the world via perception of it and another quite self-standing state motivates our actions on the world. Apprehension and motivation are not two radically separated directions from which we relate to the world, apprehension going from the world to us and motivation going from us to the world.[12] Rather, the very fact that it is something like *values* that we are perceiving in the world makes it clear that the perceptions themselves are motivating. It is not as if the perception of the threat in an impending storm and the feelings of vulnerability in the Bangladeshi fisherman, which prompts him – as a practical agent – to seek protection, are two states of mind with two different directions in their relation to the world. To have perceived the threat *is* to have felt vulnerable and vice versa.

This has implications for an entire family of states of mind. Desires, reconfigured in this way as relating in such a bi-directional form to desirabilities and undesirabilities in the world (i.e., relating to a world described and understood in evaluative terms – threats, cruelties, needs), are just one central case of mental states of this kind. Emotions too, very often, are to be conceived in just these terms. And once they are, a common and long-standing misconception about their place in practical human agency stands corrected. These points can do with some elaboration.

Too often emotions are thought of as gumming up the works of deliberation in practical life, and in politics and morals in particular. Practical reason or rationality is frequently described almost entirely in deliberative terms of rational inference, and emotions are seen by contrast as conflicting with and spoiling the deliverances of reason, so conceived (see Elster 1996). Though that no doubt happens sometimes, it is occasional and cannot plausibly be built into characterizing the nature of emotions. If we see emotions along the lines I present later – as of a piece with the conception of desires presented earlier – we can see why.

[12] This is sometimes described by the phrase "different directions of fit." To express the denial of different directions of fit for beliefs and desires, Altham coined the neologism "besires" for states that have both directions of fit at once.

On this reconceptualization, emotions, like desires, are also a mode of perception. How so? Perhaps a good way to begin to convey this is to look at what such a reconceptualization looks like in the case of physical pain. A plausible conception of pain might go this way. Take a toothache. We can perceive our teeth in the standard ways. I can put my forefinger to my tooth and perceive it tactually. I can go to a mirror, unfurl my lip, and perceive my tooth visually. But I can also, more internally, more involuntarily, perceive my tooth by – and here we run out of the right "logical grammar," to use Wittgenstein's term – by paining it. A toothache, thus, is a way of perceiving my tooth, and physical pains generally are internal modes of perception of parts of one's body. Emotions too are modes of perception of this kind, though not of one's body.[13] What, then, are they modes of perception of?

In more than one place, Aristotle writes of anger ("rage" is actually the right translation of his particular example),[14] saying that it relates to belief as follows. If I am angry with a person, that *presupposes* a certain sort of belief, for instance the belief that that person has done me harm. But this does not quite capture the right relation between belief and anger. To show why, I have deliberately emphasized "presupposes." In many cases, that seems the wrong way to think of the relationship between emotions and a belief about the world. The relationship of presupposition here would suggest that the belief is *all in place first* and *then* the anger wells up. But that does not always capture the phenomenology of anger and indeed perhaps it only seldom captures it. Often, my anger is a way of *perceiving* that someone has done me harm. It's not as if the belief is all calmly acquired and gives rise to the anger on reflection (not that this does not sometimes happen). My anger is very often my conduit to, a perception of, the fact that he has done me harm. The perception of something (value-laden) in the world and the emotion are not separable, just as I was suggesting about desire and the perception of the value-laden world.

If this is right, if emotions are ways of perceiving and forming perceptual beliefs about the (value aspects of the) world, then it cannot possibly be right to say that emotions gum up the works of rationality. In fact, far from gumming up the works, they *are* the works. The beliefs that go into rational deliberation are often the deliverances of emotions, conceived in

[13] Spinoza, motivated by his metaphysics, thought of them as being just that. See also Damasio 2004. The view presented here, motivated by an account of value and agency, is quite different.

[14] See *De Anima* 1.1.403a16–32 and *The Rhetoric* 2.2.1378a31. Scholars differ on how to read these passages, but this view is taken to be intuitively plausible by many (Elster, for instance), at least as a first thought about the relationship between emotions and beliefs.

this way as modes of perception. This point is not to be confused with the oft-made point that emotions have a propositional content – what is sometimes called the "cognitive" account of emotions. The point is not that emotions have the same form as beliefs, a propositional form; the point is that they are a path, a perceptual path to belief formation. And, as the previous paragraph makes clear, this conception of emotions cannot even be formulated if one does not view the world as containing value properties. But once one views the world that way, such a conception of emotions is entirely and naturally of a piece with doing so, as is the conception of desires presented earlier.

Desires (and emotions), I have said, are modes of perception of the value properties of the world, and *in being so*, they are *intrinsically* capable of motivating our practical agency. They do the double duty of taking in the world in its aspect of value even as they, thereby, motivate our agency to action. Now, if they are perceptual in this way, they can, of course, sometimes get the world (in its value aspect) wrong. But that is true of all perception. There can be value illusions just like there can be illusions, in general, about the nonvalue aspects of the world. This should not cause either surprise or concern. What does seem to cause some concern is the fact that just as we can have value illusions, we can have, as I admitted earlier, *differential* perception of value properties. The very fact of there being differential or conflicting perception of value properties in the world prompts the doubt that there really *are* such properties in the world. But that doubt is based on a non sequitur. For one thing, there is frequently differential perception of other properties in the world, the physical properties that natural sciences study – for instance, when we have internalized different physical theories about one or another physical phenomenon in the world. This is just a familiar consequence of what is often described as "the theory-ladenness of observation." But no one, no one sensible anyway, is tempted to conclude from this that there are no physical properties in the world. In general, it does not follow from the fact that there is some property in the world that there cannot be differential perception of it; equally it does not follow that if there is differential perception of the property that these are not really perceptions of a real property but rather, as has been suggested by some Humeans in the case of values, a subjective derivation of them from our states of minds such as desires and moral sentiments that is then illicitly "projected" onto the world.

A related tendency that is also based on a confusion is to think that because value properties in the world, by their very nature, are *intrinsically* related to the fact that those who are capable of perceiving them *as what they are* are responsive to their normative demands with practical agency, then it must be that these are (unlike other perceptible properties in the world) not real properties after all. There is no plausible inference from the fact that we understand something (value) in the world as being related intrinsically to our capacity for agentive responsiveness to it, to the conclusion that we must somehow be *making up* values all on our own and projecting them illicitly onto the world. For one thing, it is a familiar thought since Locke and Boyle at least, if not since Galileo, that color properties in the world are partly characterized in terms of *our* visual sensibilities (frogs, for instance, do not perceive color properties). But it is quite wrong to conclude from this (not that it has not been done) that the table on which my keyboard presently sits is not brown, nor any other color. I repeat that it is a non sequitur to go from the idea that some property that is perceptible may require a certain sort of subject (one with our sort of practical agency in the case of value, or one with our sort of visual sensibility in the case of color[15]) to the idea that that property does not exist in the world at all, that the subject somehow generates it from his own mentality and projects it onto the world.

I conclude with one final caveat about the nature of these visible value properties that the world (including nature) contains. In elaborating the link between them and our practical agency, I have said that our perception of these properties takes the form of perception of the *normative demands* they make on us and to which our practical agency responds. Now, the expression "normative demand" is, of course, *a metaphor*. It is *not literally* a normative demand made by the perceptible features of the world. In insisting on this point, I am declaring that, in subscribing to the view that there are value properties in the world (including nature), I am not subscribing to any sort of intentional vitalism that attributes intentionality to nature and the world. And by saying this, I am disavowing any commitment to the sort of position taken by Bruno Latour and Jane Bennett and others who think there is some quite literal form of "actants" (I assume that by this term they have made current, they mean *agents* of a kind broader than human agents) that populate nature and the world we inhabit, agents who literally address us with normative demands. My claim

[15] The analogy between color and value (first formulated by McDowell [1985]), is imperfect. There are disanalogies too, since color may quite properly be thought to be supervenient on properties studied by fundamental physics, whereas it is not at all obvious, as I said earlier, that it even makes sense to think that values are similarly dependent.

that there are value properties in the world is a much more innocuous claim than this. I am only saying that there are value properties in the world (including nature), and I am happily admitting that the idea that they make normative demands on us is a metaphor. No doubt this will seem like a copout to those who think we need a bolder commitment to vitalism (see Latour 2004a; Bennett 2010; 2018).

But it is nothing of the sort. In fact, a proper, by which I mean sober, understanding of the nature of value shows such a vitalism to be quite unnecessary for the important issues at stake. Why do I rest with the thought that the expression "normative demands" is a metaphor? Why am I not moved by the doubt that the force of a normative demand is lost if the expression "normative demand" in these uses of it (unlike when we use it to say that I, a human subject, make a normative demand of you) is not literal but metaphorical? Surprising as this may sound, the answer is that I am not moved by this doubt because of the *nature of metaphor*.

It is widely (and surely rightly) said of metaphors that they, at least the good and apt ones, are not paraphraseable away into literal statements. This, as just stated, is, and is intended to be, a claim about the nature of (metaphorical) language. But it cannot *just* be a thesis about language. Though true as a thesis about language, its significance would be limited if it were just a claim in linguistics (pragmatics) or the philosophy of language. As a thesis, its full significance only comes into view if we notice that its truth has a counterpart in a metaphysical thesis, an *extra*-linguistic claim. What is that counterpart metaphysical claim to match the claim of unparaphraseability as a claim in the philosophy of language? It is this: If a metaphor is not paraphraseable away into literal statements or propositions, what that very thesis shows is that there is a fragment or aspect of *reality* that cannot be captured by any expression but *that* metaphor. It is striking that philosophers who have made the linguistic claim don't make this metaphysical counterpart claim explicit.[16] But once made explicit, it becomes clear that there is no loss of force in saying that value properties make normative demands on us, just because the expression "normative demands" is said metaphorically. Without any commitment to intentional vitalism (a commitment that would only be generated if we insisted on some sort of *literal* deployment of the expression "normative demands"), one has said what needs to be said; one has (with a metaphor) captured something real, a fragment of reality and its unique and intrinsic relation to

[16] For instance, Davidson (1978), who makes the claim more vividly than many others, fails to draw the metaphysical significance of it.

our practical engagement. A scientism that has long denied this presence of value in the perceptible world we inhabit is thereby laid to rest without any overreaching into an implausible and unnecessary vitalism.

Latour, however, has insisted on an important point: There had better be a *politics of things* if we are to emerge with some sanity, indeed with some humanity (paradoxical as that may sound), from our destructively human-centered conception of politics (and political economy). Though I have said nothing here about what such a politics would look like, I have tried to do the philosophical ground clearing for it by providing a meta-ethical foundation for a politics of things; and have done so without any implausible commitment to the idea that things, like human subjects, possess intentionality.

This chapter has been about how a scientistic claim, via an illicit extrapolation, swept away value properties from the world (including nature) with the same brush that it swept away sacral properties from it. The claim literally renders these properties invisible. I end by noting that if the chapter's argument carries conviction, we are at least *poised* to pursue a point (on some other occasion) that has implications for politics. Usually, when one speaks of invisibility, the interest is to alert us to the fact that things are below the surface of visibility and need to be unearthed. Sometimes however – as in the theme of this chapter – things that *are* on the surface and plainly visible to us are denied that visibility due to one or other distortion of our ways of "being-in-the-world," which philosophers first perpetrate by overextending the authority of science, but which then, through the exercise of more worldly forces (the worldly forces I cited earlier were the alliances formed between the Royal Society, high Anglicanism, and commercial interests) gets dispersed into the zeitgeist as a pervasive assumption of our time. That assumption might properly be called a superstition of modernity (which is exactly what scientism is), and this chapter has tried to present the philosophy needed (an argument) to exorcise it. But a philosophical exorcism of the particular superstition that I have focused on in this chapter, even if successful, would be an arcane achievement if one did not also see through to the details of a democratic conception of politics that included "a politics of nature," a politics on which we do not have even a preliminary grip, leave alone a sense of its details. But this should not surprise us, considering the long centuries it took for human beings to develop (and to *this day* it has not been fully developed) the details of a democratic politics that included every human being.

PART II

Inspirations

Philosophies of Science and Its Social Role

CHAPTER 5

Scientific Humanisms

Sarton, Reichenbach, and the Crisis of Western Science after World War I

Alan Richardson

> Nature is blind, and faith in its benevolence and meaning is unfounded. But it is approached ever more closely through unceasing effort, surrounded and caught in the mesh of the finest conceptual net – in the sure consciousness that a knowledge of nature is one of the highest goods man is capable of attaining.
>
> Hans Reichenbach, "The World View of the Exact Sciences"

One feature of the historiography both of and by those trained in analytic philosophy, the main tradition in Anglo-American academic philosophy in the second half of the twentieth century, is the degree to which it resists using certain framing terms that are otherwise quite ubiquitous in intellectual history. It would be hard to imagine a history of twentieth-century literature, for example, that did not use – even if it ultimately rejected – the term "modernism." That term, despite being fairly straightforwardly applicable to certain aspects of the nascent project of analytic philosophy in the 1920s and 1930s, is only very rarely deployed in writing about the philosophy of the twentieth century, however.

There are reasons for this absence, of course. Two that deserve mention are, first, that historians of analytic philosophy tend to use their actors' categories as their analytic categories. Thus, if Bertrand Russell calls himself at some time slice a "logical atomist," historians of analytic philosophy will write about his philosophy at that stage as "logical atomism." Even their higher-order framing notions – such as "analytic philosophy" itself – tend to be used only because the historical actors (eventually) used them also. A second reason for this category modesty reflects the degree to which the historical writing in history of analytic philosophy is informed by the sensibilities of analytic philosophy itself. To use "modernism" as a term of art to discuss specific types of twentieth-century philosophy would be to understand "modernism" itself as a term that makes good philosophical sense. But to someone of analytic sensibilities, this is a dubious proposition at best – for what sort of philosophical sense could be made of a term that does not seem metaphysical, epistemological, logical, methodological, semantic, ethical, or even clearly aesthetic?

Compare, on these scores, a term that is in common usage in the history of analytic philosophy: "naturalism." "Naturalism" is a term that is important to many of the historical actors in that history – one cannot understand, for example, W. V. Quine's or Ernest Nagel's philosophy without understanding that they meant to endorse naturalism. Moreover, while "naturalism" as it is understood in analytic philosophy is not obviously a phrase that is limited to any one of the subdisciplines of philosophy, it can be and routinely is relativized to those subdisciplines. That is, when we are asked to get clear on the commitments of naturalism, we carve it into ontological naturalism, ethical naturalism, epistemological naturalism, methodological naturalism, etc. While these clarifications might not seem terribly promising as slogans – "ontological naturalism assumes only natural and no supernatural objects" seems analytic or circular – one can hope that a more serious consideration of these elements of naturalism will yield a set of substantive commitments. On this score, "modernism" seems less promising. What would ontological or methodological modernism even be? One scarcely knows where to begin.

My concern in this chapter is not modernism but humanism. "Humanism" occupies a middle ground in the historiography of analytic philosophy. Some historical analytic philosophers have used the term – almost always as a term of commendation. But it has not occupied much historical attention. I think this is unfortunate, because, it could be argued, much of the reception of philosophy, analytic and not, in the twentieth century (and now into the twenty-first) is bound up with concerns perhaps best expressed in terms of humanism. I gesture at that argument, but it is too large for one short chapter. My attention here is more circumscribed. Here I want merely to sketch a few episodes in the history of something I call "scientific humanism" (what I call by that term importantly does not accord with the views of some of my historical actors) and say something about why those episodes mattered to those who participated in them and why they ought to matter to us.

Varieties of Scientific Humanism

Among historians of philosophy of science, one statement of the core commitments of a view its author calls "scientific humanism" will be especially familiar, perhaps. In his "Intellectual Autobiography," Rudolph Carnap says this about some of the core commitments of the famous group of logicians, philosophers, and scientists he belonged to in the 1920s and 1930s – the Vienna Circle:

> I think that nearly all of us shared the following three views as a matter of course which hardly needed any discussion. The first is the view that man has no supernatural protectors or enemies and that therefore whatever can be done to improve life is the task of man himself. Second, we had the conviction that mankind is able to change the conditions of life in such a way that many of the sufferings of today may be avoided and that the external and the internal situation of life for the individual, the community, and finally for humanity will be essentially improved. The third is the view that all deliberate action presupposes knowledge of the world, that the scientific method is the best method of acquiring knowledge and that therefore science must be regarded as one of the most valuable instruments for the improvement of life. In Vienna we had no names for these views; if we look for a brief designation in American terminology for the combination of these three convictions, the best would seem to be "scientific humanism." (Carnap 1963a: 83)

The three views that Carnap outlines are clear enough and we need not belabor them. Moreover, it would not be appropriate to argue with Carnap about the usage of words, and thus I grant to him that the views he presents can be called "scientific humanism." When I enunciate later a somewhat different constellation of views and wish to call them by the same term, we will thus differentiate them from what I call "Carnapian scientific humanism."

To find an alternative version of scientific humanism to this one, let us draw out briefly the core commitment of Carnap's views and the place of both science and humanity within them. There are no higher powers that can solve human problems. Thus, the solution to human problems is our own task. The solution to human problems depends on achieving reliable knowledge. Science is a source of reliable knowledge. Thus, science can be used as a tool – an "instrument" – in the solution to human problems and the improvement of the human estate. This set of views is ultimately one form of a pragmatic justification for science – science is humanly valuable because it can help us reliably solve our problems.

The details of this view can be filled in in a variety of ways. Leaving aside the "no supernatural protectors" part of the view (Carnap's own version of this rejection is that supposed claims about supernatural protectors are meaningless claims and have no place in knowledge or rational decision making), we are still owed something in the way of what "improvement in the conditions of life" of some or all humans amounts to, and an account of the role of scientific knowledge in rational decision making. Carnap himself has a long answer to the latter issue, embedded in his project of the logical foundations of probability (Carnap 1950b), but we shall not concern ourselves with that account here. The former question gets us closer to our ultimate concern: There are at least two dimensions along which answers to the question of the articulation of problems and solutions, as problems and solutions, are relevant. First, there is the question of how to articulate and individuate problems. In particular, how far removed from scientific knowledge-making will the language of the articulation of problems be? Second, what normative language is necessary for the articulation of problems as problems, and of solutions as improvements in the human estate?

On the last issue, Carnap will part company from many American pragmatists who might otherwise be sympathetic to his scientific humanism. For unlike the pragmatists (see Chapters 6 and 8), Carnap will not be able to rely on either a philosophical or a scientific account of human nature and flourishing to point to an objective problem in need of an objective solution. Instead, Carnap thought all valuational claims express subjective values and desires and this expressivism in ethical judgment points to a different concern – the problems that occur in the frustration of the pursuit of subjective value. How exactly subjective values are aggregated into the social problems demanding policy solutions is a difficult, though perhaps not intractable, problem. It too is, fortunately, beyond our concern here.

The fact remains that the view fits firmly into a pragmatic justificatory stance: Science is good as a tool for the solution to problems. And it is in opposition to this pragmatic stance that I wish to articulate a different view that could also with right be called "scientific humanism." My alternative owes more to classical rationalism than to pragmatism. Its vision of scientific humanism is (with details to be filled in in various ways) this: Science is not humanistically valuable merely as an instrument for the solution of human problems; it is also valuable in itself as an intellectual pursuit of humans. Science is an expression, indeed perhaps the most central expression, of humanity; it fulfills our rational demand for the acquisition of knowledge, both of the world and of ourselves as knowers of the world. This might sound like an embarrassingly moralistic vision to those of a pragmatic bent, but humanism in most of its forms surely has a moralizing or edifying point.

It is perhaps unnecessary to give historical examples of people who have held versions of this view. It is a fairly common view – especially among scientists – even to today. But I would like to give a pair of historical examples to show its historical importance in the modern era, and to show that it can be developed in importantly different directions. Within the seventeenth-century context the person with whom I most associate this view is Baruch Spinoza. Spinoza's *Ethics* (2018/1677) is a long argument for exactly this view: The expression of human conatus (human essence in its active striving) is found in the acquisition of articulated causal knowledge of the world. Those who pursue the life of acquisition of such knowledge live from reason as free persons (as much as is humanly possible) and come closest in the mortal realm to the infinite and eternal intellectual love of God. This high rationalist variety of scientific humanism continues right through the Enlightenment of course, finding perhaps fullest expression in the Marquis de Condorcet and his view of the infinite perfectibility of humanity (Condorcet 1955/1795). In Condorcet, too, scientific knowledge is not merely a tool to be used in this perfectibility; the pursuit of knowledge is itself a crucial dimension of the increasing perfection of humanity.

In the German Enlightenment, the view also took on a more subtle tone and perhaps darker hue in the work of Immanuel Kant. For Kant, the pursuit of knowledge is an intrinsic project of reason, an expression of the highest of the human faculties. It is also a task that cannot be completed and any attempt to complete it depends on an unwarranted and faulty attempt to use reason beyond the bounds of experience. Moreover, in and of itself the pursuit of scientific knowledge is not quite an expression of moral improvement. However, the critique of reason that his transcendental philosophy provides reveals how scientific knowledge is possible and at the same time also reveals that we can think of ourselves as free. Thus, for Kant, the coming to self-consciousness of the epistemic subject reveals how we can and must embrace the moral vocation of reason. Similarly, in lieu of Condorcet's vision of the infinite perfectibility of humanity, Kant can only offer the demand of reason in its practical employment to attempt as much as possible to bring the conditions of the Kingdom of Ends (where everyone can act from moral duty) into existence on Earth – again, a task that cannot be completed but is binding on humanity.

Of course, we need not mine the early modern period for elite examples of the view. Cruder versions of the view have been a part of most public defenses of science by scientists and other public intellectuals for the past sixty years or more. It is not too much of a distortion to say that C. P. Snow's (in)famous remarks on "the two cultures" – scientific and literary – offered a form of scientific humanism. His aim was precisely to claim that "literary intellectuals" had missed half of culture, with a strong suggestion that it was the better half that they had missed. This is not scientific humanism at its subtlest or most edifying, but it is proud and it is unrelenting. Whatever philistinism scientists might perpetuate upon humanities and the arts, the reverse was much, much worse: Literary intellectuals were, by and large, scientifically illiterate to the point of being prehistoric. Having noted that literature professors could not explain the second law of thermodynamics, Snow raised the stakes:

> I now believe that if I had asked an even simpler question – such as, What do you mean by mass, or acceleration, which is the scientific equivalent of saying, Can you read? – not more than one in ten of the highly educated would have felt that I was speaking the same language. So the great edifice of modern physics goes up, and the majority of the cleverest people in the western world have about as much insight into it as their neolithic ancestors would have had. (1959: 16)

One can find echoes of Snow's contempt for literature professors all the way up to the present day in the work of public intellectuals who are scientists or who fancy that science needs them to defend it from the relativists, the irrationalists, the feminists, and the social justice warriors.

A more generous form of scientific humanism animated the work of Snow's contemporary, Jacob Bronowski. Bronowski's 1973 television series and book, *The Ascent of Man*, was not simply about what science could tell us about the ascent of humanity from the veldt to modern times; the pursuit of scientific knowledge was itself the greatest form of ascent humans had achieved. Bronowski was, he told us, interested in telling us a story of the human and the personal in the realm of scientific ideas, but he wanted also to do more than that. For Bronowski, the form of rational knowledge that was found in science was the vocation of humanity, its highest expression and its conscious embrace of human nature. The series ends on precisely this point:

> And I am infinitely saddened to find myself surrounded in the west by a terrible loss of nerve, a retreat from knowledge into – into what? Into Zen Buddhism; into falsely profound questions about, Are we not really just animals at bottom; into extra-sensory perception and mystery. They do not lie along the line of what we are now able to know if we devote ourselves to it: an understanding of man himself. We are nature's unique experiment to make rational intelligence prove itself sounder than the reflex. Knowledge is our destiny. Self-knowledge, at last bringing together the experience of the arts, and the explanations of science, waits ahead of us. (1973: 268)

That final gesture of synthesis and reconciliation should not confuse us. If knowledge is our destiny then the explanations of science, by virtue of being in the realm of knowledge, must ultimately weigh more and be more expressive of humanity than are the experiences of the arts.

World War I and the Crisis of the European Sciences

I mention Bronowski because his work brings me close to the topic of the rest of this chapter. Bronowski's ascent of man was largely a history of science, but he tells us in the introduction to the book that he intended it as a "philosophy" (Bronowski 1973: 11) – a philosophy of nature, including human nature. Bronowski was a mathematician, a biologist, an author of literature, a polymath, but he was not a professional historian or philosopher of science. Indeed, by 1973 he might have been quite suspicious of what professional history and philosophy of science had become – a highly professionalized endeavor with little interest in edification and arguing over accounts of the development of science that were not likely to support his own vision. But I want to argue that professionalized history and philosophy of science as it developed after World War I was in fact deeply influenced by the form of scientific humanism that I have enunciated, more so than by Carnap's form of it. I illustrate these claims with two main cases. The great founder of professional history of science, George Sarton, argued explicitly for a robust form of scientific humanism that depended for its expression precisely on rigorous and encyclopedic history of science. My second case is more subtle. I argue that some forms of logical empiricist philosophy of science, a logic-based form of philosophy of science associated with the Vienna Circle and colleagues in various other parts of the world, were more attached to my form of scientific humanism than to Carnap's. I argue the case for Hans Reichenbach here.

Scientific humanism in my formulation (henceforth, "scientific humanism") was not a dead letter in the 1920s when Carnap met with the Vienna Circle and found there a shared "Carnapian scientific humanism." It was, however, under considerable pressure and, if ever it could be taken for granted, it certainly had lost any claim to self-evidence by November 1918. The reason for this is simply stated: World War I was for the countries of the West an unprecedented human and cultural disaster; yet it had been fought among the most scientifically advanced of countries, using the most technically advanced equipment and weapons. For the confident expression of the inevitable improvement of humanity through scientific advancement, World War I was a fundamental crisis.

There were myriad ways to reject scientific humanism after World War I. Among the more famous and least subtle ways was Oswald Spengler's declinist project (Spengler 1991/1922/1918). Armed with an allegedly objective view of the morphology of history, Spengler argued that the West was in inevitable decline. From this it follows that the forms of scientific knowledge enshrined in the West were no bulwark against decline. Various forms of *Lebensphilosophie* (philosophy of life) and philosophy of existence also arose after the war and were by and large forms of anti-scientific humanism and sometimes forms of anti-scientific anti-humanism. The threat that was so keenly felt by the intellectual classes – and perhaps most strongly in the defeated countries of the Germanophone world – was a form of nihilism arising from the destruction of their worlds by the very tools they had thought had made their civilization the greatest in world history. The problem is expressed by the narrator of Erich Maria Remarque's wrenching war novel, *All Quiet on the Western Front*:

> I am young, I am twenty years old; yet I know nothing of life but despair, death, fear, and fatuous superficiality cast over an abyss of sorrow. I see how peoples are set against one another, and in silence, unknowingly, foolishly, obediently, innocently slay one another. I see that the keenest brains of the world invent weapons and words to make it yet more refined and enduring. And all men of my age, here and over there, throughout the whole world see these things; all my generation is experiencing these things with me. What would our fathers do if we suddenly stood up and came before them and proffered our account? What do they expect of us if a time ever comes when the war is over? Through the years our business has been killing; –it was our first calling in life. Our knowledge of life is limited to death. What will happen afterwards? And what shall come out of us? (1982/1928: 263)

This nihilism resonated with readers of the novel who had served in the war. Here is how Carl Zuckmayer (1994/1929: 23) expressed the point in his review of the book. (This passage begins with the more literal translation of the title: "In the West, Nothing New.")

> Nothing new. Except for a few hundred thousand people the world was collapsing, along with everything that until then had fulfilled and enlivened them; except that they did not know whether it was now the void, the end, a complete dissolution that would swallow them up – or the whirlpool and obscurity of a new creation. Yes, that they did not even ask, nor had any idea whether they were the plow or the earth, the axe or the wood, seed grain or a rotting carcass.

Within this context, German and Austrian professors saw a form of anti-science taking hold in their students during and after the war. Where a life dedicated to the pursuit of knowledge had once seemed like a noble and responsible goal of the student, professors now saw their students seeking something else, often enough expressed in language such as "the search for authentic experience and meaning."

Many who saw this threat took the opportunity to endorse scientific humanism, but often of a new and more subtle kind than the classic versions scouted earlier. One very interesting case is Max Weber in his famous lecture, "Science as Vocation." In the lecture, Weber is withering on the striving of young people in 1917 for "experience" and "personality." He ascribes this attitude at least in part to an anti-scientific point of view. He gives a reading of Plato's Cave allegory in which "the sun is the truth of science, which alone does not snatch at illusions and shadows but seeks only true being." Having raised this vision of Plato, Weber asks:

> Well, who regards science in this light today? Nowadays, the general feeling, particularly among young people, is the opposite if anything. The ideas of science appear to be an otherworldly realm of artificial abstractions that strive to capture the blood and sap of real life in their scrawny hands without ever managing to do so. Here in life, in what Plato calls the shadow theatre of the cave, we feel the pulse of authentic reality; in science we have derivative, lifeless will-o'-the-wisps and nothing else. (Weber 2004/1917: 14)

Weber rejects this view, not in order to go back to Plato, but rather to reject the mutual presupposition of both positions: that there is some ultimate meaning to life or the world, whether it is to be found in the pursuit of scientific knowledge or in one's own most inner and authentic experience. In a figure of speech that recalls Kant's claim that enlightenment is humanity's release from self-imposed nonage, Weber ascribes the belief in such meanings to childishness:

> Apart from the overgrown children who can still be found in the natural sciences, who imagines nowadays that a knowledge of astronomy or physics or chemistry could teach us anything about the meaning of the world? How might we even begin to track down such a "meaning," if indeed it exists? If anything at all, the natural sciences are more likely to ensure that the belief that the world has a "meaning" will wither at the root! (2004/1917: 16)

Where does Weber leave us in light of this? Essentially with the view that the value in science cannot be found in existing meaningful things that are studied in science or any other realm of knowledge – the world, God, happiness, being, what have you. It can only be found internally as a presupposition of the practice of science. That the truths of chemistry are worth knowing is a presupposition of chemistry but not one it can defend against attacks from those who do not similarly presuppose it. Whether we accept or reject this presupposition is, he says, an expression of "our ultimate attitude toward life" (2004/1917: 18).

This is, I would argue, a form of scientific humanism, but a form quite different from Spinoza's or Kant's or indeed Bronowski's. It does not answer – indeed rejects as meaningless the demand for an answer to – the question of the ultimate meaning of human life or of scientific work. Nevertheless, the scientific vocation bears within itself an answer sufficient for itself of whether it is worthwhile to pursue science. Weber speaks as a scientist about the virtues of science. If you are unmoved, then you lack the vocation for science. There is no triumphal "ascent of man" here – no insistence that the ultimate telos of humanity is self-knowledge through understanding what science reveals about the extra-human world. Weber is even more minor key than Kant – science isn't even a vocation of reason anymore in the Kantian sense of a vocation of any finite rational being.

There is, however, a social role for the scientific vocation that is more than merely self-certifying and that genuinely counts as an ethical good embedded within value-free science. For Weber, the vocation of science includes the demand of teaching, and it is in their role as teachers that scientists play a social role that cannot be played by any other profession. Only the disinterested and dispassionate teacher can confront their students with the proper set of "inconvenient facts" (Weber 2004/1917: 22) – inconvenient, that is, to the moral and political interests and values of the students. Weber concludes: "I believe that when the university teacher makes his listeners accustom themselves to such facts, his achievement is more than merely intellectual. I would be immodest enough to describe it as an "ethical achievement," although this may be too emotive a term for something that is so self-evident" (Weber 2004/1917: 22).

So, on Weber's view, there is a value to doing science as a scientist that you cannot expect to convince those without the vocation for science to agree with. There is also a social value in the teaching of science in bringing inconvenient facts before all members of the society. Because of the universalism of science (both in its topics and its pursuit), some projects associated with other humanisms Weber sets aside. For example, in elaborating his view, Weber at one point eschews using his perspective to evaluate human cultural achievement, saying (Weber 2004/1917: 23): "I do not know how you would go about deciding 'scientifically' between the value of French and German culture." If we are briefly scouting alternative humanisms available after World War I, it is useful perhaps to note that there were various forms of "new humanism" being developed that had no trouble with that question. Indeed, there were some postwar new humanisms that claimed to be aspects of German culture. A representative case is found in Paul Hensel's 1921 essay, “The New Humanism.” He begins the essay by comparing the new humanism of the twentieth century with Renaissance humanism that sought to celebrate the works of pagan antiquity, saying that there were of course similarities but that:

> The difference is more important, and it is this that is our concern above all things here. It is striking that new humanism is not a general European presence, but rather predominately a German concern and thus it cannot be as thoroughly carried through outside Germany as that has been in the land of its origin, no matter how great its effects there later became. (1930/1921: 272, translations from Hensel are my own)

Hensel attempts to explain the astonishing fact that new humanism is predominantly a German affair by in part claiming that there was a greater degree of scholarly concern among the Germans than among the French from the eighteenth century onward on the works and culture of Greek antiquity. Concerning themselves overly with the Romans, by the mid-eighteenth century the French had had enough and issued the call “back to nature,” whereas the Germans were truer to the humanist calling and instead said “back to the Greeks”! Hensel’s view seems to be that only the Germans carried forward the true spirit of Renaissance humanism whereas the French changed that project into a form of naturalism.

For Hensel, ultimately, given the history he provides, the crucial figure for new humanism is Goethe, who somehow was able to combine in one person the figures of the philosopher, the poet, and the scientist. For this reason, Hensel (1930/1921: 277) ends his remarks by saying that: “anyone’s stance toward new humanism depends on what Goethe means to them. And, thus, new humanism is above all a German concern.”

Hensel was not a major philosophical figure in 1921 and is not one now. I briefly raise his remarks here to indicate that it certainly was not the case that “humanism” in Europe after the war was inevitably going to be any form of scientific humanism or any form of political project that would aim for a unity of humanity or universal human flourishing. I mention Hensel for another, more extrinsic reason: He was in 1915 one of two directors – with the mathematician Max Noether – of Hans Reichenbach’s dissertation, and Reichenbach’s views do interest us here.

Scientific Humanism after World War I in History and Philosophy of Science

The early twentieth century was a time for the consolidation of history of science and philosophy of science as distinct professional practices. The single most important figure in the professionalization of history of science in this time period was George Sarton. As we shall see, his vision of the intellectual import of history of science was an extremely robust form of scientific humanism. Within the newly professionalizing vision of philosophy of science, scientific humanism was neither as explicit nor as robust, but I argue that in at least one version of logical empiricism – that of Hans Reichenbach – it can be uncovered.

Sarton began presenting his vision for history of science as "the new humanism" in 1918 in a paper published in French. He presented substantially the same case in 1922 at a talk before the American Association for the Advancement of Science in Boston and on the pages of his new history of science journal, *Isis*, in 1924. In the 1924 paper he presents the program of New Humanism in three principles. First, "human progress is essentially a function of the advance of positive knowledge" (Sarton 1924: 9). Second, "the progress of each branch of knowledge is a function of the progress of the other branches" (Sarton 1924: 10). He dubs this principle "the unity of knowledge." Third, "the progress of science is not due to the isolated efforts of a single people but to the combined efforts of all peoples" (Sarton 1924: 11). This principle he calls "the unity of mankind." The reason why this becomes a project in the history of science, in the first instance, is that because they are theses about the progress of science, the principles of the New Humanism can only be demonstrated historically. He pursued this project through his entire, unbelievably productive career. For example, the "unity of mankind" thesis is illustrated and motivated in dozens of essays he wrote about texts and artifacts from around the world and across historical epochs. His interests spanned material from Babylon, Egypt, India, China, the Islamic world, and much else. This was not mere eclecticism but a concerted effort to detail and appreciate some of the contributions to knowledge of all people and all civilizations.

Sarton is very clear that the New Humanism is most importantly trained on the progress of positive knowledge, of science, precisely because a science has a uniquely progressive historical development. Within the history of human endeavor, he tells us:

> The history of science would be, of course, the central history, for it would be, among these three [the other two being the history of religion and of art], the only one evidencing a continual accumulation and improvement. In spite of a few momentary regressions the history of science is, indeed, essentially a tale of progress, of conquest; the progress is slow but sure, the conquests are inalienable; man cannot tell another tale of such greatness. It is unique. This is especially obvious if the history of thought is truly encyclopedic and oecumenic, for peoples or races may degenerate or

> disappear, or some branches of science may be temporarily neglected, but if one takes a broad view of the whole tree of knowledge, deriving its substance from the whole world, the growth may be sometimes irregular, it is never interrupted. (1924: 31)

Sarton argues that this vision of the history of science allows a new philosophy – New Humanism itself – to be drawn from the study of history. This work goes forward in the spirit of more traditional humanisms but with a different and more synoptic vision:

> The New Humanism is a revival of the knowledge patiently elaborated and accumulated for many centuries by men of science, but neglected and despised by men of letters and educators, – its integration with the rest of our culture; its main spring is the history of science. It undertakes to bring together for the first time, scientists, historians, philosophers, sociologists; to coordinate and harmonize their points of view; to broaden their horizon without lessening the accuracy of their thought; to make the accomplishment of their higher task easier in spite of the increasing wealth of knowledge. (1924: 32)

While this may seem a form of triumphalist scientific humanism that my claim about the crisis induced by World War I should rule out, upon closer examination one sees within the project a clear acknowledgment of the crisis both intellectually and materially and a deeper way in which the project requires an encyclopedic vision of the history of science. Sarton argues that the unities of knowledge and humanity embedded in the New Humanism are real but hidden. They are under threat, especially in the early postwar years, from a more superficial but more self-evident social and political disunity:

> This enables us finally to solve another paradox: how can one reconcile the unity of mankind, which I postulated, with a chronic state of distrust, of discord and war, alas! but too obvious? Quite simply: the unity is hidden but deep-seated; the disunity, widespread but superficial. The unity is felt and expressed primarily by the few men of all nations whose aims are not selfish, or provincial, nationalistic, racial or sectarian in any other way, but largely human, the very few men upon whom has devolved the fulfillment of mankind's purpose; the disunity, the antagonism of interests, is felt and expressed by an overwhelming majority of other men. (1924: 15)

It is not too much to say that one educational goal of the history of science is to increase substantially the number of people who appreciate the deeper unity of humanity and, thus, to work against distrust, discord, and war.

This can be seen practically in the curious way that the 1924 essay ends – with two surprising appendices. The first appendix, spanning several pages, is "an urgent appeal to American scholars" (Sarton 1924: 35) for subscriptions to *Isis*, which had been founded in 1913 but was forced to have a hiatus from publication from 1914 to 1919. The appeal is directed at American scholars because of the ruination of the European economy by the war. Sarton was at pains to argue that the important work of the journal and the New Humanism itself depended on subscription money. If the direct appeal for money in order to advance the New Humanism was not persuasive, Sarton provided a second appendix, this one listing the names of authors and patrons of *Isis* in its first five years of production (one before and four after the war). The list includes over ninety persons distributed in over a dozen countries, including India and Japan. There are many luminaries in the world of science and history of science, including Svante Arrhenius, Pierre Boutroux, Émile Durkheim, P. E. B. Jourdain, Jacques Loeb, Hélène Metzger, Wilhelm Ostwald, Henri Poincaré, and Abel Rey. The point was no doubt to impress upon the reader that the vision for the New Humanism was not idiosyncratic to Sarton, and that a major international intellectual undertaking was indeed under way.

There is a more direct way in which Sarton saw his New Humanism as being able to begin to heal the wounds of previous wars and dissuade nations from entering new ones. He argued that the arguments nations typically gave for wars were self-interested and fraudulent – an attempt to raise narrow political or commercial interests to the standard of truth. By fostering scientific ways of thinking and standards of truth, the New Humanism would induce a higher honesty:

> Even as no honest man would care to obtain advantages for himself or his family by misrepresentation, no honest country would attempt to magnify itself by force or fraud at the expense of others. If the truth standard of politicians and diplomatists was the scientific standard instead of the legal or commercial, our international ideal would be accomplished without any difficulty. (1924: 27)

The result would be what Sarton calls "true internationalism" (1924: 26).

Some of the themes we have just scouted in Sarton are reminiscent of some of the larger elements of the logical empiricist project. Most well known, perhaps, is their insistence on the unity of science, which seems a close cousin to Sarton's unity of positive knowledge. There is more than a hint of the "unity of mankind" in the unity of science project, also, which very often emphasized its internationalism. We have already seen, however, Carnap expressing a quite different view from Sarton's regarding scientific humanism. So, our question here is, was there in logical empiricism a form of the more robust humanism we have found in common among Spinoza, Sarton, and Bronowski, and more subtly and problematically also in Kant and Weber? In accordance with practice in current literature on logical empiricism, I refrain from talking generally about the project and argue that there is such a theme in a single, more specific exemplar.

Hans Reichenbach was five years younger than Sarton and had served in the war, despite his commitment to pacificist principles, in the German army wireless telegraphy unit. In the between-war period his more popular work (which was extensive) repeatedly pointed to the importance of science and philosophy of science in solving a specific problem of the early twentieth century: he saw his culture riven by a deep divide between the everyday world of life and the modern world of science. He began his 1930 essay, "The Philosophical Significance of Modern Physics," on just this point: (Reichenbach 1978/1930b: 304): "Alienation between the world of science and the world of everyday life has emerged in our time with a force never known before." The principal guilt for this unsatisfactory state of affairs he lays at the feet of the academic philosophy of his day, which he sees as enforcing this split, assigning to Einsteinian or quantum physics its realm but claiming they amount to conceptual fictions from the point of view of everyday life. The split was troubling, not merely because it amounted to an attempt to live "a double life" (1978/1930b: 304), but also because it prevented the proper lessons of contemporary science from informing the life of the present. This lends a specific cultural significance in 1930 to the philosophy of science:

> Here, then, lies the source of that unfortunate rift, and, with all the diligence in the world, the scientifically untutored will not be able to bridge it unless philosophy, on its part, shows the way to unification. Thus we view the work of present-day philosophy of science not only from the standpoint of its scholarly significance, as a clarification of basic scientific concepts, but also from the standpoint of society. Seen from this vantage-point, a clarification of basic concepts is at the same time a reinterpretation of outmoded philosophical ideas, and only the disclosure of the continuity between the workaday world and the scientific world will be capable of carrying out that incorporation of the cultural fruits of science. (1978/1930: 305)

Now, this problem – the integration of everyday and scientific world views – might seem to be chiefly due to the progress of esoteric science; who would expect the average person to understand relativity theory or quantum mechanics? This requires, on Reichenbach's behalf, two responses. More superficially, Reichenbach was not, of course, trying to convert each person into another Albert Einstein or Niels Bohr. But his long engagement in the public understanding of science, which led, among other things, to his books *From Copernicus to Einstein* (1942/1927) and *Atom and Cosmos* (1932/1930), did presume that the concepts of science were not wholly beyond the ken of the average person and that an understanding of both the world view of science and how it was achieved was available to the average citizen. Following from this, and more importantly for understanding Reichenbach's project, he, as we have seen, wanted to blame educated people and especially philosophers for the split between the everyday world and the world of science. He argued that philosophy in Germany at the time was a fundamentally conservative activity that tried to wall off the scope of science so that far more traditional philosophical concepts could be seen as fundamental to everyday and social life. This was an intellectual project of the philosophy of his times but not the only role philosophy could play, nor the one it should play.

The chief lesson, then, of taking the world of contemporary science seriously was a rejection of philosophical doctrines, especially calcified Kantian doctrines of the a priori as determined by a rigid set of concepts that could not be rejected or replaced and in terms of which the world had to be understood. He returned again and again to this rejection of the a priori as necessary and immutable and underscored continually the liberating nature of the new sciences of nature on precisely this score. The essay I just quoted ends on exactly this point:

> It is, perhaps, to be regarded as the most significant result of modern scientific knowledge that the picture of the world to which it has led has at the same time brought to light a new vision of man as a thinking mind, for science has shown us that reason is no rigid scaffolding of logical pigeonholes, that thinking does not consist in the endless repetition of outmoded norms. She has taught us instead that man grows with his knowledge, that he carries within him the capacity for forms of thought of which he could not so much as conceive at earlier stages of his existence. (1978/1930: 322)

In other writing from this period, Reichenbach goes beyond this and argues that the liberation of thought in the development of modern science stands in close kinship to the ways of thought and action found in modernizing trends in contemporary society. In this way, modern science stands actually in quite close relation to some social and political aspects of contemporary life – just as they both stand opposed to rigid systems of concepts and values found in academic philosophy. He writes in another popular essay of the period that:

> It has become ever more obvious that decisive new insights into the meaning of life, be it visions of human society, or of the relationship between the sexes, or of education of children and adolescents, or of the distribution of work and leisure in daily life, are not found by speculative philosophers but by people in practical life who discover new values in their activities and are able to make them acceptable to others through the impact of their personalities. Specifically, it is the academic philosopher's alienation from the revolutionary social processes of our time which explains why so much that is said in academic quarters about these matters seems strangely sterile and remote from life. (1978/1931: 386)

So, Reichenbach's project was to configure a new philosophy of nature that was precisely the counter to the conservative academic philosophy of his times. One final theme that connects that new philosophy of nature – and of the human knower of nature – to the postwar crisis of the twentieth century and also to Weber's version of scientific humanism is this: In rejecting the old "reason and the understanding have only a single eternal form" view of the a priori, Reichenbach was rejecting neither the a priori itself nor, certainly, the need for a conceptual understanding of nature. Instead, he was insisting that in the realms both of concepts and values, there is ultimately human choice and, thus, human responsibility. Reichenbach's view was that his new account of the knowledge of knowledge indicated the ineliminable role of the will in knowledge and also in social life. But there are no external or internal guarantees of correctness of volition. All that he can offer instead are two things: resoluteness of will and willingness to live in a society that aims for the consilience of wills. He never gave up this view and puts the points this way toward the end of his final book, *The Rise of Scientific Philosophy* (1951: 300, 301):

> We try to pursue our own volition ends, not with the fanaticism of the prophet of an absolute truth, but with the firmness of the man who trusts his own will. We do not know whether we shall reach our aim. Like the problem of a prediction of the future, the problem of moral action cannot be solved by the construction of rules that guarantee success. There are no such rules.
>
> If a person knows that moral rules are of a volitional nature, he will be ready to change his goals to some extent if he sees that otherwise he cannot get along with other persons. Adaptation of goals to those of other persons is the essence of social education.

It is not too much to say that throughout his career Reichenbach pursued a philosophy of modern science that was also a philosophy of responsible social life.

Reichenbach's vision lacks the encyclopedic scope and the sheer grandeur of teleological ambition of Sarton's, of course. But it is a synthetic vision in its own right – for the way in which modern science has rejected traditional epistemological and meta-ethical positions is not, for Reichenbach, something the working scientist is well placed to argue. For such tasks, Reichenbach argues for a specialized group of scientifically and philosophically trained specialists, a research community in philosophy. Scientific philosophy has its own task in the modernizing projects of the twentieth century – and it is a task very much within the spirit of scientific humanism.

Coda

I have in this chapter sought to demonstrate that at the originary moment for professionalized history of science and philosophy of science in the 1920s and 1930s, there were robust scientific humanist visions embedded in some prominent exemplars of that work. This is dead obvious, if currently underappreciated, in the case of Sarton. Reichenbach's scientific humanism is less obvious, less expansive, less optimistic and teleological, but no less real.

Almost a hundred years on, things have clearly changed. Within history of science, in a variety of idioms from postmodernism to Latourian nonmodernism (Latour 1993) to a turn to social and cultural history, itself replete with what Steven Shapin has called "tone-lowering" gestures (Shapin 2010), the scientific humanism of Sarton has all but vanished from sight. The more subtle humanisms of logical empiricist and other early twentieth-century philosophy of science are more robustly found in analytic philosophy of science, although perhaps too often expressed in the mode of being a foot soldier in the science wars that raged in the 1990s and blamed sociology, history, and philosophy of science for decreased trust in science (and flare up occasionally still). These days the main and very vocal proponents of scientific humanism tend to be public scientists, and their vision seems less humane than triumphalist and more in need of than informed by serious work in history or philosophy of science.

This intellectual situation seems unsatisfying and dysfunctional. I am less interested in raising the tone than in understanding the stakes. Crises press in upon us from all sides – climate disasters, technology pressed into oppressive economic and social agendas, political extremism. We may no longer think of scientific progress as the master narrative of modernity, but we certainly need scientific and technological progress (as well as moral clarity, political will, social solidarity, critical thinking, etc.) to help us with many problems, including those caused by foolish past and current uses of science and technology. The popular scientific humanism of today seems scarcely up to the task that confronts us. We need humanities scholars – in history and philosophy of science and in many other fields – to help us achieve a scientific humanism or an alternative to scientific humanism capable of joining with the urgent integrated political action that alone can see humanity through its current crises.

CHAPTER 6

John Dewey, Humanism, and the Value of Science

Aleksandra Hernandez

> What Humanism means to me is *an expansion, not a contraction, of human life, an expansion in which nature and the science of nature are made the willing servants of human good.*
>
> John Dewey, "What Humanism Means to Me"

Thus wrote John Dewey, whose progressive pragmatist philosophy stands as one of the most important articulations of humanism in the twentieth century. Humanism in the United States, however, came in many flavors, with public intellectuals, activists, and artists from all sides of the political spectrum claiming it as the doctrine best suited to advance their vision of human flourishing. The literary humanists led by literary critics Paul Elmer More and Irving Babbitt,[1] for example, clung to a version of humanism reaching back to Plato, which held mind–body dualism to be the key ingredient to promoting the good life. For these conservative cultural critics, the wholesale embrace of science by what they called "humanitarians" and "evolutionists" such as Dewey represented a kind of regressive barbarism, which, in their view, undermined a conception of the human as distinct from other animals. Babbitt, More, and their followers believed that to keep humans' animalistic, "wild" appetites in check, we must cultivate in students what they referred to as "the inner check" or "higher will."

Many progressive intellectuals and social reformers, by contrast, espoused versions of humanism distanced from what they believed was the corrupting influence of religion, and especially Christianity, which, rather than fulfilling its promise to promote the flourishing of all human beings, was used to support economic practices that exploit children and the poor and uphold racist and sexist values (Cameron 2023). Unlike the

[1] The literary humanists were later known as the New Humanists, but to avoid confusion with the very different New Humanism of the historian of science, George Sarton, I use the term "literary humanism" to refer to the former throughout this chapter.

literary humanists, prominent sociological humanists such as W. E. Du Bois and Jane Addams adopted a naturalistic, Darwinian worldview, and turned to science as a means by which subjugated communities – African Americans, women, the working class, etc. – might achieve political, industrial, and social equality (Cameron 2023; Deagan 1988a; Early 2006; Morgan 2016). The progressive humanism of Dewey shared many of the assumptions of the sociological humanists about human nature. Dewey also believed that rather than focusing on cultivating a controlling higher will by means of an education centered around the classics, we must promote an imaginative and experimental educational approach.

The aim of this chapter is to spell out the intertwining of humanism and science across Dewey's work and, more specifically, to anchor his account of the value of science in his philosophy of humanism. I begin by situating Dewey's pragmatic humanism within a culture war in the 1910s and 1920s concerning what human nature is and whether science should guide our efforts to address social ills and promote human flourishing. I then argue that although he agreed with the literary humanists that education needed to be reformed, Dewey insisted in *Democracy and Education* (1923/1916) and *Human Nature and Conduct* (1922), among other works, that to make real social progress, we must cultivate a scientific disposition, a taste for excellence, and flexible cognitive habits to better equip future generations to meet the challenges of the changing conditions of human experience. I conclude by supplementing Dewey, vis-à-vis Addams (2002/1902), with preliminary thoughts on the role that caring about others ought to play in helping us produce knowledge that can be used to promote the common good.

The Literary Humanist Quarrel with Science

The 1920s was an era of profound cultural turmoil and social unrest in the United States. Culminating in the 1929 stock market crash, many citizens were troubled by what they perceived as a dramatic cultural decline. In a poem published in 1923, American poet William Carlos Williams perhaps best expressed a collective anxiety about the lack of human involvement in directing the course of human history, with "No one / to witness / and adjust, no one to drive the car" (1923). The public intellectuals who rose to prominence at the end of the decade, the literary humanists, likewise worried about the materialism of everyday life, relativistic and naturalistic tendencies in philosophy, and the romantic individualism they believed was being promoted in literature and the arts. Babbitt, More, and their followers embarked on a cultural crusade against what they thought was a misguided entanglement of humanism with science in academia, which, they feared, was contaminating the moral fabric of the community. The source of the problem for the literary humanists was humanity's lack of self-control. Having thus diagnosed the problem, they reasoned that humans must separate themselves from their natural selves by disciplining their animal impulses and passions, and by cultivating the higher ethical self or will as the means to discover universal values and standards against which to measure their conduct. These universal standards are required to ground the diversity of human experience in a common element, a norm to which we must all aspire (Babbitt 1930: 28).

Comprised mainly of literary critics, the literary humanists leveled many of their critiques at the Rousseauvian romantics or so-called emotional naturalists, who believed that social institutions have a corrupting influence on humans. Whereas Rousseauvian romanticism exalted primitivism, the view that humans are naturally good in a state of nature, literary humanism celebrated the ascendency of humans over nature. And whereas romantic ethics reduced morality to "an expansive sympathy," the literary humanists believed morality entails a "restraint on passion" (Hoeveler 1977: 45). The pragmatists, and "scientific" philosophers more generally, were also among their favorite targets, as they endorsed a conception of human nature as being entirely a product of a material and social environment, and based their philosophy of life on the contingent, organic element of human experience. But no group was more reviled than the so-called humanitarians, who applied the scientific method to the study of human behavior and conduct, and who neglected the fixed, spiritual "center" shared by all humans and which differentiated them from other animals (Babbitt 1930).

Although many of their portrayals of their opponents' views were simplistic and inaccurate, the literary humanists offer an important window into a culture war over whether science should play any role in our understandings of what human nature is and how humans should conduct themselves. Here I focus on Louis Trenchard More, Paul Elmer More's younger brother, whose critical attitude toward the sciences served to bolster the literary humanist critiques of naturalist thinkers and social reformers. We will see that More was precisely the man of science the literary humanists needed to legitimize their attacks on what they believed was a harmful approach to understanding the nature of human experience and ameliorating the human condition.

Trained as a physicist, More set out to delimit what should be within the purview of scientific investigation in *The Limitations of Science* (1915), and later in "The Pretensions of Science," which appeared in one of the most important contributions to the literary humanist movement, Norman Foerster's edited collection *Humanism and America* (1930), and also in *The Dogma of Evolution* (More 1925), which purports to unmask the extent to which Darwinian evolutionary theory was infecting modern thought. More agreed that science is a valuable tool for helping us to predict the future, for diminishing superstition, and for allowing us greater control of our environments (1915: 187). Yet, despite science's valuable contributions to the advancement of human civilization, he claimed that it leads us astray when scientists create "a fictitious world of the imagination made out of æthers, electrons, mathematical symbols, and have confused it in their own and others' minds with the sensible world of brute fact" (1930: 3, see also 16; and More 1915: 188). In other words, when scientists dabble in the world of metaphysics, they indirectly cause social harm, as they lend authority to what More refers to as the "pseudo-sciences" of psychology and sociology (More 1930: 4). Arguing that we need a separate method from that of the "objective" sciences to study the nature of human consciousness and behavior, More contended that any field that investigates subjective phenomena scientifically is a pseudo-science.

More fervently believed that we should be wary of the psychologists and sociologists, "far greater in number than the two descendants of James" (1930: 3)[2] and far more dangerous than the metaphysically inclined scientists, as they cause harm directly by making false and misleading claims about human nature and conduct (1930: 4). In *The Dogma of Evolution* and "The Pretensions of Science," More specifies that biologists, and especially Thomas Huxley (1930: 4–5), are to blame for popularizing the misconceptions that the world is in flux and that man is an organic machine subject to physical laws. Worried that, in his words, "an increasingly large number are . . . turning to scientific doctrines in the hope that a deeper experimental knowledge of the laws of man's individuality, of his social relations, and of his environment" will solve social problems, More insisted that naturalism is not an appropriate guide for "evolving a society nearer to the ideally good" (1915: 214). His diagnosis of "these new systems of scientific ethics," however, is based on a fundamental misunderstanding of the morally inflected scientific naturalism of some of their

[2] This is likely a reference to John Dewey and George Herbert Mead.

popularizers, especially Dewey, who, even while arguing that we should cultivate in children a scientific attitude, believed that laws derived from empirical observation are not by themselves appropriate guides to moral conduct.[3]

More's simplistic understanding of how naturalist philosophers and social reformers conceived the value of science led him to distort some his opponents' naturalistic conceptions of human nature. This was frustrating to many, including George Sarton, who wrote in a review of Foerster's collection, *Humanism and America*, "[o]ne wishes one could send some of these 'humanists' back to school that they may learn the A B C of science, and be taught how dishonest it is to condemn things of which one knows nothing" (Sarton 1930: 448). As an example of this, consider some of More's attacks, which misrepresented scientific social reform projects as "eugenics":[4]

> A third class of eugenicists consists mostly of the hysterical elements of social workers who sob over the sins of society and sob over the innate purity of the harlot, who weep over the heartlessness of the law-abiding and weep over the innate nobility of the criminal. So far as one can make out from their incoherent utterances, they wish to put all the sins of the individual on society, without comprehending that society is made up of individuals. Whatever good they may accomplish, no one in the least conversant with science will concede that they are advancing an ethics in conformity with scientific methods; for if science makes any one thing clear, it is that the actions of the individual must bring their reactions also on the same individual. (1915: 254–255)

More was not alone in his distaste for and fear of the theoretical commitments of naturalist philosophers and social reformers. His colleague, Babbitt, worried about the philosophical commitments of the naturalists (to whom he also referred as "the Baconians"): "The Baconian has inclined from the outset to substitute an outer for an inner working – the effort of the individual upon himself – that religion has, in some form or another, always required" (1930: 34). Attributing all manner of social ills to his opponents, Babbitt blamed them for encouraging "the acquisitive life and also the pursuit of material instead of spiritual 'comfort'" (1930: 34). This was due to their supposed allegiance to Rousseauvian romanticism: "the upshot of this myth of man's natural goodness has been to discredit the

[3] See Rømer 2012 for an interpretation that challenges the widespread understanding of Dewey as a naturalist.

[4] Chief among them was Addams, who did much to improve the conditions of the working class in the suburbs of Chicago – among many other social achievements.

traditional controls, both humanistic and religious" (1930: 35). More agreed and added, in *The Limitations of Science*, that "[s]ide by side with the doctrine that human sympathy is the controlling factor in ethics, and this belief is evidently the basis of eugenics, there has always persisted the contrasted doctrine that the state of man is one of warfare, a survival of the fit" (1915: 258).

It is tempting to point out the many fallacious, uncharitable, and incorrect claims made by the literary humanists, dismiss their arguments as pseudo-philosophy produced by scholars with only a superficial understanding of both philosophy and science, and move on. It is worth pausing here, however, to break down the source of their worries and why it matters. At heart, what the literary humanists were pointing to was a gradual erosion of traditional values, the emergence of schools of thought that challenged their belief that humans are cognitively and morally distinct from other animals, and educational approaches that they feared did not cultivate in pupils their capacity for moral agency. They believed that a naturalistic understanding of the human encouraged citizens to pursue their first-order desires. And they doubted the efficacy of educational and social reform programs built on what they believed were shaky theoretical foundations.[5] On their "correct" version of humanism, the study of human nature ought to be based on "intuition" (Babbitt 1930) and imaginative apprehension of some common element shared by all humans (Babbitt 1919). Literature, and in particular time-tested literature untainted by the romantic celebration of individual experience, was valuable as a source of knowledge of the subjective elements of experience. It is literary humanism cordoned off from scientific psychology, then, that should serve as a guide to human conduct.[6]

Although literary humanist efforts to inject fear and suspicion of the encroachment of science into all domains of human life in the public sphere did not find a large audience, their attacks on the authority of science nevertheless succeeded in the decades to come. Several critics at the end of the 1930s charged that faith in science was misguided and undermined the very foundations of democratic thought in the United States (Jewett 2020). In what follows, we will see that, as the literary humanists feared, Dewey's progressive humanism relocated the source of value to the

[5] Jewett (2020) gives an overview of the literary humanist cultural critiques and contextualizes them within larger debates about biological, social, and cultural determinism stemming from Russell, Krutch, Watson, and Freud. He also discusses the influence of these debates on the public understanding of science in the present.

[6] See for example Paul Elmer More 1928.

domain of human experience. We will also see that though their attacks against progressive humanism were off the mark, they nonetheless testified to a burgeoning fear of the authority of science, which helped to shape the contours of Dewey's positive proposal. As a pragmatist, he did not merely see himself as intervening in narrow, academic, philosophical debates in ethics or education. Despite their abstract quality, Dewey's philosophical contributions were a direct response to the changing intellectual landscape and social conditions of his time. His philosophy was systematic and holistic, theoretical and pragmatic, and can be understood as a philosophy of humanism.[7]

Dewey's Philosophy of Humanism

Dewey is perhaps best known for championing an instrumentalist conception of science, or the view that science is valuable as a means to helping us accomplish human ends. Science is *always* value-laden, according to Dewey, as the aims of science are tied to what we value. This could be anything from removing sexist biases from science to valuing objectivity in science. Given that for Dewey, scientific investigation is inextricably bound up with our values – in fact, as Matthew Brown (2020) has recently argued, even scientific facts, when employed as means to solve a problem, are selected for their *value* to help us accomplish our ends in view – it is surprising that not much has been written about the close entwinement of science and humanism in Dewey's thought. Perhaps this scholarly gap is the result of our contemporary academic environment, which, with notable exceptions, tends to cordon off the sciences from the humanities. Dewey believed, however, that a proper understanding of human nature and conduct was necessary not merely for designing an educational curriculum that might better serve our democratic aims but also for reenvisioning the role that science might play in advancing progressive social aims. Thus, although he was critical of the tendency of humans to employ science to advance the interests of industry, he was also hopeful that we could cultivate in new generations imaginative and critical habits of mind and a social consciousness that would enable us to harness our collective resources toward solving social problems and ameliorating the human condition.

[7] Schulenberg (2021) discusses Dewey within the context of pragmatist humanism. His interest lies, however, in the neglect by most pragmatists of the significance of aesthetic form, which I do not consider here.

Like the literary humanists, Dewey worried about the materialism of the age and the lack of a critical attitude in the public. He differed, however, in how to go about addressing these problems. First, he was critical of metaphysical discussions in ethics, which offer accounts of moral motivation disconnected from the complexity of everyday situations (Brinkmann 2013; Dewey and Tufts 1932/1908). Second, he rejected the dualistic conception of human nature put forth by the literary humanists according to which the animal self is distinct and separate from the reflective, ethical self. Third, since like other animals, we are creatures of habit, Dewey believed that rather than cultivate an inner check, which, as we will see, produces internal disharmony, we must reimagine and artfully cultivate *new* habits of thought and behavior (Dewey 1922; McClelland 2005). Dewey recognized that intelligent habits are difficult to acquire, however, as social environments tend to encourage mechanical habits of thought and behavior.[8] For this reason, he thought that we ought to concentrate our efforts on cultivating somatic awareness and cognitive flexibility in the young (Dewey 1923/1916; 1922; Westbrook 1991).[9] By adopting an experimental, or as Dewey liked to call it, a "scientific" disposition, citizens will be better equipped to meet the various ethical and social challenges of the future.

Although the literary humanists were right that the progressive humanists sought to apply the scientific method to human experience, and that many progressive thinkers adopted a naturalistic, Darwinian conception of human nature and conduct, they were entirely wrong to claim that "evolutionism," as they liked to call it, implies social Darwinism or essentialism about human nature. For Dewey, Darwin's evolutionary theory shifts our conception of the human, not as essentially distinct from other animals, but as an organism that is not only shaped by but capable of shaping its material and social environment. This new understanding propelled him to examine human valuation practices and inspired him to reenvision the conditions that need to be in place for humans to employ their capacity for moral agency. The literary humanists believed that to resist our first-order desires to consume – to reject materialism as a standard mode of conduct, in other words – we need to turn to the past and derive ideals of conduct to aspire to, but Dewey worried that this

[8] For Dewey, habits are not acquired through rote repetition and are thereby not tied to action; rather, habits are modes of responding to certain salient features of the environment. When intelligent and *thoughtful*, habits produce "creative response[s] to a changing environment and emergent impulses" (Westbrook 1991: 290, see also 293).

[9] See Shusterman 2012 for a more recent elaboration of Dewey's insight.

approach would leave false beliefs and assumptions unexamined. For example, the classics may propagate falsehoods, uninformed by the findings of the latest science, such as the belief that Black people are essentially intellectually inferior to white people. He also worried that this approach would impair our ability to think flexibly and critically when faced with particularly vexing social problems. Rather than look to the past, we should employ standards appropriate to the specific situation at hand. This is because the material and social environment changes from generation to generation, and we need to apply new solutions to ever-evolving social problems. For Dewey, cultivating an inner check encourages us either to blindly follow fixed rules and take judgment out of the equation, or to suffer because of conflicting intuitions about what we should do, where instead we could be reflecting on the value of traditional standards for solving practical problems.

Key to Dewey's proposal is an understanding of human nature and conduct as malleable and responsive to the pressures of the material and social environment (Dewey 1922; Brinkmann 2013; McClelland 2005; Westbrook 1991). If his proposed understanding of human nature and conduct is true (an understanding that, he would agree, needs to be verified by means of empirical inquiry), then it follows that flexible, or *intelligent*, habits will counteract our tendency to settle into mechanical and unproductive habits. Now, just how, precisely, are we supposed to accomplish this?

As early as the 1890s, and with the support of the journalist Franklin Ford, Dewey became enthusiastic about the prospect of engineering an epistemic environment in which knowledge is more equally distributed. Frustrated by the tendency in philosophy to produce theoretical knowledge disconnected from the problems of everyday life, Dewey and others decided to start a newspaper, which was to be called *Thought News* and distributed across southern Michigan (Westbrook 1991: 55). The aim of the paper was to spread ideas of democracy and to develop in the public habits of inquiry and a heightened awareness of their social "function" in an "interdependent community" (Westbrook 1991: 53). Dewey thought that by making philosophy accessible to the community, it would become valuable as a tool for injecting new life into what he and the other prospective founders of the newspaper referred to as the "social organism." The project fell through in part because of a fallout between Dewey and Ford, whose vision for the paper was much more ambitious that Dewey's in its scope: Where Dewey wanted to inject new life into philosophy and bridge the gap between the masses and the educated elite, Ford wanted to radicalize journalism and use the newspaper as a vehicle for studying the social organism "like a steam engine" (Ford quoted in Westbrook 1991: 56).[10] Despite its failure, *Thought News* remains an important touchstone for understanding Dewey's future scholarly pursuits, which increasingly focused on the need to develop a scientific disposition in citizens to better equip them to solve the problems facing their communities. We will see in what follows that Dewey became invested in reenvisioning childhood education, to cultivate intelligent habits at a stage when minds are most flexible, but let me focus here on just one aspect of how Dewey thought students should be educated.

Much has been written about the importance of experiential learning and, especially, of exposing students to concrete situations so they can learn to appreciate the worth of their experiences firsthand.[11] But more needs to be said about the importance of these firsthand experiences for cultivating intelligent habits.[12] "The formation of habits," wrote Dewey,

> is a purely mechanical thing unless habits are also *tastes* – habitual modes of preference and esteem, an effective sense of excellence. There are adequate grounds for asserting that the premium so often put in schools upon external 'discipline,' and upon marks and rewards, upon promotion and keeping back, are the obverse of the lack of attention given to life situations in which the meaning of facts, ideas, principles, and problems is vitally brought home. (1923/1916: 276)

Tastes, however, cannot be developed merely by teaching aesthetic, ethical, and epistemic standards secondhand. Dewey gives the example of a music student who learns that certain formal features of classical music are esteemed by the experts, and hence, that he, as a student of music, should also appreciate those features; this student can even come to believe that his own standards correspond to the conventional standards of what counts as great music. But if he has most enjoyed ragtime in the past, "his active or working measures of valuation are fixed on the ragtime level. The appeal actually made to him in his own personal realization fixes his attitude much more deeply than what he has been taught as the proper

[10] Other figures who were involved in planning the launch of the newspaper included Corydon Ford, Franklin Ford's brother, and Robert Park. See Matthews 1977 for a discussion of the importance of *Thought News*, and especially Park's relationship with Dewey and Ford for shaping Park's theoretical commitments.

[11] Some scholars describe firsthand experience as "aesthetic" experience. See Johnston 2002 and Rømer 2012, among others.

[12] For related discussions about the regulative role pre-reflective qualitative experience plays in initiating inquiry, see McClelland 2005 and Stuhr 1997. See also Rømer 2012, for an analysis of Dewey's concept of intelligence.

thing to say; his habitual disposition thus fixed forms his real 'norm' of valuation in subsequent musical experiences" (1923/1916: 275).

Dewey argues that the same principle applies to moral and epistemic judgments, in which "vital appreciation" comes to play a much bigger role in impressing upon us "the measure of the worth of the generous treatment of others," for example (1923/1916: 275). When we teach secondhand values – as the literary humanists recommended we do – we create a "split" between our habitual inclinations and the theoretical standards we were taught in school and by our parents. If one grows up in a society where slaves are considered property that ought to be returned to their masters – as Mark Twain's iconic character Huck Finn's nineteenth-century contemporaries believed one should – one will experience an internal conflict and perhaps even suffer when one's habitual inclinations go in the opposite direction. Huck's adventures with the runaway slave Jim enabled Huck to experience the value of freedom, even though he believed he would go to hell if he did not follow the rules and return Jim to his slaveowners (Twain 2014/1884). This, according to Dewey, creates a "kind of hypocrisy of consciousness, an instability of disposition" (1923/1916: 275). Similarly, a student can be taught to perform certain analytical moves and acquire information by means of "mechanical rehearsal," but unless "it somehow comes home to him at some point as an appreciation of his own," the significance of the norms will not impress themselves as standards "which can be depended upon" (1923/1916: 276). Dewey refers to "appreciation" experiences as personal responses involving the imaginative apprehension of their worth, and he emphasizes that "appreciation value" is to be found in all fields of study, not just in the realm of literature and other arts. "The imagination," wrote Dewey, "is the medium of appreciation in every field. The engagement of the imagination is the only thing that makes any activity more than mechanical" (1923/1916: 276).[13]

Dewey defined "appreciation" as "an enlarged, intensified prizing ... [an] enhancement of the qualities which make any ordinary experience appealing, appropriable – capable of full assimilation – and enjoyable" (1923/1916: 278). Significantly, Dewey thought that the fine arts have an important role to play in education, even though they are not "the exclusive agencies of appreciation." This is because they are not only "intrinsically enjoyable" but also serve the instrumental function of "fixing

[13] Dewey's theory of the role the imagination plays in moral deliberation has been discussed in Fesmire 2003 and Pappas 2008, among others.

taste, in forming standards for the worth of later experiences" by creating "a demand" or an appetite for elevating everything that we do to "their own level" (1923/1916: 279):

> They [the fine arts] reveal a depth and range of meaning in experiences which otherwise might be mediocre and trivial. They supply, that is, organs of vision. Moreover, in their fullness they represent the concentration and consummation of elements of good which are otherwise scattered and incomplete. They select and focus elements of enjoyable worth which make any experience directly enjoyable. They are not luxuries of education, but empathetic expressions of that which makes any education worthwhile. (1923/1916: 279)

For both Dewey and the literary humanists, the fine arts are instruments to be employed for finding standards by which to measure our conduct in other domains of life. For Dewey, however, they are much more than instruments for moral cultivation – they are means by which we acquire a *taste* for excellence and a love for doing and enjoying *all* human activities for their own sake, and not merely for the instrumental benefits that they may afford.

Whereas appreciative experiences of fine art are valuable in that they furnish us with the habits of mind necessary for properly measuring the worth of other, future activities, a scientific attitude inclines us to confront head-on what Brown refers to as the "contingencies" that inevitably arise when we engage in inquiry. Brown defines "contingencies" or "perplexities" as "any moves or moments in inquiry that are genuinely open, where reasonable inquirers could disagree about the way to proceed" (2020: 64). For Brown, as for Dewey, processes of inquiry that settle questions without any forethought or deliberation are not genuinely experimental or scientific, as a scientific attitude requires that we actively evaluate the problems to be solved, that we purposely determine the value of the evidence before us for solving the problem at hand, and that we measure the worth of our interpretations of the evidence. I would add that for Dewey, humans ought to undertake inquiry not merely for instrumental ends but for its own sake,[14] and like any other human activity, inquiring practices are most meaningful when we do them well. Doing things well also has the effect of developing in us a taste for *excellence*. Measured by the consequences of our practical judgments, excellence

[14] In support of this Dewey writes: "In so far as any study has a unique or irreplaceable function in experience, in so far as it marks a characteristic enrichment of life, its worth is intrinsic or incomparable" (1923/1916: 281).

further enlarges the meaning of human life and affords us the opportunity to appreciate our accomplishments – whether collective or personal – in the same way that we might prize and value a work of art.

Science and Humanism

Science, according to Dewey, is a means to human ends, and an end is to be appreciated on its own terms. Accordingly, for something to be a means rather than merely a tool, it must be part of some coordinated, intelligent activity, and fulfill an aspect of our organic need to resolve disturbances or disharmonies that inevitably arise during our transactions with the environment. To perform one of our main life functions – absorbing oxygen – our bodies mechanically employ our lungs. Every time we breathe, we experience temporary relief from the lack of oxygen and the need to take a breath. And the same is the case for every other life function our bodies automatically perform. Our bodily organs are not in themselves means, however, as they have not been intelligently employed as part of a coordinated activity to accomplish an end-in-view. It is only when our minds are focused on the rate and depth of our breath that our lungs become part of the coordinated activity of practicing mindfulness for the purposes of easing anxiety and other types of somatic disturbances. Similarly, a hammer is not a means until we use it in conjunction with other tools to hammer in nails and shape wood into a box:

> They are actual means only when brought in conjunction with eye, arm, and hand in some specific operation. And eye, arm and hand are, correspondingly, means proper only when they are in active operation. And whenever they are in action they are cooperating with external materials and energies. Without support from beyond themselves the eye stares blankly and the hand moves fumblingly. They are means only when they enter into organization with things which independently accomplish definite results. These organizations are habits. (Dewey 1922: 22)

When we first hammer a nail into a piece of wood, we must be careful not to get our fingers caught. But the longer we practice, hammering nails becomes a habit waiting to be used to build the frame of a house or repair a broken door. Likewise, our practice of mindfulness eventually becomes habitual, as we internalize the skill of slowing down our heart rate by deepening our breaths.

Some habits for Dewey are more malleable than others and can be improved with practice. Only when we have experienced standing straight in a yoga class, for example, are we able to form the idea in reflective experience of how to correctly stand straight and break bad habits outside of the studio. Only then, in other words, are we able to learn the habit of standing straight without "fiats of will" (Dewey 1922: 22–25). Even our ability to discern different colors is the product of "skilled analysis": "A moderate amount of observation of a child will suffice to reveal that even such gross discriminations as black, white, red, green, are the result of some years of active dealings with things in the course of which habits have been set up. It is not such a simple matter to have a clearcut sensation. The latter is a sign of training, skill, habit" (Dewey 1922: 25). Since skills are subject to improvement, the child could become more skilled at differentiating hues of green by engaging in the practice of realist painting and capturing the many varieties of green of a tree. This understanding of the ways that prior habits influence our ideas applies also to other types of activities, including scientific inquiry: "distinct and independent sensory qualities, far from being original elements, are the products of highly skilled analysis which disposes of immense technical scientific resources. To be able to single out a definite sensory element in any field is evidence of a high degree of previous training" (Dewey 1922: 25).

But what of habits of thought and feeling? On this point, Dewey seems more pessimistic than the "radical reformers," who put their faith in rapid institutional change, as if quickly changing institutions could change our customary habits of thought and feeling. Institutions, according to Dewey, generally embody our collective habits of thought and feeling; when we attempt merely to change the former, we leave intact the latter, which makes it very unlikely that social change will follow: "Actual social change is never so great as is apparent change. Ways of belief, of expectation, of judgement and attendant emotional dispositions of like and dislike, are not easily modified after they have once taken shape" (1922: 77). This is because, first, as we saw earlier with the example of standing straight, secondhand ideas cannot easily change dispositions to which we have been habituated by firsthand experience. Thus, if we are accustomed to experiencing firsthand a legal system that benefits some members of the community at the expense of others, it will be extremely difficult for some to give up their special privileges. And second, our cognitive habits and dispositions are shaped by our interactions with others. A child that shares her toys with other children will be praised by her teachers and parents. Indicative of a character trait customarily prized in the girl's social environment, repeated social approbation of similar behaviors serves the function of shaping the child's future disposition to behave selflessly. This suggests that if we do not change our collective attitudes toward historically marginalized groups, social change will not follow.

It follows that to change embodied habits, Dewey suggests, we must change what is collectively *valued.* And the only way to produce new, collective values, according to Dewey, is to adopt a "truly humane education," consisting of "an intelligent direction of native activities in light of the possibilities and necessities of the social situation" (1922: 70). Worried that minds have become inflexible and dependent on fixed belief and the authority of others, Dewey writes in *How We Think*, for example, that "it is its [education's] business to cultivate deep-seated and effective habits of discriminating tested beliefs from mere assertions, guesses, and opinions; to develop a lively, sincere, and open-minded preference for conclusions that are properly grounded, and to ingrain into the individual's working habits methods of inquiry and reasoning appropriate to the various problems that present themselves" (1910: 28). Rather than train children by means of "premature mechanization of impulsive activity after the fixed pattern of adult habits of thought and affection," children must be educated to form new habits of inquiry – they must be trained to think critically – to be "serviceable under novel conditions" (Dewey 1922: 75).

A truly humane education must also draw its lessons from "the elaborate systems of science [which] are born *not of reason*" but of the impulse to hunt, combine things, and communicate with others, methodically organized into "the dispositions of inquiry, development and testing" (Dewey 1922: 136, emphasis added).[15] Yet, above all, it must exploit children's natural tendency to be curious about the world: "[t]raining is such development of curiosity, suggestion, and habits of exploring and testing, as increases their scope and efficiency. A subject – any subject – is intellectual in the degree in which *with any given person* it succeeds in effecting this growth" (Dewey 1910: 46). In sharp contrast to the literary humanists, who, we may recall, recommended that we cultivate in children a higher ethical will to control their base, animal instincts, a truly humane education cultivates a scientific disposition in future citizens, not merely for slowing down the impulsive drive to hurriedly accomplish one's goals, but also for avoiding becoming overly interested in reflection disconnected from everyday experience: "We may become so curious about remote and

[15] On this point, Hickman (1990) argues that "[i]f by 'scientism' it is meant that the methods of experimentation have proved so successful in the domains in which they have been developed and applied that they ought to be utilized and further developed in areas where they have not been tried, then the term is applicable to Dewey's instrumentalist program" (quoted in Johnston 2002: 4).

abstract matters that we give only a begrudged, impatient attention to the things right about us. We may fancy we are glorifying the love of truth for its own sake when we are only indulging a pet occupation and slighting demands of the immediate situation" (Dewey 1910: 137). Both the tendency to hurriedly accomplish one's goals and to become overly interested in reflection, Dewey writes, are irrational to the extent that "the foresight of consequences is warped to include only what furthers execution of predetermined bias" (1910: 138).

The hallmark of a humanistic education is the inculcation of a habitual disposition to inquire into, deliberate about, and test the efficacy of traditional principles under new conditions.[16] Progressive humanism does not reject conventional principles outright, as some principles have been tested and proven to be efficacious instruments for ameliorating the human condition, such as the principle of respecting a person's freedom. Some other principles, such as certain legal principles, may need to be modified so that, in Dewey's words, they can become "more effectual instruments in judging new cases" (1910: 165). Some traditional principles in fact contribute toward social inequities and must be rejected on the grounds that they do not cohere with the principle of respect for a person's freedom, as we saw in the example of Huck Finn. But the issue cuts deeper than that. According to Dewey, blindly adhering to old principles is simply "another manifestation of the desire to escape the strain of the actual moral situation, its genuine uncertainty of possibilities and consequences" (1910: 166). A scientific disposition toward moral inquiry takes older principles to be *hypotheses* to be tested in the imagination against concrete situations (1910: 167).

Dewey gives the example of a young person who has repeatedly experienced the consequences of being kind to others. These experiences culminate in the disposition of kindliness (1923/1916: 275), which, in addition to acquiring appreciation value (by which he means the value an activity acquires while we are enjoying, prizing or appreciating it),[17]

[16] The role that science might play in our "valuation" practices, or, as Dewey also refers to them, "evaluation" practices, is controversial. See Johnston 2002 for an overview of the debate. See also Waks 1998 and McCarthy 1999 for enthusiastic endorsements of Dewey's scientistic theory of value judgments, and Boisvert 1998, Hickman 1990, Garrison 1997, and Johnston 2002, which emphasize, to different degrees, the aesthetic dimension of human valuation practices.

[17] Stroud understands Dewey's use of "appreciation" as synonymous with "taste" (Stroud 2011: 41). I see a connection, however, between what Dewey calls "appreciation value" and Stroud's concept of "immediate value." Stroud writes: "When Dewey explicitly links his notion of immediate good to intrinsic value . . . he is careful to note that the *immediacy* of value is what is intrinsic, and not some sort of value that has an essentialist primacy" (Stroud 2011: 44).

reliably produces good consequences in experience. A disposition to approach moral situations as occasions for inquiry will yield reliable predictions of good outcomes that may or may not conform with conventional morality, as can be observed in Huck Finn's taste for the value of freedom and his intuition that it also applies to Jim's situation. Although it is true that Huck's belief that he will go to hell if he does not tell on Jim somewhat limits his moral development – insofar as Huck fails to generalize from Jim's situation to the plight of all slaves – this is not because of the inherently biased nature of empathy, as Alan Goldman (2010) argues, but rather because Huck is surrounded by adults who dogmatically impress their corrupt moral values on children and steer them in the wrong direction.

A humane education would have afforded Huck the possibility of greater moral development, provided he had internalized the notion that conventional principles are not infallible. In a different social environment, Huck would not have experienced internal conflict but would have rather been encouraged to take in "the full scope" (Goldman 2010: 276) of the situation, test the outcomes in his imagination, and derive the correct principle from the ground up. This is not to say that it is always inadvisable to apply a rule to a situation. But Dewey warns in "The Logic of Judgments of Practice" (1915) that adhering to an ideal or a standard involves no judgment. He gives the example of being faced with the situation of whether or not to buy a suit. If the operating principle is that you already have a suit in mind, then you are not really selecting a suit through a valuation process, as you have already prejudged the situation (1915: 518). It is the oppositive of taking a scientific approach to the process of figuring out what you should do with the outcome of your judgment in view.

A scientific disposition is a cognitive habit that can serve as a means to solving practical problems beyond the domain of what is narrowly referred to as "science." What we call "science," according to Dewey, is a sociological artifact, which more than anything tells us about our conventional styles of organizing experience. It also reflects what in his mind is an unproductive bifurcation of the humanities from the sciences in education, which he takes to be largely responsible for impairing our ability to solve problems jointly. In *Democracy and Education* (1923/1916), Dewey argues that we have arbitrarily assigned values to various fields, as if aesthetic value belonged only to the domain of literature, and epistemic value, to science. Science, he argues, can have many different values, depending on the problem to which it is applied, as a means:

> [T]he attempt to distribute distinct sorts of value among different studies is a misguided one, in spite of the amount of time recently devoted to the undertaking. Science for example may have any kind of value, depending upon the situation into which it enters as a means. To some the value of science may be military; it may be an instrument in strengthening means of offense or defense; it may be technological, a tool for engineering; or it may be commercial – an aid in the successful conduct of business; under other conditions, its worth may be philanthropic – the service it renders in relieving human suffering; or again it may be quite conventional – of value in establishing one's social status as an "educated" person. As matter of fact, science serves all these purposes, and it would be an arbitrary task to try to fix upon one of them as its "real" end. All that we can be sure of educationally is that science should be taught so as to be an end in itself in the lives of students – something worthwhile on account of its own unique intrinsic contribution to the experience of life. (Dewey 1923/1916: 282)

Although the instrumental value of science is undeniably far more consequential, given the many uses to which scientific knowledge can be put, we cannot neglect the importance of enjoying or valuing the process of doing science *well*. In other words, although science is instrumental in helping us solve problems, we stand a better chance of ameliorating the human condition when we cultivate scientific curiosity combined with a taste for both excellence and goodness in future generations. This will guarantee better tools at our disposal for solving problems.

Conclusion

At this point, it is worth returning to the problem that Dewey started with: How might we employ science in the service of values that better serve our communities? Recall that he believed cultivating in children a scientific attitude would better equip them to meet the challenges of the future. He also emphasized the importance of firsthand experiences for deriving norms that are personally meaningful and that have been tested in experience for their reliability and success at producing good (or valuable) consequences. But how is Deweyan valuation supposed to work in practice? And how do we guarantee that by cultivating in children a scientific disposition, they will arrive at values that serve the common good? How do we ensure, in other words, that these children of the future, equipped with flexible habits of mind and a taste for the value of excellence and goodness, are going to employ science to serve social rather than selfish, economic interests?

To answer this question, let me first touch on the concept of sympathetic understanding in the work Jane Addams, a prominent humanist sociologist and social reformer in the early twentieth century who influenced the emancipatory focus of pragmatism, including Deweyan ethics, through her exemplary work with poor immigrants in the suburbs of Chicago (Seigfried 1999). Like Dewey, for whom moral deliberation is a dramatic, artful, and *caring* process involving play acting, perspective-taking, and anticipation of the consequences of our judgments (Caspary 2000; Fesmire 1995; Goldblatt 2006; Hamington 2010), Addams worried about values that were increasingly becoming more influential – mechanization and materialism – and believed that cultivating the proper uses of the imagination could enable citizens to discover new values (1930: 124) and redeem industry from the role it had played in augmenting social "evil" and "distress" (1930: 28). For Addams, ethical agency is realized when we deliberately choose to expose ourselves to other values firsthand, as we cannot discover new values when we are isolated from other human beings (2002/1902). For this reason, she lived among the poor immigrants of the suburbs of Chicago, and in 1889 cofounded with Ellen Gates Starr the socialist settlement, Hull-House – a thirteen-building complex equipped with a daycare, dining, and other types of gathering spaces where the middle-class residents would learn from the community about its needs. Through that work, she came to appreciate the inherent dignity of all persons and was inspired to mobilize legal reforms to address various social injustices, including women's oppression, child labor, and the exploitation of laborers.

Addams' theoretical and practical work sheds light on the importance of cultivating personally meaningful and caring relationships with others for not only experiencing but also internalizing other values by means of sympathetic understanding (2002/1902). The core idea behind sympathetic understanding, as Charlene Haddock Seigfried explains, is that of reciprocity, in which we recognize both our responsibility toward others and our dependency on them (2002/1902: xx–xxi). But perhaps even more important is that sympathetic understanding is a mode of attention that opens space to "the viewpoint, values, and goals of others" (2002/1902: xxi). These values and goals impress themselves on us and become part of "moral deliberation and social transformation" (2002/1902: xxi). Addams' social ethics suggests that moral deliberation cannot simply presuppose care for the welfare of and a sense of responsibility for others; rather, these things arise from a process in which we internalize the goals and values of others – which will sometimes conflict with our own.[18]

[18] See Helm (2010), whose account of caring about others as "persons" serves as a useful supplement to Dewey's and Addams' ideas.

Dewey was not always explicit about the importance of the value of caring about others to his theory of valuation, perhaps because he worried about the limitations of our capacity to sympathize with others. In *Ethics*, which Dewey cowrote with his colleague from the University of Chicago, James Tufts, though he follows Darwin in claiming that our success as a species lies in our capacity to sympathize and cooperate with one another, Dewey nonetheless warns that sympathy "rarely extends beyond those near to us, members of our own family and our friends. It rarely operates with reference to those out of sight or to strangers, certainly not to enemies" (Dewey and Tufts 1932/1908: 261). For this reason, he draws on Adam Smith in posing the figure of the "ideal spectator" to take the place of the social group. By imagining not whether our peers would approve of our actions but whether the ideal spectator would, we stand a better chance of executing moral judgments that "merit approbation because their execution will conduce to the general wellbeing" (Dewey and Tufts 1932/1908: 270). Dewey thus emphasized that actions that merit approbation are those in which we voluntarily make a choice to "bring good to others" (Dewey and Tufts 1932/1908: 272).

Despite his reservations about relying on our natural sympathy when we are making moral judgments, Dewey also claimed that "sympathy is the animating mold of moral judgement ... because it furnishes the most efficacious *intellectual* standpoint. It is the tool, *par excellence*, for resolving complex situations" (Dewey and Tufts 1932/1908: 298). When sympathy is mechanical and the controlling factor in our actions, it is likely to produce actions that benefit us and those who are in our circles of concern at the expense of distant others. Sympathy combined with refection, or what Dewey and Tufts in the *Ethics* refer to as "intelligent sympathy," by contrast, is bound to produce actions that merit approbation from the ideal observer, who upholds the standards of beneficence and social welfare. From this perspective, sympathy is not merely the animating mold of moral judgment but also the pillar of a humanist science. For science to be *humane*, caring about others must influence the values we bring to bear when we make epistemic judgments.[19]

Following Addams, who "studied the everyday world ... [and] connected this analysis to the political and economic conditions that generated the mundane and oppressive reality" (Deagan 1988b: 255), for Dewey, a humane science must be similarly employed to serve human rather than

[19] See Kitcher 2011 for an account of how "broad," "cognitive," and "probative" schemes of values interact with one another to produce practical judgments.

industry interests.[20] The great irony here is that, given the shared aims of the literary humanists and the pragmatists, men and women such as Dewey and Addams were one of the primary targets of the literary humanists, whose skepticism about the metaphysical foundations of humanism was completely unwarranted. Whereas the literary humanists dealt armchair critiques, the progressive humanists were on the ground, successfully employing science to relieve the plight of the working class, women, and children, and advancing the cause of racial equality. When faced with questions about the role that science might play in ameliorating the human condition, one need only read about this period in American history and derive lessons about what the human spirit is capable of when armed with the right imaginative, aesthetic, and scientific resources. The solution, it seems clear, is not to cordon off the sciences from the humanities. If humanism is to ameliorate the human condition, it is as a science of experience.

[20] See Scimecca and Goodwin (2003), who discuss Addams' humanist sociology within the context of American pragmatism.

CHAPTER 7

Sociopolitical Engagement and Scientific Value Freedom

The View from the Left Vienna Circle

Thomas Uebel

Introduction

The reevaluation of the philosophies of science of logical empiricism has been underway for several decades among historians of analytic philosophy and philosophy of science.[1] Increasingly it has interested not only contemporary anti-metaphysicians but also feminist and anti-racist philosophers.[2] What attracts them is what has taken historians the longest to recover and impress upon the philosophical public at large (where it still has not resonated fully). This is the fact that by some of the members of the Vienna Circle their philosophy of science was regarded as closely related to ongoing struggles for the social, economic, and political transformation of society.[3] In later years, versions of this engaged perspective were also promoted under the heading of "scientific humanism."

Unsurprisingly, this recent reappreciation of Vienna Circle philosophy has not been wholesale. One doctrine commonly attributed to logical empiricists has proven particularly rebarbative: scientific value freedom, often summarized as intending to safeguard objectivity by the demand that "social, ethical and political values should have no influence over the reasoning of scientists" (Douglas 2009: 1). (Epistemic values such as truth, coherence, and explanatory power are viewed as presupposed by science and as such uncontested.)[4] Consequently, the Circle's left wing, which pressed the politically critical and transformative agenda, stands accused of

[1] Logical empiricism (aka logical positivism or neopositivism) was a philosophical movement that originated in 1920s Austria (the Vienna Circle around Moritz Schlick) and Germany (the Berlin Group around Hans Reichenbach). In its Anglophone exile from Nazism, it laid much of the foundation for post–World War II analytical philosophy of science but fell out of favor in the 1960s.

[2] See, e.g., Okruhlik 2004, Longino 2006, Yap 2010, Bright 2017, Dutilh Novaes 2020, and LaVine 2020. The misconceptions of S. Richardson 2009 have been addressed in Uebel 2010 and Romizi 2012 and are not revisited here.

[3] Most prominently so by its so-called left wing: Rudolf Carnap, Philipp Frank, Hans Hahn, and Otto Neurath.

[4] The radical challenge to that distinction is briefly considered below.

doctrinal inconsistency. As described, value freedom proscribes the type of value-laden engagement that is demanded by contemporary feminists and anti-racists.

Investigating the matter demands close attention to the content of the doctrine of scientific value freedom in context and its understanding by Rudolf Carnap and Otto Neurath. What is at issue is not, of course, whether science can or should be used to broadly speaking political ends by building on or implementing its results in policies or administrative measures: It obviously can, has, and will be. The question is whether we should conceive of considerations pertaining to its (potential) use as external or internal to science.

The standard way to think of value freedom is to locate any concern with values other than epistemic ones outside of science itself. This proscribes as unscientific all investigations that take account of nonepistemic values both in the evaluation of hypotheses and with regard to their potential applicability. A different way of conceiving of value freedom focuses solely on the results of scientific investigations. It proscribes certain types of statement being issued as scientific ones, that is, as justified by scientific reasoning. As explained later, this approach to value freedom is considerably less restrictive. Note also what informs the standard view of value freedom. The demand that scientific activities remain uninfluenced by nonepistemic values is meant to ensure that science remains unbiased by perspectival partialities, renders reality unfiltered by subjectivity, and stays "objective." According to this view, science does not "do" subjective perspectives but seeks a "view from nowhere."[5] Also at issue for the standard view therefore is whether this traditional conception of objectivity ought to be upheld. By contrast the narrower version of value freedom has no commitments of this sort.

Unless further specified, denials of scientific value freedom could mean the denial of either of these versions or both, but it is the former version that is commonly under discussion in the Anglo-American literature. Evaluations of the Vienna Circle's position on the matter (especially that of its left wing) have long suffered from inattention to the differences involved. My discussion draws on the distinctions just made and explores an overlooked combination of positions. I begin by detailing the apparent dilemma faced by the Vienna Circle advocates of scientific humanism and then ask whether Neurath offers a promising way out. This leads to specifying his and Carnap's distinctive understanding of value freedom

[5] See Nagel 1979 and relatedly Williams 1978; for cogent opposition, see Fine 1998.

and then to investigating whether their noncognitivism is as detrimental to their project as many have claimed it is. My point is that it isn't. The meta-ethical differences between many current feminist and anti-racist theorists and the proponents of the left Vienna Circle's "scientific world-conception" do not condemn activism of the sort advocated by Neurath and Carnap to incoherence.

A Promising Program Threatened

In his autobiography Carnap reported:

> All of us in the Vienna Circle took a strong interest in the political events in our country, in Europe, and in the world. These problems were discussed privately, not in the Circle which was devoted to theoretical questions. I think that nearly all of us shared the following three views as a matter of course which hardly needed any discussion. The first is the view that man has no supernatural protectors or enemies and that therefore whatever can be done to improve life is the task of man himself. Second, we had the conviction that mankind is able to change the conditions of life in such a way that many of the sufferings of today may be avoided and that the external and the internal situation of life for the individual, the community, and finally for humanity will be essentially improved. The third is the view that all deliberate action presupposes knowledge of the world, that the scientific method is the best method of acquiring knowledge and that therefore science must be regarded as one of the most valuable instruments for the improvement of life. In Vienna we had no names for these views; if we look for a brief designation in American terminology for the combination of these three convictions, the best would seem to be "scientific humanism." (1963b: 82)

Most readers of the present volume will regard the points made here as rather obvious, but not perhaps Carnap's elaboration:

> It was and still is my conviction that the great problems of the organization of economy and the organization of the world at the present time, in the era of industrialization, cannot possibly be solved by "the free interplay of forces," but require rational planning. For the organization of economy this means socialism in some form; for the organization of the world it means a gradual development toward a world government. (1963b: 83)

Yet even Herbert Feigl – not a member of the left wing but more closely associated with the liberal Moritz Schlick – stressed the need for intervention in his own post–World War II manifesto for scientific humanism: "Cooperative planning on the basis of the best and fullest knowledge available is the only path left to an awakened humanity that has embarked on the adventure of science and civilization." (1981/1949: 377) Yet whether they differed over the kind of intervention needed, all Circle members presumably agreed with Feigl's conclusion:

> [S]cience, properly interpreted, is not dependent on any sort of metaphysics . . . a mature humanism requires no longer a theological or metaphysical frame either. Human nature and human history become progressively understood in the light of advancing science. It is therefore no longer justifiable to speak of science *versus* the humanities. Naturalism *and* humanism should be our maxim in philosophy and in education. A Scientific Humanism emerges as a philosophy holding considerable promise for mankind – *if* mankind will at all succeed in growing up. (1981/1949: 377)

Note that what Carnap and Feigl called "scientific humanism" is clearly an expression of values. While Carnap separated such concerns from the "theoretical" discussions in the Circle meetings (and Feigl most likely followed him in this), it is questionable whether the scientific humanist stance could remain a wholly "private" matter.

Consider Carnap's own Preface to the *Aufbau* and the collaborative pamphlet "The Scientific World-Conception: The Vienna Circle," two publications from the late 1920s, and Feigl's own manifesto of 1949. While the *Aufbau* celebrates "an inner kinship" "between the attitude on which our philosophical work is founded and the intellectual attitude which presently manifests itself in entirely different walks of life," including "movements which strive for meaningful forms of personal and collective life, of education, and of external organization in general" (Carnap 1967/1928: xviii), the collaborative pamphlet speaks of an "inner link" between "attitudes toward questions of life" and the "scientific world-conception" of the Vienna Circle, with the former including "endeavors toward a new organization of economic and social relations, toward the unification of mankind, toward a reform of school and education" (Verein Ernst Mach 2012/1929: 80–81). Not only Neurath (see Neurath 1928; 1931; 1932a) but also Carnap advertised "the struggle *we wage* against superstition, theology, metaphysics, traditional morality, *capitalistic exploitation of workers*, etc." (2013/1934: 177, emphasis added, my translation). More obliquely, Feigl's North American manifesto – uniting pragmatists, naturalistic realists, scientific empiricists, and others – signaled its social relevance by its historical reference: "All these trends of thought and many others converge in a broad movement that one may well be tempted to regard as the twentieth-century sequel to the Enlightenment of the eighteenth century" (1981/1949: 367). What characterized the

Enlightenment, of course, was the ambition to bring the advance of theoretical knowledge to bear on liberating people from the shackles of traditional prejudice, religious dogma, and political tyranny.[6]

It is possible to discern in the "inner link" or "kinship" that Carnap and Neurath perceived between their philosophy and contemporary movements for social and economic change a purely epistemic ideal, what I have elsewhere dubbed "intersubjective accountability" (Uebel 2020): assertions about what is and could be the case have to be backed by intersubjectively available evidence. It is also possible to regard Feigl's new enlightenment in related terms. It is inconceivable, however, that Carnap, Neurath, or Feigl would not have noted that using science to improve human life conditions in support of the movements mentioned earlier also requires nonepistemic judgments of value to be made about what should be the case. So the question arises: Is making such judgments a proper part of science and its philosophy?

Despite their sympathy for his scientific humanism, contemporary activist scholars would judge Feigl's conception of it to be invalidated by the standard picture of value-free science. For Feigl, all value involvement appears to be on the side of applied science. "Scientific knowledge itself," he declared, "is socially and morally neutral." (1981/1949: 375). Carnap also, on first impressions, only offered the standard noncognitivist, neopositivist diagnosis. Value questions are external to science and of an entirely different type. Unlike statements of fact, value statements are devoid of "cognitive" meaning, that is, they are not truth-valuable or truth-apt.

> As regards superstition, theoretical questions are at issue. It is possible to disprove by scientific means the assumption that prayers or charms can prevent hail storms or railway accidents. However, whether somebody is in favor of or against cremations, in favor of or against democracy, in favor of or against socialism, that is an issue of adopting a practical attitude, not of theoretical proof. By theoretical means one can only determine here that this or that institution brings with it these or those hygienic, economic or cultural consequences. ... *Scientific considerations do not determine the goal, but only ever the pathway to the goal adopted.* (Carnap 2013/1934: 177–178, emphasis in original, my translation)[7]

[6] For recognition of the need of Enlightenment thought to renew itself with every generation, see Frank 1949/1917.

[7] Carnap's remarks in his first Bauhaus lecture of October 1929 are fully consonant ("Wissenschaft und Leben," RC 110-07-49, Archive of Scientific Philosophy, Hillman Library, University of Pittsburgh).

It seems that for Carnap, too, the shared enlightenment perspective which his talk of the "inner kinship" designated was limited to the epistemic ideal of intersubjective accountability. Did he also then subscribe to the standard conception of value freedom which bars nonepistemic values from science entirely?

Neurath to the Rescue?

It is at this juncture that one is advised to consult Neurath's philosophical work.[8] Like Carnap's, it combines anti-foundationalist holist empiricism and metatheoretical constructivism, but it also promises to give political engagement in science a clean bill of health.[9] Quite apart from his nonreductive naturalism,[10] what is particularly attractive is Neurath's pioneering work on argumentation, now standard in the feminist literature, that legitimates appeal to "contextual," that is, nonepistemic, values in central areas of science.

Powerfully proposed in the post-positivist literature by Helen Longino (1990) and widely adopted since, it was previously employed by members of the left Vienna Circle, as early as Neurath (1928) and as late as Frank (1957). Known as the argument from underdetermination, it builds on Neurath's and later Quine's generalization of a conclusion of Pierre Duhem's to all suitably abstract scientific theories. This was the idea that the theories of theoretical physics are underdetermined by empirical evidence: Testing them requires auxiliary theories that themselves resist direct testing. It follows that alternative but logically incompatible theories are able to account for the very same data. It is in the gap between evidence and theory this opens up that Neurath's and Frank's logical empiricism and Longino's "contextual empiricism" locate the logical space that allows scientists to appeal to tiebreakers that are nonepistemic in the following sense. Their employment allows the scientists to settle on a theory or hypothesis to work with but does not distinguish them as epistemically superior to its competitors.

8 That Carnap and Neurath are of one mind on this issue is argued in Uebel 2005. On Neurath's philosophy in the round, see Cartwright et al. 1996, Cat 2019, and Howard 2019.

9 Contrary to popular misconceptions of logical positivist philosophy, secure atomistic foundations for scientific knowledge were never sought by Neurath, nor by Carnap after his long debate with Neurath in the early 1930s. Given their holistic fallibilism of Quinean proportions (see Carnap 1937/1934: 318–319), the tools to reflect on our knowledge claims in metatheory were also not given but had to be constructed for the purpose.

10 Neurath's "physicalism," far from asserting a crude materialism, only served to bar dualist speculations; see Uebel 2019.

Traditional philosophy of science informed by such Duhemian holism had recognized the gap and sought to address it by employing various background theories (e.g., of perception and scientific instruments) and background assumptions that allow for the assessment and legitimation of evidential relations between the data at hand and a given theory (e.g., the data's relevance and strength). Yet contextual empiricism also recognizes that among those "auxiliary" or background assumptions, nonepistemic ones figure, sometimes long unnoticed or neglected: Here "contextual values" enter. To safeguard the probity of scientific reasoning, therefore, Longino proposes a procedural conception of objectivity. It is not its production of supposedly perspectiveless representation that distinguishes scientific inquiry as objective, but rather the fact that its knowledge claims are subject to comprehensive criticism of all their presuppositions and assumptions, both evidential and conceptual. Transformative criticism has the task to uncover, uproot, and replace previously unnoticed, unwarranted assumptions.[11]

Neurath's views on how theories are chosen are highly congenial: "Poincaré, Duhem and others have adequately shown that even if we have agreed on the protocol statements,[12] there is an unlimited number of equally applicable, possible systems of hypotheses. ... We select one of the systems of statements that are in competition with each other. The system thus selected is not, however, logically distinguished" (1983/1934: 105, translation amended). Neurath remained unspecific about the means by which such a choice is made. It is tempting therefore to invoke a notion he introduced when he criticized Descartes' sharp distinction between foundationally grounded theoretical and pragmatically oriented practical thinking. For Neurath, the distinction between abstract and action-oriented thinking was not an epistemologically categorical one:

> We have seen that in many cases, by considering different possibilities of action, a man cannot reach a result. If he nevertheless singles out one of them to put into operation, and in doing so makes use of a principle of a more general kind, we want to call the motive thus created, which *has nothing to do with the concrete aims in question*, the auxiliary motive, because it is an aid to the vacillating, so to speak. (1983/1913: 4, emphasis added)

[11] Whereas Longino's discussions tend to focus on assumptions of evidential relevance higher up in chain of reasoning leading to the acceptance of hypotheses, Neurath focused on the considerations governing the admittance of observational data; see Neurath 1983/1932b and Uebel 2009. Both types of scrutiny are required.

[12] Protocol statements are statements of evidence, typically of intersubjectively observable states of affairs.

Neurath's point was that, given that "the differences between thinking and action are only of degrees, not kind," that both abstract and action-oriented thought must proceed from uncertain ground, it follows that "thinking too needs provisional rules," that abstract thought also needs rules "which have to be applied as long as one has not reached complete insight" (1983/1913: 2–3). He concluded that scientific thinking is clarified by recognition of the notion of auxiliary motives.[13] As he noted, the simplest form of an auxiliary motive is to have one's action decided by drawing lots, but his more general formulation deserves notice: Auxiliary motives have "nothing to do with the concrete aims in question." Transposed from practical to theoretical thought, this means that an auxiliary motive does not, in and of itself, make it more likely that the theoretical aim of thought, truth, is realized. Adopting an auxiliary motive allows a decision to be taken in virtue of its singling out one utility (one particular type of information wanted about the issue at hand) as determining how the inquiry will proceed. In consequence, both epistemic virtues (coherence, simplicity) and nonepistemic criteria (practical utility) are there to be invoked to select one among the empirically equivalent theories.[14]

For Neurath and Frank, the gap argument was the point at which their logical empiricist epistemology joined forces with John Dewey's pragmatist attack on spectator conceptions of knowledge (thinking of knowledge as faithful copying). Their conception of how to accommodate nonepistemic values in scientific theorizing has considerable appeal to contemporary theorists who also, however, question whether Neurath and Frank went far enough. For many activist scholars, the gap argument is but the first step toward their rejection of the idea of value-free science: They also embrace the "entanglement" of fact and value (Putnam 2002). Rejecting the principled separation of fact and value and claiming their categorical indistinguishability on epistemic grounds is said to allow for the full truth-valuability of value statements.[15] Precisely due to this entanglement, science in conditions of underdetermination and partial ignorance is said to be unable to avoid value questions when decisions about hypothesis acceptance must be taken. In challenging the distinction between

[13] Uebel 1996: 135–136. This reading of Neurath's auxiliary motives is employed also by Okruhlik 2004, Howard 2009 and 2019, Stuchlik 2011. Frank went further than Neurath in explicitly noting values to be involved in theory choice (Frank 1957: 354); for discussion of the opposition he faced see Howard 2003: 61–63.

[14] Cf. Longino 2016.

[15] For different versions of such value cognitivism see, e.g., Anderson 2004 and Brown 2013.

epistemic and nonepistemic values the entanglement thesis undermines all conceptions of scientific value freedom, but whether this further challenge must be granted is itself highly questionable.[16]

Yet consider how Neurath's and Frank's position looks from another radical variant of philosophy of science that, it has been noted, has begun to merge with empiricist feminism (like contextual empiricism), but has its own controversial history: feminist standpoint theory.[17] Where the former can be regarded as originally concerned simply with providing a framework the acceptance of which would make for better and truly objective science (overcoming biases undetected by standard accounts of objectivity), the latter was formulated as a political theory aiming to legitimate interventions in and disruptions of "business as usual." Here let's adopt a formulation of standpoint theory by Alison Wylie that renders earlier controversies irrelevant:

> It is an explicitly political as well as social epistemology characterized by the thesis that those who are marginalized or oppressed under conditions of systemic inequity may, in fact, be better knowers, in a number of respects, than those who are socially or economically privileged. Their epistemic advantage arises from the kinds of experience they are likely to have, situated as they are, and the resources available to them for understanding this experience. Feminist standpoint theorists argue that gender is one dimension of social differentiation that makes such an epistemic difference. (2012: 47)

Standpoint theory starts from a normative position: It provides an epistemology for social cognition that contests the findings and theoretical presuppositions of traditional value-free philosophy of science *as part of* a general struggle for justice and equality.

Both Kathleen Okruhlik (2004) and Don Howard (2019) plausibly identify Neurath as a standpoint theorist of an older variant, namely of Marxist persuasion. Both cite his "Personal Life and Class Struggle":

> The workers who lack a rich bourgeois education, can become superior to the bourgeois precisely in the field of social life in that they have a greater understanding for social connections and can apply even a smaller amount of knowledge more significantly. Marxism shows the proletarians who are engaged in the class struggle what is especially important to know; and it preserves adherents from the often disorganized educational endeavour of

[16] For defenses of the distinction of epistemic and nonepistemic values, see Steel 2010 and Lacey 2017; see also Blackburn 2013 on disentangling thick concepts.

[17] For further details see Intemann (2010), whose suggestion of large-scale convergence appears widely accepted now.

> bourgeois enlightenment, which from the outset sees in merely increasing knowledge something worth striving for as such. (1973/1928: 292–293)

Howard aptly comments:

> It is precisely the oppressed status of the working classes that affords them a privileged epistemic status, more clearly grasping social relations and seeing the lie in rationalizations of bourgeois privilege, rationalizations the falsity of which bourgeois thinkers cannot see as lies because their class status places them in an epistemically disadvantaged state. They cannot see through those lies because their doing so would undermine the power and prerogatives of their own class. Neurath's philosophy of science in action thus paints a picture of politically engaged, indeed revolutionary science in service to the achievement of justice. (2019: 51)

Neurath's commitment to the cause is unquestionable. (If Carnap had still been inclined to writing polemics engaging with issues of the day – as he did early on[18] – this would be equally evident in his case.) But is Neurath's position more consistent than Carnap's appeared to us earlier? Moreover, is his own position up to the task?

Okruhlik voices concern about Neurath's appeal to auxiliary motives. Appreciative of the fact that they allow value-driven decisions inside of science, she worries whether this construction is robust enough to sustain the value commitments it facilitates – and whether it takes the values in question seriously enough. She points to the role of auxiliary motives as "randomizing devices" (suggested by Neurath's talk of casting lots and rolling dice) and contrasts that with decisions taken "non-randomly" by activist scientists who do as activist scholars deem fit (Okruhlik 2004: 63). As we will see, Okruhlik's is not merely a difference of emphasis. There is a further worry. Auxiliary motives seem to be the wrong vehicles altogether to facilitate nonepistemic value input into science:

> [T]hose elements of Neurath's social science that seem to us most overtly political or value-laden do not arise from employment of auxiliary motives but from Neurath's version of Marxist standpoint epistemology. Standpoint theory and the auxiliary motive do not yield to easy assimilation because auxiliary motives come from (and remain) outside science, while it appears that, for Neurath, Marxist social science just is the form that the scientific world-conception takes in the social sciences. (Okruhlik 2004: 64)

To be sure, Neurath's Marxist social science was not grounded by an auxiliary motive but by his belief that "[o]f all the attempts at creating a

[18] See, e.g., Carnap 2022/1918 discussed in Uebel 2012 and Damböck 2022.

strictly scientific unmetaphysical physicalist sociology, Marxism is the most complete" (1973/1931: 349). For Neurath, doing Marxist sociology followed from the ideal of anti-metaphysical social science itself. Yet this does not show that the auxiliary motive does not come into play at other junctures in social scientific reasoning, for instance when cases of underdetermination need resolving. Neurath's one example of this suggested opting for one of the empirically equivalent hypotheses or prognoses about the historical situation faced on the grounds that doing so provided the broadest base for collective action (1973/1928: 293). Here strategic class war considerations served as an auxiliary motive: an interpretation that was not only plausible on its own but also acceptable to the comrades was what was required. Okruhlik's conclusion that "Neurath's standpoint theory is not really a departure from or a rival to [his] empiricism" (2004: 64) nevertheless suggests a major drawback. Neurath "did hold to the empiricist dogma that puts values outside the domain of meaningful discussion. It is this dogma that may constitute the biggest difference between Neurath and feminist philosophy of science" (Okruhlik 2004: 67). The spoiler is a dogma of empiricism that even Quine shared: ethical noncognitivism.

By contrast, Howard sees no problem with the way Neurath resolves the gap argument. "For the purposes of understanding Neurath's philosophy of science in action, what is most important is his argument about the role of the auxiliary motive, for this is what provides legitimation for Neurath's politically engaged science" (Howard 2019: 64). Rather than see in it a coded stance on meta-ethical matters, he reads it as a description of all too commonly misunderstood problem situations:

> It is noteworthy that Neurath terms these factors auxiliary *motives*, not reasons. He means deliberately to make this an issue about the psychology of judgment and not pure reason alone. Neurath's epistemology of science is a kind of naturalistic epistemology. What he gives us here are supposed to be psychological and, thus, scientific facts about how reason operates, not a priori norms. Still, our recognizing the role of auxiliary motives has normative implications because of the widespread failure to discern or admit the work that such motives do. (Howard 2019: 53)

Howard's endorsement of Neurath's model of politically engaged science does not speak to the charge that noncognitivism undermines the rationality of his political engagement. Instead, Howard stresses that "since, in the end, we must choose on the basis of nonempirical factors, we enhance the intellectual integrity of science by frankly asserting the agendas that motivate science in action" (2019: 54). This leaves Okruhlik's challenge open – for from Neurath's understanding of Marxism no rejection of noncognitivism follows.[19]

Like Carnap, Neurath dismissed normative ethics as cognitively meaningless. Already very early on he declared that "a moral demand can never be proved" (1973/1912: 119): He agreed with Hume's denial that norms follow logically from facts. He also argued against utilitarianism as a general principle of social organization for, without an arbiter or dictator, "it is not possible to create an order of life which takes account of different *views* as to the best distribution of pleasures, as would have to be the case with the *pleasures* of each in a purely utilitarian world" (1973/1912: 122, emphasis in 1912 original). Kantian deontology fared no better, with the categorical imperative a ready object of scorn: "how should we demarcate a discipline as 'ethics' if God is eliminated? Can we make a meaningful transition to a 'command in itself', to the 'categorical imperative'? We could just as well introduce a 'neighbor-in-himself without a neighbor'" (Neurath 1983/1932a: 79).

Neurath then was in no better position to argue for socialism than Carnap – if socialism was understood as an ethical position. But as he did not understand it so, no contradiction obtained for him. Yet Okruhlik's challenge remains alive as a pragmatical one. As a social scientist, Neurath could argue the case that a radical reorganization of socioeconomic relations is more likely to improve the lot of the proletariat than a continuation of business as usual and therefore is to be recommended if such improvement is desired – but not that it should be desired. This may appear too weak a stance. Is noncognitivism then as detrimental to political activism as many contemporary critics, such as standpoint theorists, claim when they charge it with putting values "outside the domain of meaningful discussion"?

The Appropriation of Weberian Value Neutrality and Value Relevance

The most commonly discussed version of the doctrine of value freedom forbids taking account of nonepistemic values in science generally, especially in theory evaluation. It is also this version that is under attack in the currently most commonly discussed counterargument to the doctrine, the argument from inductive risk.[20] Roughly, accepting a finding or theory

19 Neurath rejected the philosophical, expressly ethical dimension that Max Adler tried to impress on Austro-Marxism (see 1973/1928: 297).

20 For a recent installment of the debate see Douglas 2017 and Betz 2017.

means to certify it as reliable for use by third parties, yet since virtually all findings are only ever reached on a balance of probabilities, their acceptance involves a judgment that the risk of harm caused by inductive failure is low enough to be neglected. Unsurprisingly, no consensus regarding this argument has been reached, resistance turning on whether risk assessment properly falls to the scientist investigating a hypothesis or to the agents or agencies seeking to make use of the findings. Yet like the gap argument, albeit along a different route, the argument from inductive risk seeks to show the entanglement of science with value questions.

For better or worse, Neurath and Carnap appear to have neglected the argument from inductive risk.[21] They were exercised by the possibility of unchecked intrusion of political-ideological values into science. With this concern and the very broad outlines of their response they agreed with Max Weber. Now relations between members of the left Vienna Circle and Weber and his legacy (he died in 1921) were very complex. As economists, Neurath and Weber sparred repeatedly in the *Verein für Sozialpolitik*, jointly attended the 1917 meeting of the German Youth Movement at Burg Lauenstein as critical "elder statesmen," and encountered each other again during Neurath's trial in postrevolutionary Munich in 1919. As philosophers, Neurath and Weber took contrary stands on the materialist conception of history and in the socialist calculation debate (about whether rational economic planning is possible in a socialist commonwealth); Neurath also remained opposed to Weber's interpretive sociology, forever suspicious of seemingly idealistic tendencies.[22] Given furthermore that Neurath was concerned with what the conception of scientific value freedom provided freedom for, as opposed to Weber's concern with what it proscribed, it is perhaps not surprising that Neurath did not advertise his understanding of value freedom as a version of Weber's – especially as he also had to cleanse it of metaphysical accretions. Carnap fell in with Neurath's take on the matter.[23]

Weber's version of value freedom concerns the results of scientific investigations: It bars a certain type of value statement from being issued as justified by scientific reasoning. Importantly, Weber did not forbid all

[21] As noted in Magnus 2013, the argument is not original to Rudner 1953 but goes back to James 1896.

[22] See Uebel 2018 and 2022.

[23] Unlike Neurath, Carnap never referred to Weber in any capacity, but in October 1928 his list of literature read includes Weber's *Wissenschaft als Beruf* (2012/1919; see Carnap 2022: 748). As Weber there restated his doctrine of value freedom in popular form it is not surprising that Carnap's first Bauhaus lecture (see note 7) shows striking similarities.

value statements but only unconditional ones – in all modalities, be they purely descriptive ("*x* is good") or prescriptive ("*x* should be the case") or expressing commands ("do *x*!") – and he left conditional ones untouched. Phrased differently, Weber barred categorical imperatives from science but not hypothetical ones. Neurath's and Carnap's agreement with Weber on this point is seldom recognized, but the distinction between conditional and unconditional value statements was equally central to the Circle's noncognitivism – and their version of value freedom – as is clearly documented in Carnap's autobiography:

> In our discussions in the Vienna Circle we were much concerned with clarifying the logical nature of value statements. We distinguished between absolute or unconditional value statements, e.g., one that says that a certain action is morally good in itself, and relative or conditional value statements, e.g., one saying that an action is good in the sense of being conducive toward reaching certain aims. Statements of the latter kind are obviously empirical, even though they may contain value terms like "good." On the other hand, absolute value statements that speak only about what ought to be done are devoid of cognitive meaning according to the empiricist criterion of significance. They certainly possess noncognitive meaning components, especially emotive or motivating ones, and their effect in education, admonition, political appeal, etc., is based on these components. But, since they are not cognitive, they cannot be interpreted as assertions. (1963b: 81)

Carnap equated the distinction between conditional and unconditional value statements with the distinction between cognitively meaningful and cognitively meaningless ones. For Weber unconditional statements were unscientific, but he did not deny their truth-valuability. This illustrates that one need not be a noncognitivist to accept Weber's demand for value freedom (he wasn't one).

Weber held that in issuing unconditional value statements science overreached itself. He could have but did not appeal to Hume or argue explicitly against the naturalistic fallacy of "deriving an ought from an is." But neither did he merely claim that "it can never be the task of a science of empirical experience to determine binding norms and ideals from which practical prescriptions may then be deduced" (Weber 2012/1904: 101–102):

> [T]he problem of establishing facts, demonstrating what is true in mathematics or logic, or uncovering the internal structure of cultural values is entirely *heterogeneous* from the problem of furnishing an answer to the question of [what] is the *value* of culture and of its individual elements, and how one should accordingly act within the cultural community and political groupings. (2012/1919: 346, emphasis in original; cf. 2012/1904: 103)

What made these two sets of problems so different was the fact of moral and political "value pluralism." What Weber noted as a striking and novel fact of "modernity," we take for granted as a fact of "multiculturalism." "The 'scientific' advocacy of practical standpoints is impossible . . . (except in cases where one is discussing the means for achieving a goal that is presupposed as a fixed *given*). It is meaningless in principle, because the different value orders of the world are in irresolvable conflict with each other" (Weber 2012/1919: 347, emphasis in original). There is, Weber took it, no evidential standard for which of the many conflicting value judgments should prevail in society. (Neurath's judgment on utilitarianism as a social philosophy, outlined earlier, converges with this.) For unconditional value statements it is impossible to establish the type of evidence base that is required to sustain claims to objectivity. The question of which social values were to be realized was one to be decided not by science but by civic society and depended on the active engagement of the citizens.

What use then was there for science, indeed social science? Weber's answer (of which we heard echoes in Carnap earlier) is as follows:

> [A]ll that an *empirical* discipline can demonstrate with the means at its disposal are the following: (1) the unavoidable means [to effect a certain goal]; (2) the unavoidable side effects [of doing so]; (3) the resulting competition between a number of different *possible* valuations [on the basis of] their practical consequences. . . . But the question: (1) to what extent a goal may justify the unavoidable means; (2) or to what extent the unwanted side effects may be acceptable, let alone: (3) how to resolve conflicts between a number of goals that one has set for oneself or that are regarded as obligatory, and that collide in the concrete case – even such simple questions are entirely matters of choice or compromise. No (rational or empirical) scientific procedure of any kind whatsoever can decide them. *Our* strictly empirical science can least of all presume to relieve the individual of [the burden of] his choice. (2012/1917: 315, emphasis in original; cf. 2012/1919: 349–350)

Furthermore, scientific policy advice had to respect the same strictures as purely theoretical science: Unconditional value judgments were barred. Any advice was to be formulated in terms of conditionals which asserted means–ends relations: These are bona fide empirical statements, legitimated by intersubjectively available evidence (Weber 2012/1904: 102). (From here on I distinguish Weber's and the Circle's versions of value freedom as "value neutrality.")

To see these ideas implemented in a social science context, consider Neurath's contribution to the *Werturteilsstreit* in an internal discussion document for the *Verein für Sozialpolitik* dedicated to addressing Weber's challenge to social scientific value discourse. (Neurath's use of "pleasure" and "pain" as generic terms – "*Lust*" and "*Unlust*" – does not indicate a sensualist understanding of utility.)

> 7. Moral judgments can impinge on the discipline of economics at *two* points. (a) In the investigation of concrete relations of pleasure and pain. The pleasure or pain resulting from an individual's moral evaluation is co-ordinated to the pleasure and pain which is caused by clothing, food, accommodation, works of art, etc. (b) In the evaluation of a concrete system of institutions which causes pleasure and pain. I can state, for instance, that some order of things conditioned by a certain institution and causing a particular distribution of wealth is of lower moral value for me than some other order of things. *In this case what is evaluated morally is the order of things, whereas in the first case the moral evaluation itself was part of this order.*
>
> 8. Moral evaluation can be considered as a manifestation of pleasure and pain in every concrete investigation, for instance by also taking account of the moral indignation caused by servitude in some region, besides taking account of the lack of food that comes along with the servitude in that region.
>
> 9. The moral evaluation of systems of wealth distribution, say the free market or some other system, is amenable to scientific formulation once one has agreed on the principle serving as basis for the moral evaluation. One can raise the question: which of the orders A, B, C, . . ., N accord best with principle X? Whether an answer can be always given, or even a univocal one, is another matter. (Neurath 2004/1913: 297–298, emphasis in original)

Note that the two occurrences of value statements specified in §7 are illustrated in §8 and §9 respectively. Value judgments may become a datum for empirical behavioral science (as in §7a and §8). Yet value judgments can also be passed within empirical science (as in §7b), but only under one condition: that the standard of evaluation be agreed, that is, made explicit (as in §9). In other words, conditional value judgments about matters investigated in empirical social science are permissible.

So Weber held that the value pluralism of modernity prevents unconditional value judgments from commanding universal consent and therefore excluded them from empirical science. Neurath and Carnap excluded unconditional value judgments from science because of their verificationism according to which statements must, at least in principle, be testable by reference to intersubjectively available evidence to be cognitively significant. Since the practical outcomes remain the same, one may wonder whether Neurath's and Carnap's version adds anything significant to Weber's value neutrality. The answer is that, importantly, it subtracts something. Given Weber's repeatedly advertised allegiance to the neo-Kantian Heinrich Rickert's idealist value theory, empiricists could not but reject Weber's original version. For Rickert's philosophy of value, modernity's value pluralism was simply a mistaken illusion of the age; whether he himself agreed with this or not, Weber limited his prohibitions to the realm of empirical science. Since Weber also denied, like Rickert, the unity of science thesis – a core doctrine of logical empiricism which disputed a special status for the human sciences – Neurath and Carnap had to transpose Weber's conception of value neutrality from a neo-Kantian to a naturalistic setting.

Consider that Weber spoke as if "value relations" constituted the sole objects of "the cultural sciences," that is, social science, whose "transcendental precondition" was "that we *are* cultural *beings*, endowed with the capacity and the will to adopt a deliberate *position* with respect to the world, and to bestow *meaning* on it" (2012/1904: 119, emphasis in original).

> The concept of culture is a *value concept*. Empirical reality *is* "culture" for us because, and to the extent that, we relate it to value ideas; it comprises those, and *only* those, elements of reality that acquire *significance* for us because of that relation. Only a tiny part of the individual reality that we observe at a given time is coloured by our interest, which is conditioned by those value ideas, and that part alone has significance for us; it has significance because certain of its relations are *important* to us by virtue of their connection to value ideas. (2012/1904: 116, emphasis in original)

This is the neo-Kantianism that the Circle theorists were unable to accept. Fortunately, it was possible to rescue something tangible, as Weber himself once hinted at.

> As for the meaning of the term "value relation" ... suffice it to recall that [it] simply represents the philosophical interpretation of that specifically scientific "*interest*" which governs the selection and formation of the object of an empirical inquiry... even purely empirical scientific research is *guided* by cultural interests – that is to say: value interests. (2012/1917: 317, emphasis in original)

Detranscendentalize and demetaphysicalize Weber's value talk and what you get is the simple recognition that the pursuit of social science is guided in the choice of its subjects and in the determination of its research agendas by the interests of its researchers – and that there is nothing wrong with this. Indeed, as has often been noted (e.g., Nagel 1960: 486), what's also called "value relevance" is not the sole property of social science at all but extends across all disciplines (saving the unity of science).

Another difference between Weber and Neurath also deserves notice. Value neutrality on its own does not address worries about biased procedures in the gathering of data, the generalization of hypotheses, and the evaluation of theories by peer groups.[24] It must be complemented by an argument that recognizes and regiments nonepistemic value choices in these respects. It may not have been a coincidence that Neurath's gap argument also makes room to consider these matters so as to complement his adoption of Weberian value neutrality and value relevance.

Neurath's Noncognitivist Standpoint Theory

For Weber, value neutrality came combined with value relevance which Neurath and Carnap separated from the idealist philosophy with which he had associated it. This allows for the partisan choice of research projects but forbids partisan formulations of research findings. Importantly then, it allows for a transformative agenda quite independently of the value considerations legitimated by the gap argument. Neurath's socialist economics, in particular his radical proposals for the socialization of entire national economies in the wake of World War I, also express this stance. Depending on whether he was speaking as a scientist or citizen advocate, we can find fiery speeches and propaganda among his output, but also scientifically neutral discussions of the conceptual frameworks required to develop such schemes for social transformation.[25]

Yet Neurath's transformation of Weberian value-neutrality-cum-value-relevance stands in a challenging relation to standpoint theory. One might wonder whether the description of Neurath as a standpoint epistemologist is felicitous: Without affirmation of nonepistemic values, standpoint theory may feel like *Hamlet* without the Prince of Denmark. The puzzlement is understandable, but two further questions arise. First, whether pursuing research programs that are informed by political agendas demand for their success that the unconditional value statements that inform their adoption be proclaimed alongside and on par with their results. Second, what the role of unconditional value statements is for standpoint theory

[24] On the latter see, e.g., Rollin 2017 and Intemann 2017.

[25] Compare Neurath's 2004/1920a in an academic journal with his 2004/1920b, a freestanding propagandistic pamphlet.

and what that tells us about the relation between standpoint theory and general epistemology and philosophy of science.

The first question must not be misunderstood. The issue concerns neither scientists hiding the value commitments of their research programs nor strategies of obtaining and dispensing research funding. The question is rather: What is lost in terms of research output if the prescription of value neutrality is followed? I submit that it is nothing that is of strictly scientific value. To be sure, the public persona of scientist activists may be less headline-grabbing than if they were moral crusaders. But note that value neutrality does not prohibit the very same scientists from being passionate advocates of the agendas their research is meant to further – but this they would do as only citizens in the civic arena, not as expert scientists: *vide* Neurath! They can even use their scientific results to bolster their political argument (present the facts of deprivation, say, and likely means of alleviation). The only thing they cannot do is claim that science gives unconditional backing to their agenda (here of providing alleviation of the deprivation).

Answering the second question is more complex. With standpoint theory regarded as a normative political theory, the role of unconditional nonepistemic value statements is plain: they state its basic value axioms and are thus indispensable. With standpoint theory regarded as epistemology, it is not clear what role they have to play. What is clear, however, is that under the heading of standpoint theory, both normative proposals and descriptive theses have been put forward.[26] Standpoint theory, we saw, emerged as a normative political theory to articulate anti-discriminatory demands and overturn androcentric bias in traditional epistemology and philosophy of science and found application in the sciences and in the provision of health, social care, and law across society generally. Yet standpoint theory is not only about advocacy (especially in criticizing undesirable practices), however important that is; it also made significant contributions to epistemology itself. For instance, it has challenged what Okruhlik called "the dogma of the intersubstitutability of epistemic subjects" (2004: 67) – that epistemology be blind to their social situatedness[27] – and from this recognition of a desirable pluralism of perspectives follow consequences for how to think about objectivity and question the ideal of the "view from nowhere."

[26] See Wylie (2012), who summarizes its history before defining standpoint theory as quoted earlier.

[27] For the political valence of the concept of situated knowledge see Haraway 1988.

Given that standpoint theory is both politically and epistemically normative, its stand on unconditional nonepistemic values can be a differentiated one. It would of course be nonsensical to bar it, as a political theory, from asserting unconditional value statements, but it is not at all clear why, as an epistemology, it should insist on issuing them. Prohibiting them would not rob standpoint epistemology of its critical bite, given the transformative agenda of its political wing, but only distinguish between the roles of engaged advocate in the civic arena (with scientific malpractice in view) and the role of epistemologist (parallel with first-order scientists). The move to procedural objectivity away from the view from nowhere conception would not be endangered.

Let me stress that it is not my business to suggest how feminist epistemologists should go about theirs. What is my business, however, is to argue that value neutrality is much less detrimental than it may at first appear. To ask the question, seemingly so absurd, of what standpoint theory would lose if it were to renounce unconditional nonepistemic value statements, is to clarify in what sense it is appropriate to think of Neurath as a standpoint theorist in pursuit of a transformative agenda. My answer is that he can count as one if we allow for a noncognitivist version of standpoint *epistemology* (and set him to work further on the situatedness of cognition which he only began to consider). Cognitivist and noncognitivist standpoint epistemologists can speak as one as civic actors; they agree in their politics, after all. Only their activism as scholars and scientists proceeds in different voices – but this does not change any potentially transformative results of their theorizing.

Discussing Standpoints with Carnap

Yet is this all there is to the issue of cognitivism versus noncognitivism? Carnap's remark (1963b: 82) that one's meta-ethics rarely if ever determines behavior may well be true, but there remains Okruhlik's worry, prompted perhaps by unduly strident talk in Carnap's London lectures (1935: 23), which may suggest that noncognitivism "puts values outside the domain of meaningful discussion."

Elsewhere, however, Carnap had already clarified that "the exclusion from the domain of *theoretical judgments* does not relieve us of the ability, even the duty to adopt a *practical attitude*. There is a *fundamental difference* between both, however, which we must come to understand" (2013/1934: 176, emphasis in original, my translation). Here we touch on the all too often neglected positive part of the message of Carnap's noncognitivism:

his recognition of the "other" of scientific reason and the indispensable complementation of reason by this other, the will and willing. All action requires decision, and this demands that we "adopt a practical attitude" on what's at issue – and that includes value questions.

> By theoretical means one can only determine here that this or that institution brings with it these or those hygienic, economic or cultural consequences. This is a very important preparation for our adoption of an attitude, but it does not render this adoption otiose. We must decide whether we are in favor of or against the consequences which theoretical investigation has established will follow (e.g., the elimination of economic crises and unemployment). It is on this that, guided by theoretical insight, our action depends. (2013/1934: 177, my translation)

"Adopt[ing] a practical attitude," taking a stance, is what agents do. (One is tempted to say that is what makes for an agent.) Carnap's terms are striking: "*Pflicht der praktischen Stellungnahme*" (duty of adopting a practical attitude) and "*Sache der praktischen Stellungnahme*" (matter of adopting a practical attitude), the former denoting the normative, the latter the descriptive dimension of exercises of the will. The same duality applies to assuming, taking, and adopting a "standpoint." ("*Praktischer Standpunkt*" is a close cognate of "*praktische Stellungname.*")

But what, in Carnap's hands, makes for a responsible *Stellungnahme* that is within the means, intellectual and affective, of the agent? Elsewhere I discussed the recognition of cognitive autonomy and reflexivity as required for rational action by Neurath (2004/1913); here I turn to Carnap's later analysis which illuminates their recognition of the all-too-human condition of having to adopt practical attitudes:

> This result of a logical analysis of value statements and the controversies concerning them may appear as a purely academic matter without any practical interest. But I have found that the lack of distinction between factual questions and pure value questions leads to confusions and misunderstandings in discussions of moral problems in personal life or of political decisions. If the distinction is clearly made, the discussion will be more fruitful, because with respect to the two fundamentally different kinds of questions the approach most appropriate to each will be used; thus for factual questions arguments of factual evidence will be offered; whereas persuasion, educational influence, appeal, and the like will be brought to bear upon decisions concerning pure value questions. (1963b: 81)

Carnap offered the fact–value distinction as a basis for an "explication of value statements" (1963b: 1009). The distinction is an analytic one made for pragmatic purposes: It cannot be overstressed that it is *not* an ontological distinction (1963b: 1003).

This is also true of noncognitivism: It provides a framework for analysis. Assume a list of (1) "statements connected with values or valuations" (behavioral descriptions, means–ends and utility claims) and a list of (2) statements connected with values or valuations that are "clearly analytic if true, otherwise contradictory" (statements whose truth is intelligible given only the meaning of the terms used: logical truths, T-sentences, conceptual explications). Now the "thesis of noncognitivism" can be stated in a conditional form: If a statement on values is neither factual (belongs to category [1]) nor analytic (belongs to category [2]), then it is noncognitive. This is consistent with some value statements being factual and "rejects only those conceptions which regard knowledge of values as a knowledge sui generis, essentially different from factual and logical knowledge" (Carnap 1963b: 999–1000). Next, define "optative" as "a general kind of meaning common to all statements expressing a wish, a proposal, a request, a demand, a command, a prohibition, a permission, a will, a decision, an approval, a disapproval, a preference, or the like, whether or not they also contain meaning components referring to matters of fact." Any sentence that "has a meaning component of this sort" is an "optative sentence." Now noncognitivism asserts unconditionally: "There are pure optatives" (Carnap 1963b: 1001). So even pure optatives are far from meaningless, but their type of meaning is not descriptive: The direction of fit does not go from world to mind but from mind to world. (Their acceptability to a subject is determined by whether they correctly express the way she wants the world to be and whether they are consistent with her other value commitments.) Noncognitivism only holds that there are statements that do not describe and cannot be true or false since they instead express that something should be the case.

In a recently discovered fragment, Carnap called such statements "value functions" and integrated them in a Bayesian decision-theoretical framework. (Given a credence function, a body of evidence, and a set of possible actions, it can be defined what a "rational action" is, namely an action for which there exists no alternative that is preferred by the agent in that situation.) People possess many different partial value functions; importantly, however, Carnap allowed that "there is also a comprehensive value function" which "comprises all aspects" of what a person values and "in which the relative weight of each aspect in any possible overall situation finds expression – aspects that are sometimes in mutual conflict" (2017: 192). Carnap affirmed that there are "standards of rationality for value functions" and made some proposals, but noted that they would not rule out as irrational value functions that "would be considered by most people, perhaps all, as completely wrong and immoral" (2017: 193). As elements of a decision-theoretic calculus, Carnap's value functions were judged only for their formal fit. No "purely valuational criteria" to feed into the comprehensive value function were mentioned by him, so the nature of its "weighting" of individual value functions is left undetermined.

The significance of his decision-theoretical calculus for our concerns is that it shows that, his noncognitivism notwithstanding, Carnap took practical reasoning very seriously (the model links up with his long work on inductive logic). Moreover, by analyzing "complex value statements" into components which are either purely factual or purely optative it becomes possible to exhibit the value commitments expressed by the complex statement (Carnap 1963b: 1009–1011). This provides an example of the first of the modes in which "a scientific treatment of value-judgments" may proceed, according to Weber. It can "help the striving person to reflect on the ultimate axioms that form the basis of what he is striving for, on the ultimate value standards that he applies or that he should apply in order to be consistent" (Weber 2012/1904: 103). Significantly, it is science – here formal science: logic in the broad sense – that provides this clarification. Add to this what, as Weber already noted, the empirical sciences can offer regarding practical value questions: consideration of "(1) the unavoidable means; (2) the unavoidable side effects; (3) the resulting competition between a number of different *possible* valuations [on the basis of] their practical consequences" (Weber 2012/1917: 315, emphasis in original). Together, the logical and the consequential analyses of value statements – that is, analyses addressing questions of logical consistency and dependence and questions about means–ends relations and resources – provide endless material for discussions that inform decisions. However, what "the cognitive" cannot do for us – and on this point all three, Weber, Neurath, and Carnap, are uncompromising – is what only the will can do: make the decision.

> What can be proven theoretically is that philosophical and religious metaphysics is a potentially dangerous narcotic that damages reason. We reject this narcotic. If others love its use, we cannot refute them theoretically. This does not mean at all, however, that we must be unconcerned about how people decide on this point. We can give theoretical information on the origin and the effect of this narcotic. We can also work on people's practical decision of the matter by exhortation, education, example. But we must in this be clear that this work lies outside of the theoretical field of science." (Carnap 2013/1934: 179, my translation)

Here Carnap embedded what became his decision-theoretical conception in a naturalistic psychology that is open to elaboration by neighboring disciplines. Carnap's discovery of the "other of reason" does not reveal a hidden metaphysics but points to the complexity of the behavioral sciences. It also reveals a refreshing honesty about what philosophy can do: If we wanted to ennoble his common sense as "metatheoretical reflexivity," the additional adjective "deflationist" would be appropriate. (Given this analytical stance, he and Neurath need not even deny the de facto entanglement of facts and values in the wild, only that they cannot be disentangled.)

In sum, when it is alleged that noncognitivism "puts values outside the domain of meaningful discussion" it must be answered that this is false for Carnap's and, we may take it, Neurath's versions of it. They can discuss what value statements and valuations entail and presuppose logically and what practical consequences are likely to attend to action taken or not taken in their light. Thereby they can impress on agents the responsibilities they face. What they cannot do is establish the truth of unconditional value statements. I submit that cognitivists cannot do this either. (Forceful claims to truth without evidence, if repeated often enough, may prove effective in certain historical situations, but this does not make them rationally justified.) Noncognitivists are no less fit for the public contestation of values than cognitivists.

Conclusion

Needless to say, what I have defended here needs elaboration and supplementation in all sorts of ways, not being a theory in its own right but a gloss of a perspective recovered from underappreciation.[28] What prompted this investigation of the practical dimension of the metaphilosophy of the left wing of the Vienna Circle was the disquiet felt by activist theorists about the doctrine of value freedom and noncognitivist value theory. With their position on value freedom clarified as subscription to demetaphysicalized Weberian value neutrality and their position on noncognitivism identified as recognition of the other of reason (there are pure optatives, statements whose acceptability to a person is not determined by the satisfaction of truth conditions), their use of value relevance can now be

[28] Like all philosophical positions, noncognitivism faces outstanding problems; note that Carnap's to-do list is broad enough to comprehend the Frege–Geach problem which had not yet entered the literature when he wrote (the Schilpp volume was long delayed): "logical rules must be stated for the logical relations, especially for logical implication, both between value statements and between value statements and cognitive statements" (Carnap 1963b: 1013).

regarded as noncognitivist standpoint-taking within science. Other arguments may also have to be considered, but given those discussed here I conclude that activist scholars and scientists need not deny all forms of value freedom. Neurath's and Carnap's form of value neutrality, even their noncognitivism, does not prevent the epistemology of science playing its part in the moral and political struggles of the day.

CHAPTER 8

The Pragmatic and the Religious Functions of Science

Matthew J. Brown

Introduction

"Science is like religion," it is sometimes said, or perhaps, "Science is merely another religion." In popular conversations, science and religion are equated for various reasons: to point out that science involves "believing where we cannot prove" (Kitcher 1983), or that it involves "faith" in some sense. In a more extreme register, it can be a nod to relativism, an argument that scientific "knowledge" is nothing more than a set of beliefs among other competing beliefs or knowledge claims, none more valid than the rest. The science-as-religion idea is sometimes bolstered by philosophical arguments, such as a version of the thesis of underdetermination of theory by data, according to which radically different theories and assumptions can be equally well supported by the same empirical data. Those who attend to the social dynamics of science find analogies with religion as well, for instance, in the ways in which a scientist changing allegiance from one theory or research tradition to another resembles something akin to a "conversion experience" (Kuhn 1962), or in the degree to which political struggles among parties to a scientific controversy bears resemblance to disputes among members of different religious denominations. Call this idea, that science and religion are the same sort of thing, that science is a religion or like a religion, *science-as-religion*, or *religionism* for short; and the opposite view, that the two are inherently very different sorts of things, *anti-religionism*.

The religionist line of thinking runs up against a powerful objection: In a profound sense, unlike religion, *science just works*. That science is a highly pragmatically successful endeavor, enabling accurate and reliable powers of prediction and control, seems obvious. While not all sciences are equally successful in this regard, overall the scientific approach has proven to be pretty good at prediction and control of the world around us, while religion, whatever its benefits, offers no such practical track record. Science has given us incredibly precise predictions of astronomical, microphysical, and chemical phenomena, as well as medical, transportation, communication, and computational technology beyond the imaginings of prior generations. For many, this is the primary way to understand the significance of science: The primary function of scientific methods, theories, laws, models, techniques, etc. is to enable us to predict and control the parts of the world that interest us. Let us call this "the *pragmatic function* of science." That it is *an* important function of science is clear and relatively uncontroversial.

A more controversial, extreme *pragmatism* would argue that this primary function of science is also the whole story with science; according to such a pragmatist, science just *is* problem-solving inquiry that helps expand our ability to predict and control the world around us when our habits and practices fail us.[1] Even in the realm of so-called pure or basic science that seems to have little practical applicability, there is often a very high degree of precise and accurate prediction and ability to create and manipulate phenomena. While the bizarre subatomic particle behaviors that are exhibited in high-energy supercolliders seem to have little use on a practical level, the pragmatist can still insist that our theories predict their behavior with a high degree of accuracy and our experimentalists can manipulate and control that behavior in highly specific ways. Likewise, our ability to predict astronomical phenomena goes well beyond our practical needs for calendar-making or space travel into realms with no practical significance that we currently anticipate.

What the pragmatist lacks is an account of those aspects of science that are not closely connected to our practical capacity to predict and control. In other words, the pragmatist seems unable to explain both inquiry that is governed by standards and values orthogonal to predictive and experimental precision and accuracy, including the crafting of grand theories of universal scope that synthesize many of the local achievements of a scientific field. Scientific *realists* have pointed to these gaps as examples of the inadequacy of this sort of pragmatism. According to scientific realism, the aim of science is to produce a true picture of the world, and the picture of the world it has produced deserves our belief, at least in most of its details.[2] Some pragmatists have fired back against the realists

[1] Pragmatic approaches to science of various sorts are enjoying renewed attention in philosophy of science. See Kitcher 2011; 2015; Miller 2014; Douglas 2014; M. J. Brown 2020; Chang 2022; Andersen and Mitchell 2023.

[2] See Chakravartty 2017 for a thorough introduction to the varieties and complexities of scientific realism.

dismissively, arguing that the aspects of science that exceed the concerns of prediction and control are merely metaphysical or religious clap-trap, a failure of the Enlightenment to carry out the project of disenchantment of the world thoroughly enough. This "disenchantment" was theorized by Max Weber as a process that began with the tendency of Abrahamic monotheism to eliminate magic and ritual, and continued with the scientific secularization and rationalization of the world (Mishima 2020). In Weber's words: "It means that in principle, then, we are not ruled by mysterious, unpredictable forces, but that, on the contrary, we can in principle *control everything by means of calculation*" (Weber 2004/1917: 12–13, emphasis in original; see Mishima 2020).

What is at stake, then, between the scientific realist and the pragmatist are competing conceptions of humanism. The realist seeks a rational, scientific alternative to the religious-metaphysical worldview of "mysterious, unpredictable forces," gods, spirits, souls, magic, or miracles. It seeks to substitute a worldview composed of unseen laws of nature, forces, fields, fundamental particles, quarks, strings, etc. The extreme pragmatist sees this quest as itself of a piece with rather than a true overcoming of the religious-metaphysical worldview; metaphysics comes from a longing for the unseen, "really true" world in both cases, rather than a disenchanted world in which everything that *is* is in principle a potential subject of our control.

Religionists would, in a sense, agree with the pragmatist response against the scientific realists, though not necessarily with its pejorative tone: When we take science to produce a metaphysical worldview, science is playing the role of a religion. According to the religionists, the scientific realist illegitimately uses the pragmatic success of the parts of science concerned with practical prediction and control (what is often called "applied science") to argue for the status of a naturalistic, scientific worldview. But this is a kind of bait and switch; the local practices of prediction and control that are highly successful in science are only loosely connected with the grand-scale synthetic theorizing of the scientific worldview. The superiority of the scientific worldview over the religious thus cannot be defended on the basis of the pragmatic successes of science.

In this chapter, I attempt to broaden the pragmatist approach in a way that threads the needle between the different concerns of the realists and those who are for and against the thesis of science-as-religion. I argue that in addition to its pragmatic function, science also has a religious function; here I use "religious" in an entirely nonpejorative sense. This "religious function" of science explains the significance of the grand synthesizing of the scientific worldview that has no (direct) pragmatic value of its own. I attempt to provide a pragmatic defense of this aspect of science by drawing on classical pragmatist philosophies of religion. The writings of William James and John Dewey on religion give us a way of assimilating the value of the religious function of science to a broadly pragmatist philosophy of science, thus answering rather than dismissing the concerns of the realist about the completeness of a pragmatist philosophy of science. The classical pragmatists give us a way to think of the religious function of science as a positive contribution to the construction of a naturalist, humanist worldview that is desperately needed in the present era, without an illicit argument based on the empirical successes of science.

I first explore in greater detail what realist philosophies of science have found missing in narrowly pragmatist philosophies of science. I argue that there are ultimately three sorts of things for the pragmatist to worry about: (1) apparently useless science, (2) nonpragmatic epistemic criteria for evaluating scientific claims, and (3) the construction and status of scientific worldviews. Next, I explore each of these aspects of science in turn, arguing that only the third poses any real difficulty for the pragmatist. Then I explore ideas from classical pragmatist accounts of the pragmatic function of religion, which allow me to conclude by articulating a positive religious function for science in our society on pragmatist grounds.

What Needs to Be Explained

Science, in fact, gives us great powers of prediction and control. What more should we want from science? What is it, exactly, that the realist thinks science does that the pragmatist cannot explain? Scientists and philosophers of a realist bent have regularly insisted that we should want more, much more, from science than mere pragmatic success in prediction and control, and that science can or does deliver such things.

One common refrain is that what is central to scientific progress is *basic* or *pure science*, the pursuit of scientific knowledge wholly independent from our practical aims, interests, and activities. Such science pursues or arrives at *Truth* in the sense of accurate representation of *Reality*, or at least knowledge of the *deep structures* or *unobservable features* of our world. Science aims at, and its success is judged by, not only increasing success in prediction and control but also increasing our power of explanation and understanding of the world, judged according to a set of *explanatory* or *superempirical virtues* or *epistemic values*; these virtues or values guide inquiry as much as the pragmatic ones, says the realist. Ultimately, the goal is not just instrumental but also to arrive at a full *scientific world-conception* or *worldview* in which we can understand the universe and our place within it. Indeed, realists commonly argue that the high degree of predictive, experimental, and technological success science has achieved would be a *miracle* if its theories were not in fact tapping into *deep truths* about the nature of the world beyond our senses. It is generally thought that this set of values, aims, goals, and achievements cannot be accounted for by the pragmatist. Let us try to get clear about what, exactly, the lacunae are supposed to be, and then determine which pose genuine problems for the pragmatist.

As a preliminary point, while some pragmatists of the past may have been committed to philosophical views that would prevent them from acknowledging that science posits unobservable entities, the contemporary pragmatist has no such compunctions. If electrons, quarks, markets, mental states, laws, kinds, or what have you play significant roles in bodies of knowledge that enhance our abilities to predict and control, many contemporary pragmatists have no qualms about them as objects of knowledge realistically construed. There is no reason that the pragmatist need be a strict empiricist. Indeed, the classical pragmatists frequently criticized the traditional empiricists for their view of experience; rather than understanding experience as composed of atomic sense-data (a bundle of independent and simple sensations such as color and shape), the pragmatists saw experience as having depth, structure, and continuity. There's no reason that a pragmatist cannot say, first, that our main contact with electrons concerns what they can help us predict and control, and second, that on that basis we understand them as real elements of the furniture of the world.

Preliminaries out of the way, there are three challenges to the pragmatist in accounting for these aspects of science that seem to go beyond mere prediction and control.

First, there is the question of "pure" or "basic" research with no obvious or immediate applicability. In such cases, scientists are surely doing scientific inquiry, and that inquiry is aimed at expanding our powers of prediction and control; but being able to predict and control those particular phenomena serves no particular use that we can foresee. That might be because the objects of that research are distant in time (paleontology and the biology of dinosaurs) and space (astronomy and the distant stars and galaxies). It might be because there is nothing inherently interesting about the subject to anyone but the scientists who study it, perhaps because it is too removed from our common experience and no technological application has been conceived (much recent high-energy particle physics). It might be because the nature of the phenomenon forbids fine-grained prediction or any intervention whatsoever, such that no utility is on offer (physical cosmology). How can the pragmatist account for the value of such inquiry?

Second, there is the question of virtues, values, standards, or criteria for science that go beyond the pragmatic concerns of prediction and control. We can think about this challenge in a few different ways. Such superempirical epistemic standards might include things such as a scientific theory's explanatory power, simplicity, unifying ability, or fruitfulness for future research. Some have tried to give pragmatic justifications for these standards, arguing for example that they make a theory easier to use, and thus more testable. In this case we reduce the supposed nonpragmatic virtues to pragmatic ones. Some would argue that such standards come in only when the evidence has run out, when empirical and pragmatic factors underdetermine theory choice. If we hold that these superempirical standards are on a par with pragmatic and empirical criteria, then there may be contexts where we choose less accurate and less "useful" theories, because, say, they provide simpler, more unifying explanations. (If we would never do this, then those standards are not actually on a par, and the challenge to the pragmatist is minimized.) If choosing such theories is a reasonable way to proceed, the pragmatist must be able to account for it.

Third, there is the question of scientific worldviews. What I call "construction of a scientific worldview" is an important part of the creative and constructive activity of science that does not consist of empirical inquiry into specific phenomena. Scientific theorizing also involves synthesizing across a wide range of empirical inquiries, in order to provide a larger picture of the universe (and our place in it). This form of theory construction typically builds on past achievements of observational and experimental research, but it need not and often does not have much direct contact with empirical inquiry itself. The grand theorizing by figures such as Newton and Einstein sometimes have such a character; so does the work of synthetic popularization by figures such as Stephen Jay Gould, Richard Dawkins, Stephen Hawking, and Carl Sagan. Sometimes this work inspires, redirects, or guides future empirical inquiry, and so in retrospect its pragmatic value seems clear. In other cases, often the most ambitious examples of such work, the connection to particular empirical inquiries remains tenuous. The significance of the latter seems very difficult for the pragmatist to explain.

In what follows, corresponding to these three challenges, I seek to provide some basic evaluation from within a broadly pragmatist point of view of the following:

(1) Scientific inquiry that is "useless."
(2) Scientific inquiry guided or judged by standards other than prediction and control.
(3) The construction of a "scientific worldview."

In the following three sections, I explore these three topics in turn. As we will see, the first two of these challenges can be handled in a relatively straightforward way. The third challenge, however, will require us to explore in depth pragmatist views about religion and humanism in order to fully assimilate the positive significance of scientific worldviews into a pragmatist approach.

The Pragmatic Value of "Useless" Science?

In a way, ironically, those inquiries that are supposedly impractical are the easiest for the pragmatist to account for. The results of a scientific inquiry may not be "useful" in the narrow sense of immediate applicability in medicine, engineering, or policy; but nevertheless, as genuine inquiry, it might be governed by the broadly pragmatic criteria of prediction and control. Still, insofar as the pragmatist emphasizes the increase in our prediction and control of phenomena *that interest us*, there remains a problem of accounting for why we would be interested in such recherché phenomena as basic science often tackles, such as the biology of prehistoric creatures or the behavior of distant stars.

In "Genuine Problems and the Significance of Science" (Brown 2010), I worried about this issue in the context of thinking about Philip Kitcher's (2001) account of "scientific significance" and the aims of science. According to Kitcher, the "significance" of a scientific problem or scientific inquiry can be understood as its place in a network of interconnected aims and projects, and the grounding points in this network, from which all significance ultimately flows, are obvious practical uses, on the one hand, and questions of what he calls "natural curiosity," on the other.

Consider the work of an auto mechanic. A mechanic is an inquirer, engaged in problem solving in relation to the diagnosis and repair of malfunctioning vehicles. The work is not quite scientific, although it is notoriously difficult to draw such lines, but it draws on some science and engineering knowledge, design specifications of the vehicle, manuals, heuristics and rules of thumb, intuition hard-won by experience, and a good bit of guess-and-check. It is typically pretty unsystematic, ad hoc, in response to the specific case in front of the inquirer.

Suppose the auto mechanic consistently runs up against a problem that cannot be solved with the resources available to them. After reflecting on the pattern of failures, they determine that the source is not in a lack of skill or knowledge on the part of the mechanic but with some principle of engineering that they regularly rely upon. In most circumstances, this principle helps the mechanic in the repair of diverse automobile engines. But in a certain number of cases, their inquiry fails, and no successful repair can be made. Most mechanics would just accept that some cars cannot be fixed, but our mechanic is particularly dogged and becomes so consumed with the solution to the problem that they go to school for an engineering degree, hoping to determine its source, and so becomes eventually a working researcher revising the very principles of engineering that they once used as a mechanic.

Suppose our newly minted engineering researcher consistently runs into trouble when dealing with particular principles learned from basic physics. Often those principles serve them well, but on certain occasions forming a pattern, they fail to aid the engineer in their inquiry. Our engineer reads more and more about the physics involved and realizes that the ultimate source of the problem is a gap in our knowledge of physics, finally pursuing another degree in that field so as to work on revising our understanding of the laws of physics.

In this fantasy story, we see how the work of a physicist might have traceable lines to the work of the engineer, even the auto mechanic – and such lines (understood conceptually rather than embodied in a single person) are part of the story of significance for Kitcher. But other work in physics does not seem to have such clearly traceable lines. Nevertheless, the physicists working on problems without such traceable lines of connection to practical concerns are drawn to those problems as much as our fantasy mechanic-turned-engineer-turned-physicist. According to Kitcher, there are some kinds of questions about the nature of the world, life, and human nature that we are all naturally curious about. A similar story could be told in terms, not of practical inquiries, but of these questions of natural curiosity. According to Kitcher, for even seemingly abstract and technical scientific inquiries, we can trace their significance back to a combination of all the practical problems and questions of natural curiosity that they bear on in some way. The amount and strength of such connections helps us compare the significance of different scientific projects.

The concern I raised in my earlier paper in response to Kitcher was that when we start weighing the value of different projects, on Kitcher's view, and we have to rank projects that might, for example, contribute to reducing worldwide deaths from malaria (or cancer or COVID-19) with projects that mainly satisfy our "curiosity," the latter would be totally swamped. "Curiosity" seems inadequate to defend anything like a robust program of basic research whose significance is largely basic knowledge rather than practical results, given the wide range and depth of immediate practical needs that scientific inquiry might help us meet. In other words, Kitcher's account of significance seemed to me unable to provide the defense of basic research that he seemed keen to provide. So, although Kitcher's account seeks to defend basic research with little practical application on the basis of our natural curiosity, it seems like his account will systematically devalue it in favor of practically significant inquiries.

Perhaps this devaluation is the right approach, though. After all, when it comes to possibly saving human lives or satisfying our curiosity about whether megafauna from tens of millions of years ago had feathers, does it not seem inhumane to prefer the latter? When these trade off, should we not obviously prefer the former? The pragmatic point of view seems to be understood that way.

Scientists develop conceptual, material, explanatory, and methodological resources in systematic ways. They are often driven by the existence of a difficult puzzle that only a specialist can understand *as* a puzzle. Some of these have the sort of obvious lines of relevance to practical problems that I described earlier. Other puzzles may only have relevance later, when systematic generalization of the puzzle solution is achieved, and the practical payoff can be seen. Vannevar Bush made a strong claim that basic science would inevitably yield useful by-products. At the same time, he held that scientific progress depended on it being unconstrained by a focus on practical results: "Scientific progress on a broad front results from the free play of free intellects, working on subjects of their own choice, in the manner dictated by their curiosity for exploration of the unknown" (Bush 1945: 12). In this strong form that guarantees progress and applicability from unconstrained basic research, this argument is untenable, because often basic research fails to translate to application, while mission-driven, applied research is much more fruitful on its own than Bush would admit (Sarewitz 2016). Still, this is one path by which seemingly useless science sometimes proves its use, through unanticipated future application, even when Kitcher's "lines of significance" cannot be traced beforehand.

Perhaps part of the problem concerns the way that we think about what is or isn't "useful." There is a tendency, under the regime of neoliberal capitalism, to assimilate the "useful" to the economically productive; to see scientific knowledge as useful if it contributes to technology that sells or to technocratic solutions to policy problems. Here, turning to classical pragmatist philosophy and particularly Dewey's reflections on the concept "useful" provides a crucial corrective. In *Art as Experience*, Dewey writes:

> Wherever conditions are such as to prevent the act of production from being an experience in which the whole creature is alive and in which he possesses his living through enjoyment, the product will lack something of being esthetic. No matter how useful it is for special and limited ends, it will not be useful in the ultimate degree – that of contributing directly and liberally to an expanding and enriched life. The story of the severance and final sharp opposition of the useful and the fine is the history of that industrial development through which so much of production has become a form of postponed living and so much of consumption a superimposed enjoyment of the fruits of the labor of others. (1987/1934: 33–34)

And again in *Experience and Nature*:

> The existence of activities that have no immediate enjoyed intrinsic meaning is undeniable ... So we optimistically call them "useful" and let it go at that, thinking that by calling them useful we have somehow justified and explained their occurrence. If we were to ask useful for what? we should be obliged to examine their actual consequences, and when we once honestly and fully faced these consequences we should probably find ground for calling such activities detrimental rather than useful.
>
> We call them useful because we arbitrarily cut short our consideration of consequences. We bring into view simply their efficacy in bringing into existence certain commodities; we do not ask for their effect upon the quality of human life and experience. They are useful to make shoes, houses, motor cars, money, and other things which may then be put to use; here inquiry and imagination stop. What they also make by way of narrowed, embittered, and crippled life, of congested, hurried, confused and extravagant life, is left in oblivion. But to be useful is to fulfill need. The characteristic human need is for possession and appreciation of the meaning of things, and this need is ignored and unsatisfied in the traditional notion of the useful. (1988/1925: 271–272)

Here Dewey captures and responds to a common misconception of pragmatism and "instrumentalism," and helps us resolve some worries about "useless" inquiries and the value of curiosity. Dewey points out that the definition of "useful" must be situated not in the demands of capitalist modes of production but within human experience, where what is useful concerns not only survival but flourishing. In this sense, delight, wonder, the enjoyment in finding a clever solution to a difficult puzzle, are all definitely useful, perhaps even more useful than those results that "contribute" in an economic sense. In any case, insofar as scientific practices of inquiry enable forms of prediction and control that contribute to the enrichment of life and experience, our "appreciation of the meaning of things," then they are useful and significant.

Nonpragmatic Criteria in Science

Some philosophers of science insist that there are standards or criteria for hypothesis acceptance or theory appraisal that are independent from the criteria of successful prediction and control and that these nonempirical criteria may in some cases supplement or even outweigh empirical prediction and control. In some cases, it is taken as simply a brute fact about science that it answers to such nonempirical standards or "epistemic values" (Kuhn 1977). Others see these "superempirical virtues" as independent grounds for belief in the truth of a scientific theory, and thus part of a robust case for scientific realism (Churchland 1985). My response here is relatively brief: These claims are largely confused; either the criteria in question reduce to or are instrumental to prediction and control; or they are not criteria that guide genuine scientific inquiry and belief.

One version of the idea of nonpragmatic criteria for science is simply that science aims at *more* than prediction and control; it aims at *truth*. Another version is that certain superempirical epistemic standards or values are valuable *because* they are truth promoting. We could mean two things here by "truth." One thing we might mean is *true predictions* of observation and experiment. This is just to restate the claim that science aims at prediction and control, rather than being an alternative to it.[3] Or we could mean truth in a broader sense, the truth of the whole theory in all its parts, not just the truth of the predictions it makes. Truth in this sense by definition exceeds prediction and control; but also truth in this sense cannot be considered an independent *aim* or *standard* for anything. Our only means of assessing whether we have come near to the truth in this sense is our assessment of the success of theories in facilitating successful

[3] Steel (2010) for example defines epistemic values in terms of truth promotion but argues that this can simply mean accurate prediction, which is neutral regarding scientific realism. By contrast, Churchland (1985) understands there to be truth-promoting epistemic values that are orthogonal to predictive accuracy, requiring a realist framework.

prediction and control. We have no additional access to the truth, no way to aim at it, other than through our most successful science.[4]

This is not to deny that science achieves the truth sometimes; that is a separate issue. Nor is it to deny that science might aim *indirectly* at the truth, by aiming at successful prediction and control. But the idea that science *aims* at truth directly and independently as such is a misunderstanding of what an *aim* is. To be able to aim at something, one must have some sense of how the aim might be achieved, how to recognize whether the aim has been achieved, or whether one is moving close toward achieving the aim, however indirectly. In the case of truth in a sense that goes beyond accurate prediction, we cannot meet any of these requirements (Laudan 1984: 137). If truth means that our theory accurately pictures the "underlying reality" beyond our concepts and observations, then we have no way of getting outside of our experience and conceptual frameworks to compare the picture with what is pictured. Likewise, if we want to know whether certain superempirical virtues of theories are truth promoting, not in the sense of enabling better prediction and control but in the broader sense, we have no independent grounds to answer this question. At best, we can say either that theories with such virtues tend to have characteristics that match our assumptions about what the world is like, or that they are instrumental to better prediction and control of phenomena.

An argument from the pragmatist philosopher Charles S. Peirce suggests a third option for thinking about truth as an aim (or, in his terms, "regulative ideal") of science. According to Peirce, the truth is whatever belief scientific inquirers would tend to settle on in the long run of inquiry; a belief (result, hypothesis, theory) is true if it would withstand testing and evaluation in every test scientific inquirers might subject it to, without falling into doubt. This is sometimes taken to be a pragmatist definition or theory of truth, and often not considered a plausible one.[5] It is probably better understood as an elucidation of the relations between truth, belief, and inquiry (Misak 2004). This is of a piece with various other attempts by pragmatists to redefine or elucidate truth in ways that make it a tractable aim: as successful belief (James), as unrevised in the ideal limit of inquiry (Peirce), as warranted assertibility (Dewey),[6] or as ideal rational acceptability (Putnam 1981). This pragmatist line intrinsically links truth to success

[4] See Laudan 1984; Putnam 1981; Putnam 2002.

[5] See Quine 1960: chapter 1 for a sympathetic rejection.

[6] Actually, Dewey's view is more complicated. See Brown 2015.

in prediction and control, rather than treating it as an independent goal or standard.

One could also understand the long-run achievement of truth as a substantive claim rather than a definitional move: If there is a truth of the matter, then scientific inquiry, taken to indefinite lengths, would settle upon it. The long-run success of science is thus evidence that science aims at truth. This is a common claim of the scientific realist. However, truth in any of these senses is, again, not an aim that inquirers could have in view *instead of* successful prediction and control. Rather, the claim is that in thoroughly exploring improvements to successful prediction and control, inquirers get the truth, in the long run. In this way, "truth" is aimed at only indirectly, through the pragmatic aims of prediction and control. The former does not provide independent standards of evaluation.

Some philosophers of science have argued for certain nonpragmatic standards for scientific inquiry as being intrinsic scientific criteria in their own right. Thomas Kuhn, for example, articulates five criteria for good scientific theories: "accuracy, consistency [with other theories], scope [of phenomena encompassed by the theory], simplicity, and fruitfulness [for future research]" (Kuhn 1977). The first, accuracy, is just another term for predictive success. The other four are ambiguous: They might be understood purely as features of the theory or relations between theories, or they might be understood as characterizing relations between theory and evidence (Douglas 2013). Simplicity understood as a feature of a theory might be characterized ontologically (number of theoretical posits required), mathematically (as a property of the equations or models constituting the theory), computationally (in terms of the difficulty of making calculations), or otherwise. Understood as a relation between theory and data, simplicity means that the theory has a lower degree of complexity than the evidence it covers (Douglas 2013: 799). If these nonpragmatic standards (also called epistemic standards, superempirical virtues, epistemic values, or cognitive values) are ways of characterizing the relation between theory and evidence, then meeting them is either instrumental to greater success in prediction and control, or they pick out a particular type of prediction or control as particularly valuable. They do not constitute a standard *independent* from success in prediction and control. On the other hand, if we conceive of standards such as simplicity as inherent properties of theories independent of their relation to evidence, then such standards cannot be *criterial* for science at all.

To sum up what has been said so far, we have considered two types of scientific inquiry that at first glance seem not to fit the pragmatists' account of science: "useless" science that lacks immediate applicability, and science guided by nonpragmatic criteria. In both cases, we have found that the pragmatist can fully accommodate the value of the relevant science while clearing up certain misconceptions. But not all scientific activity can be understood as problem-solving inquiry directed at our capacity to predict and control. Much of the science that we find really inspiring, that informs both public understanding of science and science education, consists of attempts to synthesize and build on the results of pragmatic inquiry in order to understand *how it all fits together.* I group such attempts under the traditional heading of "the scientific worldview" or "scientific worldviews" in the plural; it is these that constitute the greatest lacuna for the type of pragmatism under discussion here.

Scientific Worldviews

Religionists are often inclined to see the tension between science and religion in terms of a clash between very different *worldviews.* "The scientific worldview" is variously depicted in terms such as materialism, mechanism, determinism, and reductionism, and as opposed to ideas such as spirituality, idealism, magic, and miracles. An early exemplar of scientific worldview building is René Descartes' treatise *The World,* written between 1629 and 1633 and published posthumously (1998/1677). The book combines epistemology, physics, biology, and metaphysics to paint a picture of an entirely mechanical understanding of the physical world encompassing the nature of matter and light, astrophysics, living organisms, and the mechanics of perception (with room, however, for God and the rational soul). Descartes sought to provide a complete and systematic alternative to the worldview late medieval philosophers had created in synthesizing Aristotle and Christianity; in the process he synthesizes new scientific research and ideas with his own creative speculations. Many scientists, philosophers, educators, and popularizers have followed in Descartes' footsteps, attempting to build on the latest science to create a comprehensive account of the nature of the world.

How are the various results of science synthesized into a scientific worldview? It is a more difficult and complex matter than it may seem. On the ground, we see diversity and disunity in science. Science involves a hodgepodge of approaches, theories, concepts, and conflicting results. A frank survey of everything *actually* going on in science shows that it provides no single map of reality, that it has little or no overall organization to its theories and methods, and that inconsistencies abound (see

Feyerabend 1999; Dupré 1993; Galison and Stump 1996; Kellert, Longino, and Waters 2006). Even fundamental physics contains different approaches that are apparently inconsistent and so far resist combination into a successful, testable theory. Beyond the realm of theoretical physics, we're in even more of a mess. On a pragmatic and contextual account of science as problem-solving inquiry, aimed at prediction and control of phenomena that interest us, this is not much of a surprise. We might well expect science to be as diverse as our interests. In everyday science, there is no need to take it all together – specialization and contextualism help us keep the mess in hand, and localized conflicts are a spur to further inquiry.

But all this plurality, as Paul Feyerabend aptly pointed out in his later work (1999), problematizes the notion of *the* scientific worldview. One way we might articulate "the scientific worldview" would be a thorough survey of this mess, a list of achievements, gaps, and internal clashes. This would be pretty convoluted, to put it mildly, not to mention self-contradictory, and it would not do the job the scientific worldview is thought to do; when we ask how it all fits together, we expect something more than *concatenation* in response. Instead, we need to ask what happens when we try to create, from a survey of this mess, a single, coherent worldview. Thanks to the messy reality of scientific practice, any attempt to craft a coherent worldview based on science is not a straightforward matter.

The construction of a scientific worldview has three features. (1) it is *selective*; it leaves a lot out, and emphasizes certain aspects of science over others. (2) it is *constructive and creative*; it stitches the remaining pieces together into a coherent and compelling story. Finally, (3) it is *philosophical*; it is part of metaphysics or ontology, not on the same footing as ordinary scientific inquiry, whose warrant is largely connected to situations of practical problem solving and successes in prediction and control. It goes beyond particular empirical problems to paint a grand picture but not a uniquely compelling one. In other words, the scientific worldview cannot be read off of science directly but must instead be constructed by creative, philosophical interpretations that select certain elements of science for emphasis. Because there is a degree of free choice in deciding how to construct a worldview from the materials of science, it is probably best to think about this in terms of multiple, potentially competing scientific worldviews in the plural.

Any worldview has consequences for our lives, hopes, and sense of purpose. In *Science and Moral Imagination*, I describe worldviews as "complex evaluative standpoints where particular valuations are tied up with more general ideals, principles, and institutions, as well as factual beliefs, theoretical claims, and metaphysical commitments" (M. J. Brown 2020: 142). This emphasizes the fact that our sense of how it all fits together is never a neutral, disinterested matter, but one that touches on questions about the meaning of human life and our place in the world. This connection should come as no surprise to those familiar with constructions of scientific worldviews. Many of the most creative articulators and passionate defenders of versions of the scientific worldview – such as Jacques Monod, Carl Sagan, Richard Dawkins, Neil deGrasse Tyson, and Jerry Coyne – have explicitly drawn out moral or political lessons, or have argued that the scientific worldview challenges not only earlier ideas but traditional *ways of life*.[7]

As such, scientific worldviews closely connect with the traditional function of myth or religion, and a scientific worldview will have to compete with other worldviews – traditional religious, speculative philosophical, and alternative interpretations of science. These worldviews, because they are so loosely connected to the pragmatic dimensions of science, will be judged less by typically scientific standards than by philosophical, aesthetic, and ethical ones.

Why do we need a scientific worldview at all? What is the pragmatic value of such a thing? Why should we not simply rest content with the various results of particular scientific inquiries and leave these further questions to theologians and metaphysicians (or perhaps just leave them alone)? I think there is a story for the naturalistic pragmatist to tell about why we should want a well-crafted scientific worldview. To tell it, I turn to what might seem like an odd source: pragmatist analyses of religion and religious experience.

Pragmatist Accounts of the Religious

Recall a part of the earlier quote from Dewey's *Experience and Nature*: "The characteristic human need is for possession and appreciation of the meaning of things" (1988/1925: 272). We can join this to a statement from Dewey's pragmatist forerunner James that "the life of religion … consists of the belief that there is an unseen order, and that our supreme good lies in harmoniously adjusting ourselves thereto" (James 1902: 53).[8]

[7] See also recent analyses of the ways in which the Vienna Circle's "Scientific World-Conception" is tied up with their political projects, such as Romizi 2012, and Chapters 5 and 7, this volume.

[8] See the interesting commentary and discussion of this quotation in Grinnell 2009: 163.

This starts to give a flavor of the classical pragmatist philosophy of religion. According to James and Dewey, religion speaks to deep needs in human life and experience, for *meaning* on the one hand and *guidance* on the other. Insofar as religion allows us to understand and appreciate the meaning of things and to adjust ourselves to the world in a way that promotes the good, it thus performs a pragmatic function.

These pragmatist thinkers draw a distinction between two aspects of religion as ordinarily understood. On the one hand, there is institutional religion. Both thinkers exclude the institutional side from their positive account of the value and function of religion. They considered organized religion a secondary development that at best did not get at what was centrally important about religion; at worst, they saw religious institutions as tied to "creeds and cults" that interfered with the expression of religious experiences and values (Dewey 1989/1934: 21). On the other hand, there is what James calls "personal religion" and what Dewey calls "the religious," which they see as playing a positive role in helping us understand the world and our place in it, thereby providing a solution to our uneasiness about ourselves and our standing in the world, as James puts it (1902: 508).

Where James and Dewey differ is on the question of whether the *supernatural* is an essential feature of religion. For James, religion requires not only some "unseen order" but also a mystical or supernatural order that resonates with the "higher part" of our own being (1902: 508). On a Jamesian account, then, it seems doubtful that a scientific worldview could perform the religious function. Dewey believes, on the contrary, that supernaturalism is an addition to the religious from the institutional side of religion, and that we can liberate the religious to better function in our lives by removing the supernatural accretion.

Supernatural belief hardens into dogma as a result of institutional forces; Dewey explores this claim at length in *A Common Faith* (1989/1934). But supernaturalism has its origins, not as a primitive attempt at science or philosophy, but in an aesthetic urge, Dewey argues in *Art as Experience*:

> Were the hold of the supernatural on human thought an exclusively – or even mainly – intellectual matter, it would be comparatively insignificant. Theologies and cosmogonies have laid hold of imagination because they have been attended with solemn processions, incense, embroidered robes, music, the radiance of colored lights, with stories that stir wonder and induce hypnotic admiration ... Most religions have identified their sacraments with the highest reaches of art, and the most authoritative beliefs have been clothed in a garb of pomp and pageantry that gives immediate

> delight to eye and ear and that evokes massive emotions of suspense, wonder, and awe. (Dewey 1987/1934: 37)

In sum, religion functions as much or more so on the aesthetic plane than on the intellectual, scientific, or philosophical. This is not to dismiss the value of religion in the least; according to Dewey, the aesthetic is also the realm of experience where meanings are at their fullest. Insofar as the role of religion is to give us and help us appreciate the meaning of things, we must ultimately operate on the plane of artistic expression as much as if not more so than the merely cognitive or intellectual.

The function of religion is not only aesthetic, however, but also moral. In *A Common Faith*, Dewey emphasizes both the continuity of humanity with nature and the continuity of the human community past, present, and future, calling this the "community of causes and consequences" that is "the widest and deepest symbol of the mysterious totality of being the imagination calls the universe" (1989/1934: 56). The interacting network of human beings with nature throughout time is the environment in which our ideals, aspirations, and values are formed by acts of *moral imagination*, which unifies the aesthetic and moral functions of religion with its attempt to wrestle with the universe as a totality. As we use our moral imagination to shape the purposes of our lives, we make our character as we make a work of art (Fesmire 2003: 107).

It is in this relation between our values and the "mysterious totality" of the universe that Dewey, contra James, gives a *naturalistic* analysis of the role of faith and of the religious in experience. That is, he attempts to give an account of faith and religious experience without any reference to supernatural entities such as spirits, souls, immaterial substances, and so on. Dewey's hope is that a naturalistic picture can help provide what traditional religion once provided in terms of social cohesion and personal meaning, but for a secular, humanistic, democratic world. One of the core concepts in Dewey's account is the concept of *natural piety*:

> The fact that human destiny is so interwoven with forces beyond human control renders it unnecessary to suppose that dependence and the humility that accompanies it have to find the particular channel that is prescribed by traditional doctrines . . . Our successes are dependent upon the cooperation of nature. The sense of the dignity of human nature is *as religious* as is the sense of awe and reverence when it rests upon a sense of human nature as a cooperating part of a larger whole. *Natural piety* is not of necessity either a fatalistic acquiescence in natural happenings or a romantic idealization of the world. It may rest upon a just sense of nature as the whole of which we are parts, while it also recognizes that we are parts that are marked by

> intelligence and purpose, having the capacity to strive by their aid to bring conditions into greater consonance with what is humanly desirable. Such piety is an inherent constituent of a just perspective in life. (1989/1934: 18, emphasis added)

To practice natural piety is to approach the world with humility and reverence but not passive fatalism. In a secular age, it may be difficult to see the value in talk of "piety." But for Dewey's part, *natural piety* is a much-needed perspective attuning us to our dependence on the world for the success of our endeavors. We are in the world, Dewey is saying, not in the way that a button is in a box, but as part of a complex network of interacting dependencies. This should be a cause for humility and reverence. We are parts of the world that can act toward a reflectively chosen end to improve our lot. We form purposes and ideals reflecting not only momentary desire but our attempt to understand what is ultimately desirable.

It is in this capacity for reflective or intelligent action that Dewey finds room for naturalistic interpretations of faith and the divine. Dewey defines *faith* as "the unification of the self through allegiance to inclusive ideal ends, which imagination presents to us and to which the human will responds as worthy of controlling our desires and choices" (1989/1934: 23). To have faith is to believe in one's ability to bring such ideals to realization through our desires and choices, even if only in the long run. Similarly, Dewey defines *God* or *divinity* in naturalistic terms as the unity of our ideal ends, "the values to which one is supremely devoted," in our imagination (1989/1934: 29). It is important to Dewey that such ideals are neither (yet) actualized, nor "mere rootless ideals, fantasies, utopias" (1989/1934: 34). Rather, these ideal ends are possibilities made coherent through action in connection with conditions in nature that promote their realization. As Dewey says, "It is this active relation between ideal and actual to which I would give the name 'God'" (1989/1934: 34).

Here, aesthetic meaning, moral values and ideals, human ingenuity and intelligence, and social cooperation come together to form secular concepts of faith and the divine (if words such as "God" and "the divine" are too inseparable for you from supernaturalism, consider instead using a term such as "the sacred" or "the spiritual"). Central, again, is the notion of natural piety, the humility and reverence for nature that makes our lives possible, as well as a faith in the human community to cooperate toward realizing those ideals.

According to Dewey, militant atheism and modern supernaturalism are allied in presenting an image of humanity in isolation from nature, and thus the negation of natural piety.[9] Against both, Dewey holds that the religious quality of our experience, our worldview, and our way of life is ineliminable but must be situated within the natural world of our experience. In the present day, it seems that natural piety and faith in our ideals and our community are precisely what we need more of. We need a common, secular, democratic faith that can support and encourage them.

The Religious Function of Science

Here I think we can begin to address the religious function of science. The synthetic and visionary parts of science associated with the articulation of a *scientific worldview* can, I believe, help fulfill the role of cultivating natural piety and providing the basis of a faith in our ability to realize our ideal values. There seems to be a deep human need to understand the nature of our world and our place in it, which has long been fulfilled by mythology and religion. This understanding is often linked with the grounding of the values of a culture. We can also point to the value of the experiences of *wonder* and *belonging* created by such an understanding, and, by extension, we can see worldview-making as an imaginative and inspirational attempt to use science to help us appreciate the wonder of the universe.

Dewey found just such a cultural role for science: "The flights of physicists and astronomers today answer to the esthetic need for satisfaction of the imagination rather than to any strict demand of unemotional evidence for rational interpretation" (Dewey 1987/1934: 37). This is the aesthetic role that was previously played by supernatural "theologies and cosmogonies." Perhaps on the somewhat darker side, in certain ways, science is also hardening into dogma in just the way that supernatural belief had done before:

> The world of physical science is no longer new and strange; to many it is now familiar; while many of those to whom it is personally unfamiliar take it for granted on authority. To a considerable extent its subject-matter is

9 One might rightfully argue that Dewey is being too quick and overgeneralizing. Perhaps, because he can stipulate the definition of "militant atheism," we can permit that generalization. But it is unclear how Dewey can account for the attitudes toward nature recommended by theists as diverse as Francis of Assisi, Pierre Teilhard de Chardin, and Pope Francis, or the wide variety of thinkers sometimes grouped under the heading "ecotheology." (My gratitude to Eric Martin on this point.) Still, I think Dewey's charge is applicable to many mainstream, modern religious traditions, and captures something important about supernaturalism as typically conceived. That some religions treat nature itself as divine, or recommend a religious reverence toward nature, shows a complexity in the concepts of "natural" and "supernatural" that Dewey missed.

> taking the place of the subject-matter of older creeds as something given ready-made, demanding unhesitating credence and passive acceptance. (Dewey 1988/1925: 185)

From our perspective today, this may seem like an exaggeration. After all, are we not inundated, especially in the United States, by those who deny the authority of science? First, no; despite a few exceptions of politicized issues, the US public still strongly trusts scientists and scientific knowledge (National Science Board 2020). Second, Dewey, as both a preeminent philosopher of education and a pioneer in empirical education research, was long interested in science education and concerned that the instructional methods common to science teaching focused exclusively on content, taught as timeless truths, rather than on scientific methods of inquiry. We can and should question whether treating science as something that should be accepted unfailingly by a passive public is what we want, though it certainly makes the religionist analogy more persuasive.

Let me instead emphasize the positive. Imaginatively constructing a scientific worldview can serve the positive religious function for the public identified by James and Dewey. It can give us an understanding of our place in the world, the meaning of it all, a pious relation to nature, and the faith in the ideals we seek to realize. "The scientific worldview" can help not only the secular public but also the scientific community; it can act as a motivation for scientists, something grand to work toward, a flag to rally to. For Dewey, "Faith in the continued disclosing of truth through directed cooperative human endeavor is more religious in quality than is any faith in a completed revelation" (Dewey 1989/1934: 18). This faith is on display in the everyday inquiry of scientists that in turn is consolidated by the worldview-builder.

Ultimately, this part of science, which is an activity genuinely connected with scientific inquiry, science education, and science communication, has more in common with religion than with experiment, inquiry, or technical application. This similarity should affect the way we approach scientific worldviews. Understanding the religious function of scientific worldviews suggests complex criteria for responsible worldview construction that pulls against many current tendencies in contemporary naturalism, humanism, and scientism. In particular, we need to think carefully about the relation between our worldviews and our values and traditional ways of life.

In his very late works, Paul Feyerabend defended a thesis that he called "Aristotle's principle," or sometimes the "Existential Criterion of Reality." The reason for the former name has to do with the way that he interprets Aristotle's response to Parmenides' monism, the philosophical theory according to which the World is one and unchanging: "Aristotle criticized Parmenides in two ways. He tried to show the mistakes in Parmenides' reasoning[,] and he pointed out that change, which Parmenides had called unreal, is important in human life" (Feyerabend 1999: 200). Feyerabend extracts a principle behind the second strategy: "real is what plays a central role in the kind of life we identify with" (1999: 201).[10] This principle is already tacitly at work in Feyerabend's work starting in 1975, where he attempts to combat overconfidence in science. Feyerabend came to see scientific realism as supporting a kind of dogmatism about science and an undeserved special role for science in society, which hurt the freedom of people in our society to pursue their own values and traditional forms of life. This was the sort of work for which Feyerabend was labeled "the worst enemy of science" (Theocharis and Psimopoulos 1987). His work in that period exemplifies the negative version of Aristotle's principle – don't treat something as real if it conflicts with the life you want to live, and don't accept pictures of reality that make that life impossible or burdensome. He thus became concerned with, as the title of one provocative essay put it, "How to Defend Society from Science" (Feyerabend 1975).[11]

The positive version of Aristotle's principle treats "real" as an honorific appended to those results of inquiry that we are willing to incorporate into our worldview, as a result of endorsing their role in our practices. In other words, we are willing to treat something as real insofar as it plays a role in our valued practices and forms of life: what we care about and identify with. This is a value-laden judgment. The principle does not license an "anything goes" attitude toward what we should regard as real, but rather links it to our cherished values and practices. What "plays a role" in our practices should be understood pragmatically, as what actually plays a role in practices that we value, that are successful and unproblematic. What's more, while our decisions about what we give the honorific "real" to makes a real difference to our practices, it is a matter of the philosophical interpretation of science, not a matter of acceptance or rejection of the science itself.[12]

10 I take no position on whether this is a good interpretation of Aristotle.

11 Thus, in this negative phase, Feyerabend claims: "Scientific results and the scientific ethos (if there is such a thing) are simply too thin a foundation for a life worth living. Many scientists agree with this judgement" (1993: 131). Thanks to Eric Martin for reminding me of this point. I think it is an open question, whether in his more positive phase, Feyerabend might be more optimistic about the value of a scientific worldview constructed along the lines described here.

12 Science itself, in its plurality, would of course remain somewhat independent from any version of the scientific worldview.

Between scientific worldviews and our values (and the practices and ways of life they are connected to) is a complex, two-way street. Of course, in various ways, values inform the results of scientific inquiry. (For studies of values in science, see Part III.) What's more, as Feyerabend argues, our values should play a role in the selective activities of worldview building. The worldviews we adopt are or imply complex evaluative standpoints, informing the ideals we pursue and the values we hold.

As I put the finishing touches on this chapter, we continue to deal with the legacy of the COVID-19 pandemic crisis, and there remains major public dissensus over the severity of the problem, the efficacy of the scientifically validated preventatives for the disease and public health policy decisions made during the crisis, and questions concerning the seriousness of "long COVID." We face increasingly severe and irreversible fallout from the climate crisis, but our elected representatives seem focused at best on half measures, where they acknowledge its reality at all. We have faced significant challenges to the institutions and the very values of our secular, pluralistic, democratic society. If ever there were a need for a secular worldview that could inspire natural piety and a faith in our ability to realize our ideals, now is the time. It is the religious function of science to provide such a worldview. Those of us committed to science and to humanism should bring our moral imagination to bear in order to better meet that need.

PART III

Interventions

Scientific Knowledge and Social Imperatives

CHAPTER 9

The Present Plight of Science, and Our Plight

Janet A. Kourany

We need the help of the sciences now more than ever, what with the various coronavirus pandemics and other global diseases; repeated economic downturns; environmental pollution and global warming; racial, ethnic, and other sources of social unrest; the ever-present threat of cyberattacks; and much, much more. Yet the sciences these days are suffering from their own set of problems, and have even contributed in significant measure to many of these problems that now beset us (cf. Chapter 10). Are the sciences, therefore, up to the job we need done right now, or can they be helped to be up to that job, and if so, how? These are serious questions that a socially relevant science studies should take up. What might be philosophy of science's role in that endeavor? This is my topic. But the scene is extremely complex. So it is best to start at the beginning.

The Way Science Was Supposed to Be

Let us begin, therefore, at the dawn of modern science. For it was then that a promise was made: If society would but support the new enterprise, society would be richly rewarded not only with unprecedented insights into the workings of the universe but also with all the benefits such insights would provide. Indeed, Francis Bacon, one of the chief architects of the new experimental science of the seventeenth century as well as one of its more exuberant press agents, promised that the knowledge science would offer would "establish and extend the power and dominion of the human race itself over the universe" for the benefit of all humankind (1960/1620: 117–119). What did Bacon mean? The problem, as he saw it, was that the human race had been thrust into "immeasurable helplessness and poverty" by the Fall from Eden and needed to be rescued. And science would be the rescuer. In other words, science would provide a solution to the plight of humankind (Bacon 1964/1603).

To explain how this would go, Bacon offered a blueprint for the new science, a blueprint that was later adopted by the Royal Society as well as other early scientific societies and that is still in effect today. In it he included illustrations of the benefits he expected from the new science. Science, Bacon suggested, would make possible the curing of diseases and the preservation and prolongation of life; science would produce the means to control plant and animal generation; science would lead to the development of new materials, including new building materials and new clothing materials; and science would provide new modes of transportation ("through the air" and "under water") and even new modes of defense (Bacon 2008/1627). In all these ways and others too, science would make humans once again the masters of nature as they had been in the Garden of Eden, and hence once again "peaceful, happy, prosperous and secure" (Bacon 1964/1603).

True, religion would have to play an important role in this achievement. In fact, Bacon emphasized the theological dimensions of the scientific activities he supported. For him the study of nature, the study that would bring all manner of practical benefits, would also be the study of the Creation, thereby increasing human knowledge and glorification of the Creator and thus adding to the justification of the study. Moreover, this study would require spiritual as well as intellectual discipline, and would involve spiritual as well as intellectual purpose. "We have certain hymns and services," Bacon had the scientists in his utopian *New Atlantis* report, "which we say daily, of Lord and thanks to God for his marvellous works: and forms of prayers, imploring his aid and blessing for the illumination of our labours, and the turning of them into good and holy uses" (2008/1627). So religion was to be a necessary complement to the new science (McKnight 2005), but a religion very much reformed – "purified" – by the dominant intellectual movement of the day: humanism. Indeed, Bacon's promise regarding what science would achieve for humanity incorporated central tenets of Renaissance humanism: that humans were essentially good, or at least deserving of the benefits that God had placed in nature for their use (the benefits that Bacon's science would uncover and further develop); that God had given humans vast intellectual and creative powers, powers that should be cultivated to the fullest (just the powers that Bacon's science would require); and that such powers should be used to improve the lot of humanity – their intellectual and physical worlds as well as their moral and social ones (which was at least a good deal of what Bacon's science was about). Without these humanist tenets, in fact, Bacon's promise would not have been nearly as compelling (see for further details Sargent 2002; 2005; 2012).

At the dawn of modern science, then, Bacon promised all manner of societal benefits if science were supported. And over the next four centuries many other distinguished representatives of the scientific establishment made that same promise. One of the most famous of these in the twentieth century was Vannevar Bush, the engineer and inventor who headed the United States Office of Scientific Research and Development during World War II. At the end of that war, Bush sent a report to President Franklin D. Roosevelt that became the basis of US science policy for much of the twentieth century. In it Bush promised that, if science is supported by society but also left free of societal control, its advances will bring

> more jobs, higher wages, shorter hours, more abundant crops, more leisure for recreation, for study, for learning how to live without the deadening drudgery which has been the burden of the common man for ages past. Advances in science will also bring higher standards of living, will lead to the prevention or cure of diseases, will promote conservation of our limited national resources, and will assure means of defense against aggression. (1945: 10)

What's more, Bush added, such advances in science will be crucial for attaining these benefits. "Without scientific progress no amount of achievement in other directions can insure our health, prosperity, and security as a nation in the modern world" (1945: 11).

So, here was Bacon's promise again. The seventeenth-century theological infusions were gone, to be sure, but so much else, including so much of Renaissance Humanism, remained. Indeed, where Bush now promised "health, prosperity, and security" for people as a result of science, Bacon had promised that they would be "peaceful, happy, prosperous and secure" as well as healthy; where Bush now promised that science would banish the "deadening drudgery" of their pre-science existence, Bacon had promised that science would end the "immeasurable helplessness and poverty" of that existence; and so on.

Bush's promise did depart from Bacon's in one respect, however. It had to do with what counted as *legitimate* science and how social benefits would arise from it. For Bacon, scientific research was all about – *should be* all about – attending to the needs of society:

> Lastly, I would address one general admonition to all – that they consider what are the true ends of knowledge, and that they seek it not either for pleasure of the mind, or for contention, or for superiority to others, or for profit, or fame, or power, or any of these inferior things, but for the benefit and use of life, and that they perfect and govern it in charity. For it was from lust of power that the angels fell, from lust of knowledge that man fell;

> but of charity there can be no excess, neither did angel or man ever come in danger by it. (1960/1620: 15–16)

If such research (inspired by humanism as well as religion) were supported, Bacon promised, science's social benefits would result. For Bush, on the other hand, the most important kind of scientific research, the kind on which other scientific research depends, was all about freely pursuing "the truth wherever it may lead." "Scientific progress on a broad front results from the free play of free intellects, working on subjects of their own choice, in the manner dictated by their curiosity for exploration of the unknown" (Bush 1945: 12). And only if society supported *that* kind of research would science's social benefits result.

By the end of the twentieth century, however, "the free play of free intellects" was no longer considered "the best precondition for maximizing the utility of science" (Rohe 2017: 745; see also Gibbons 1999; Guston 2000a; Krishna 2014; Sarewitz 2016). Science had just gotten too big and too costly, with no end in sight to its continued and ever-increasing demands for support. As a result,

> The sheer size of the system and its need for sustainable allocation of funds is finally unbalancing Bush's claim for the "free play of free intellects." … To continue feeding the science system, a broad societal consensus is needed, in which legitimization is increasingly, often tightly, linked to performance measures and other demonstrable evidence of contributions to social welfare, economic growth, and national security. (Rohe 2017: 746)

No matter. Whether the free play of free intellects was what yielded the social benefits of science (as Bush had claimed) or whether they resulted most reliably only from research explicitly aimed at them (as Bacon had suggested), Bacon's promise – that such benefits *would* result if science were supported – was still very much taken for granted.

The Way Science Is Now

Today, well into the twenty-first century, Bacon's promise has never been more important, what with the problems mentioned at the outset: global diseases such as COVID-19; repeated economic downturns; environmental pollution and global warming; racial, ethnic, and other sources of social unrest; and all the rest. And yet, the obstacles to the fulfillment of that promise have also never been greater, even with the support lavished on science by society. Of course, there have always been obstacles. Bacon himself recognized obstacles – such as the "idols of the mind," the various sorts of errors in human reasoning (whether innate or acquired) that are part and parcel of the human condition, and "the dullness, incompetency, and deception of the senses," "by far the greatest hindrance and aberrations of the human understanding" (1960/1620: 52) – and Bacon sought ways to overcome them (Sargent 2002). Still, those obstacles seem modest in comparison to the obstacles that now confront scientists. If we are ever to be "healthy, happy, prosperous, and secure" as a result of science, as Bacon promised, the current crop of obstacles must also be addressed.

Consider, then, the current obstacles to the fulfillment of Bacon's promise – or at least some of the most pressing of them – and consider, in particular, the scene in North America, the place I know best. There, during the last decade or so, those in the science and science studies communities have been anxiously discussing a variety of problems within the sciences – actually a variety of *sets* of such problems – that they say are of great consequence for society. Indeed, taken together these problems may very well undermine the possibility that science will be able to help us deal with the global challenges that now confront us.

The War on Science

Start with one of the oldest of these current sets of problems. It is said, by science journalists and even many scientists, to involve nothing less than a *war* on science, a war that has been going on for *decades*. Take, for example, Pulitzer Prize-winning *Washington Post* science reporter Chris Mooney's 2005 book *The Republican War on Science* and science writer and filmmaker Shawn Otto's 2016 book *The War on Science: Who's Waging It, Why It Matters, What We Can Do about It*. They describe the war in the United States, while science writer Chris Turner's 2013 book *The War on Science: Muzzled Scientists and Wilful Blindness in Stephen Harper's Canada* describes the war that has taken place in Canada. These books have been supplemented by documentaries on the war, such as one by the BBC in 2006 (BBC Horizon 2006) and one by CBS in 2020 (CBS News 2020), and they have been supplemented, as well, by a continuing stream of articles on the war in such venues as the *New York Times* and the *Washington Post*, *Scientific American* and the *National Geographic*, and the *Guardian* and the *Globe and Mail* – a continuing stream of articles that turned into a torrent after Donald Trump was elected.

The details contained in these war reports are jarring: how, starting in the 1980s, influential Republicans, first in the US Congress and then in the White House, joined forces with corporate interests and fundamentalist Christians to challenge scientific findings in a wide range of areas, including health, education, and the environment. Particular issues concerned, for example, the efficacy of condoms in preventing the spread of sexually transmitted diseases, the efficacy of abstinence-only sex education programs, the status of creation science and of evolution, the status of endangered species, and, of course, global warming. Their tactics included misrepresenting scientific debates to the public, exaggerating scientific uncertainty, preferring outlier scientific views to the views of recognized experts while attacking the integrity of those experts, and stacking government agencies and advisory committees with partisan individuals who could and did hold back or alter scientific reports with which they disagreed.

Not to be outdone, Canadian prime minister Stephen Harper, starting in 2011, not only engaged in these same kinds of practices but also instituted sharp cutbacks in basic research and the overall funding of climate, energy, and environmental research, leaving thousands of government research scientists out of work and hundreds of scientific research institutions and more than a dozen federal science libraries shut down. And after he took office in 2017, US president Donald Trump tried to outdo even this, with an average of two administration efforts to restrict or misuse science per week at the federal, state, and local levels – over 400 in all, as documented by the Silencing Science Tracker, a joint initiative of Columbia University's Sabin Center for Climate Change Law and the Climate Science Legal Defense Fund.[1]

Harvard University science historian Naomi Oreskes and California Institute of Technology science historian Erik Conway, in their 2010 book *Merchants of Doubt: How a Handful of Scientists Obscured the Truth on Issues from Tobacco Smoke to Global Warming*, have explained how even well-placed academic scientists have contributed to this ongoing war on science. Referring to strategies that Stanford University fellow science historian Robert Proctor called the "tobacco strategies" in an earlier war book of his own (*Cancer Wars: How Politics Shapes What We Know and Don't Know about Cancer*), Oreskes and Conway detail how these strategies were intended to produce doubt and confusion in the American public regarding such serious problems as acid rain and the hole in the ozone layer as well as global warming and secondhand tobacco smoke. The

[1] See NowThis Impact 2020 and, for further information about the Silencing Science Tracker, https://climate.law.columbia.edu/content/about-silencing-science-tracker.

strategies included: supporting decoy research to distract from critical questions, thereby "jamming the scientific airwaves"; organizing "friendly research" for publication in popular magazines and even setting up scientific "front organizations" to advocate for their friendly conclusions; producing divergent interpretations of scientific evidence and also misinterpretations as well as engaging in suppression of such evidence; forever calling for more research and more evidence and setting standards for proof so high that nothing could ever satisfy them; and exploiting or actually producing divergent expert opinions (see also Michaels 2008; 2020). The scientists involved included such luminaries as Fred Seitz, past president of the National Academy of Sciences and of Rockefeller University; Robert Jastrow, founding director of the Goddard Institute for Space Studies; William Nierenberg, past director of the Scripps Institution of Oceanography; and Fred Singer, first director of the National Weather Satellite Center and founder of the Science and Environment Policy Project in his home state of Virginia. In each case, Oreskes and Conway tell us, what motivated these scientists to work against the existing strong consensus within the international scientific community were anti-regulation, market fundamentalist political commitments rather than interests in safeguarding industry profits. Still, the activities of Seitz and the others were backed by major conservative think tanks that were, in turn, backed by the US fossil fuel industry, particularly ExxonMobil.

All of these activities constituting the ongoing war on science are only the first set of problems currently confronting science – a set of problems, to be sure, specifically confronting *North American* science. But since that science is a major part of the international scientific scene, these problems have had significant effects, as well, on the rest of the world's science. This should be quite apparent even for those not especially engaged in science-watching. Trump's denial of climate change and his dismissal of, interference with, and finally extraordinary actions to undermine US climate science, for example, together with his withdrawal from the Paris Climate Accord, put definite strains on international collaborative scientific efforts to limit climate change. Add to this Trump's attacks on the science as well as the scientists dealing with COVID-19, his strident criticism of the World Health Organization's handling of the pandemic, and his plans to withdraw its US support, and you have another example of the relevance of the North American scientific scene to the rest of the scientific establishment. As these events indicate, we are all in this together!

The Failure of Incentives

A second set of problems, currently confronting US science in particular, is newer than the first. But it is widely considered to be even more troubling right now to the scientific community – and here the relevance to the larger scientific community is even more apparent. This second set of problems has at least two components. One is what scientists have been calling the "perverse incentives" now prevalent among US scientists, perverse incentives that result from the way science is currently funded here. The story goes like this. Academic researchers in the US require outside grants (in addition, for example, to start up funds provided by their universities) to cover most of their research expenses, including even much of their salaries. But since the number of academic scientists in the US has been increasing while the supply of such grants (mostly from the federal government) has generally been decreasing, competition has become particularly intense, a situation difficult for all but especially so for younger researchers. Moreover, since the term of these grants is usually quite limited – only three or so years – scientists are discouraged from pursuing the more challenging, more significant, long-term projects that produce the big gains for science when they pay off, but which may not pay off. What the present funding situation encourages, instead, are small, safe projects that can be completed in short time spans, the kind of projects that will ensure publications, tenure, promotions, and still more grants. And success tends to be measured by quantity – the number of grants awarded, the number of publications achieved, the number of citations gained – rather than quality, depth, and rigor. In short, what is being encouraged, say scientists, is large quantities of mediocre work (Belluz, Plumer, and Resnick 2016; Roy and Edwards 2017a; 2017b; Ioannidis 2018; Boyle 2018).

This is the best-case scenario. At worst, what is being encouraged is work that cuts corners, takes liberties, and hypes up results, either consciously or unconsciously. Of course, researchers can always turn to private, for-profit sources of funding instead of the public funding that invites all these problems, but private funding comes with its own perverse incentives: conflicts of interest and pressures to deliver the kind of research and results that will be favorable to the sponsor. The cases on record of this, involving the food industry and the pharmaceutical industry, for example, are chilling (see, e.g., Welch, Schwartz, and Woloshin 2011; Dumit 2012; Moss 2013; Nestle 2018).

The second component of the current research scene that is especially troubling to scientists – one that complements the prevalent perverse incentives – is the just as prevalent *nonincentives*. Take replication: the successful reproduction of experimental results. Called the cornerstone of scientific method, it is an absolute requirement for the proper grounding of science. Yet, in recent years, even attempts at replication in science have been relatively rare.

The reasons are many. For one thing, replication studies are not normally viewed as major contributions to their fields; hence they have received less funding and less attention from both scientists and the media. What's more, they are harder to publish since journals prefer original research to replications of previous research. And they take time and resources away from other projects that reflect scientists' own original research ideas. So there has been little incentive to attempt replications. And when they *are* attempted, and especially when the results are negative, there has been little incentive to even try to publish them since journals have a strong disinclination to publish research concerning any kind of negative or failed experiments (Price 2011; Anonymous 2013a; Sheldrake 2015; Engber 2016; Hastings 2017).

A similar situation holds of peer review. On the one hand, peer review is meant to weed out poor quality work before it reaches publication, again a crucially important requirement of successful science. But on the other hand, researchers are not paid or otherwise rewarded for the time they put into reviews, and the work takes time away from their own projects (and don't forget that these researchers are also endlessly applying for grants to support that research, so they have little time to spare). The result is that researchers have not been motivated to do the really careful reviews that are needed, and to do them in a timely manner (Anonymous 2013b; Balietti 2016; Belluz, Plumer, and Resnick 2016).

All this has yielded an unsettling outcome – a current "replication crisis" across all of science, but especially psychology and biomedical research, precipitated by spectacular failures to replicate even "landmark" studies done by the best scientists using the best methods and published in the best journals (Begley and Ellis 2012; Open Science Collaboration 2015; Baker 2016; Nosek et al. 2018); and epic cases of fraud and even years-long runs of fraud wholly undetected by peer review together with epic cases of exemplary work, even Nobel Prize-winning work, that had been rejected by peer review (Altman and Broad 2005; Altman 2006; Balietti 2016; Harvey 2020). When added to the perverse incentives that, as noted earlier, also characterize science, the conclusion is particularly depressing. Arizona State University's Consortium for Science, Policy, and Outcomes codirector Daniel Sarewitz (2016: 5–6) lays it out well:

> Scientists are more productive than ever, pouring out millions of articles in thousands of journals covering an ever-expanding array of fields and phenomena. But much of this supposed knowledge is turning out to be contestable, unreliable, unusable, or flat-out wrong. From metastatic cancer to climate change to growth economics to dietary standards, science that is supposed to yield clarity and solutions is in many instances leading instead to contradiction, controversy, and confusion. Along the way it is also undermining the four-hundred-year-old idea that wise human action can be built on a foundation of independently verifiable truths.

In short, this second set of problems with science, when added to the war on science covered in the first set, suggests that science is unlikely to help us deal with the important global challenges that confront us – global diseases, repeated economic downturns, and global warming and environmental pollution.

The Taint of Social Bias

But what about the racial, ethnic, and other sources of social unrest currently rocking the US and many other regions of the world (see, e.g., Haynes 2020)? Might science yet help us deal with that? In the US, the social unrest especially concerns Black Americans and their supporters and their response to the repeated killings of Black men and women at the hands of police officers. But other factors also enter the picture and help to explain the deep anger, despair, and frustration that Black Lives Matter protests display. For one thing, Black Americans were harder hit by the coronavirus pandemic than other Americans; for example, they have been nearly three times as likely as White Americans to be infected with the virus, nearly five times as likely to be hospitalized, and more than twice as likely to die, a death rate far higher than all other racial and ethnic groups (Soucheray 2020; cf. Ford, Reber, and Reeves 2020; Gould and Wilson 2020). For another thing, Black Americans were especially hard hit by the economic downturn, harder hit than most other Americans (Coleman 2020; Hardy and Logan 2020). And then there is the continued racism that Black Americans confront on a daily basis – fewer employment opportunities than other Americans, lower pay than other Americans, poorer housing options than other Americans, less of everything than other Americans, especially respect (see, e.g., Shelby 2016; Porter 2021).

When we turn to science to help deal with the situation, however, the resources available are disappointing. Black economists have pointed out, for example, that mainstream economics (neoclassical economics) seems simply to deny that discrimination exists (see especially the analyses by Howard University economist William Spriggs appearing in the days after the murder of George Floyd, especially his 2020). That Black Americans have fewer employment opportunities or lower pay, the mainstream seems to suggest, must be a matter of Black Americans' inferior educational backgrounds or lower intelligence or greater likelihood of involvement with crime or the like, or their potential employers' statistically based understanding of Black people's lesser reliability and promise. Or it has simply to do with their potential employers' *taste* in job applicants. For no other possibility makes sense for rational, competitive employers, according to mainstream economics. So, the varieties of facts relevant to understanding and coping with discrimination against Blacks are simply not sought by mainstream economists. Between 1990 and 2018, for example, less than half of 1 percent of all peer-reviewed papers in the top five economics journals even took up the issue of race/ethnicity – that is 29 papers out of a total of 7,567 (Francis and Opoku-Agyeman 2020).

Nor have the relevant facts generally been sought by medical researchers regarding Black Americans' greater vulnerability to COVID-19 and other global diseases such as cancer and heart disease, or their lesser propensity to be helped by standard treatments. For Blacks tend to be left out of clinical trials and medical research more generally (Oh et al. 2015). Of course, there are exceptions. The gathering of facts about Black people has traditionally been extensive in some areas of medical research, such as those associated with promiscuity (including sexually acquired diseases), antisocial behavior (including drug abuse, violence, and sexual assault), and underachievement (Osborne and Feit 1992), and there are all those facts energetically gathered in other areas of science such as the psychological and genetics research associated with intelligence deficits (see, e.g., the past and present research scene detailed in Evans 2018 and Saini 2019). But such research efforts have seemed to offer little help to Blacks.

True, *Black* researchers have been seeking other sorts of facts, the sorts of facts that *are* helpful to Black people as well as other disadvantaged groups. For example, the major professional associations of Black scientists – such as the National Medical Association (formed in 1895), the Association of Black Psychologists (formed in 1968), and the Caucus of Black Economists (formed in 1969 and later renamed the National Economics Association) – have all had, as part of their mission, the production and distribution of knowledge that improves the quality of life of native and immigrant African Americans, Latinxs, and other people of color. And other organizations, such as the National Black Child

Development Institute, have been pursuing projects with the same aim – such as the "Being Black Is Not a Risk Factor: A Strengths-Based Look at the State of the Black Child." That project, for more than forty years, has focused on achieving positive outcomes for vulnerable children who suffer from the dual legacies of poverty and racial discrimination.[2]

But these Black researchers represent only a tiny proportion of their fields. For example, according to National Science Foundation figures for 2019, Black people are only 3 percent of US economists, less that 7 percent of US psychologists, and an indefinite percentage of US health researchers.[3] At the same time, Black researchers face massive amounts of discrimination in these fields. Black psychologists, for example, have reported that "[the specialty of] Black psychology was born from the struggle of Black psychologists who were constantly exposed to messages of Black deficiency, pathology, and inferiority" (Cokley 2020). And just recently the American Economic Association released a statement that said, in part: "We recognize that we have only begun to understand racism and its impact on our profession and our discipline. We have learned that our professional climate is a hostile one for Black economists" (AEA Executive Committee 2020; cf. Blanchard, Bernanke, and Yellen 2019). In short, Black researchers have had a very small voice in their disciplines, a voice not frequently listened to. Small wonder that the facts these researchers have uncovered, the facts that are so helpful to Black people, have not had a powerful effect on their fields, the media, and the social surround. This, then, is the third set of problems currently confronting science to which I want to draw attention, a set of problems of far longer duration than the other two. Of course, other marginalized groups in American society, such as Native Americans, Hispanic and Latinx Americans, and Asian Americans, face many of the same challenges as Black Americans. Science has been largely unresponsive to their needs too.

A Role for Philosophy of Science

The foregoing concerns three sets of problems currently at the forefront of discussion. These are not the only obstacles to the fulfillment of Bacon's

[2] For further information about these organizations, see their websites at www.nmanet.org (for the National Medical Association), www.neaecon.org (for the National Economics Association), https://abpsi.site-ym.com (for the Association of Black Psychologists), and www.nbcdi.org (for the National Black Child Development Institute).

[3] See NSF's figures for "Women, Minorities, and Persons with Disabilities in Science and Engineering" for 2019, table 9.6, at https://ncses.nsf.gov/pubs/nsf21321/data-tables.

promise now facing the sciences, of course. There are also the problems of androcentrism, sexism, heterosexism, and a variety of related LGBTQ issues that feminist scientists and philosophers and historians of science have been discussing for decades (e.g., Harding 1986; Creager, Lunbeck, and Schiebinger 2001; Kourany 2002; Fausto-Sterling 2020). There are the problems regarding the science carried out in the private sector – problems of so-called commercialized or commodified science – that have also been the subject of discussion for decades (e.g., Mirowski and Sent 2002; Krimsky 2003; Radder 2010). And there are the problems more recently under discussion – the now mostly unfulfilled need for interdisciplinary collaboration to solve multidisciplinary problems sometimes called the *silo problem*, the problem of so much science kept secret by government or industry or locked behind paywalls, the problems stemming from the public's distrust of science, and so on (see, e.g., Galison 2008; de Melo-Martin and Intemann 2018; Worthy and Yestrebsky 2018; Brown 2020). Still, the foregoing three sets of problems are thought by many to connect more closely than any of these others to our present most pressing global challenges, the challenges for which we need science at its best to help us (witness just the terms – the *war* on science, the (replication) *crisis* in science, the *hostile climate* of science with its messages of (Black) *deficiency, pathology, and inferiority*, etc. – used to represent these problems). And this makes our three sets of problems especially worrisome, and their resolution especially urgent. Might philosophers of science have a role to play in this effort? The problems, after all, concern threats to science as a knowledge-producing activity, threats so serious that scientists are now devoting considerable attention to them. But the focus of philosophy of science is precisely on science as a knowledge-producing activity. So, these threats to science should claim attention from philosophers as well. What contributions might we make to deal with them?

Fortunately, we don't have to start from scratch. The current discussions that take up these problems also offer solutions to them, or at least strategies to consider. Science journalist Shawn Otto, for example, ends his 2016 *War on Science* book with fourteen "battle plans" to "beat back the war." These include such initiatives as science-informed policy debates for candidates for public office, pro-science pledges for the successful candidates, religious institutions that integrate the results of scientific investigation rather than function at odds with them, and the formation of chambers of *progressive* commerce (or boards of progressive trade) for business leaders. Historians of science Naomi Oreskes and Erik Conway end their war book with what amounts to an historically informed tutorial for the public on how to recognize the legitimate scientific experts on an issue, so that the public will be able to tell whom to listen to and whom to ignore when it comes to issues such as global warming. And scientists have sought to beat back the war on science in still other ways, such as by galvanizing public sentiment and public pressure against the war. Recent examples of this strategy are impressive: the march Canadian scientists organized in 2012 that involved 2,000 scientists, a coffin, tombstones, and a mock funeral on Ottawa's Parliament Hill to commemorate, as they said, the "death of evidence" brought about by funding cutbacks and other actions of the Harper administration (for accounts of it, see Pedwell 2012 and Smith 2012), or the "March for Science" American scientists organized post-Donald Trump in 2017 that took place in Washington, DC (where 100,000 people gathered) and more than 600 other cities all across the globe – the largest science demonstration in history (March for Science 2017; Smith-Spark and Hanna 2017). Examples also include ongoing statements by the Union of Concerned Scientists and other scientific organizations, public letters of protest signed by hundreds of scientists from all over the world, lectures and interviews on the internet, and other public outreach activities by scientists, all in response to the war on science (especially memorable was the open letter to Canada's Prime Minister Stephen Harper signed by more than 800 scientists from 32 countries; see Chung 2014).

Scientists have directed their attention to the scientific community rather than the public in their response to the second and third sets of problems discussed in this chapter. Regarding the second set – more specifically the "perverse incentives" currently pervading science – scientists have suggested such possibilities as a funding system for science that, lottery style, randomly determines which of a group of acceptable proposals should be funded, or that funds particular scientists or particular labs for specified periods, perhaps especially excellent ("rigorous," "efficient," "effective" as well as "original" and "innovative") scientists or especially well-run labs, independently of their announced projects, or that privileges new fields or fields that are high risk, or that leaves it up to research to determine the best way to fund research. To combat the "replication crisis," on the other hand, scientists have suggested ways to make replication studies easier, such as by requiring authors of publishable papers to be more detailed and transparent about the methods used in their research, by encouraging them to share their data, and even by encouraging them to have engaged in at least one replication study themselves before publication. And to combat what some have called a "broken" peer review system, scientists have suggested such possibilities as posting "pre-prints" of articles to be evaluated by a wider audience before formal peer review, "post-publication" peer review to continue the peer review process on the web even after publication, and either a more anonymous system of peer review in which reviewers don't know authors or a less anonymous system in which authors also know reviewers (see, e.g., Alberts et al. 2014; Baker 2016; Belluz, Plumer, and Resnick 2016; Munafò et al. 2017; Ioannidis 2018). To these suggestions of scientists, moreover, a number of science policy analysts have added ways to steer science specifically to solve socially important problems (e.g., Sarewitz 2016, Korte 2019).

Finally, to deal with the third set of problems discussed here – the racism both in and outside of science – scientists, particularly Black scientists, have suggested such possibilities as research programs in psychology that investigate the nature of racism in all its forms, its wide-ranging effects, and the most successful methods to eradicate it,[4] funding programs in economics that commit to multiyear or recurring support for actively anti-racist science initiatives,[5] and outreach programs in biomedical research that encourage and enable those in various minority communities to join research efforts (such as clinical trials) that can improve their health and well-being.

These proposals from the science and science studies communities offer a wide range of strategies to (in the words of one of the contributors) "save science." But do they save *Bacon's* science, the specifically humanist science Bacon promised? Certainly, some of them do, or at least try to – such as the third set of proposals supporting research efforts to fight racism and increase the health and well-being of minorities, and the proposal from the second set supporting organizational efforts to resteer science more efficiently toward socially important problem solutions. But many others do not. The second set of proposals supporting such strategies as lottery-type research funding systems, greater transparency in research, and longer peer reviews, for example, may increase the reliability of research results, but they include no commitment to also promote the human flourishing Bacon promised. And similarly for the first set of proposals, the ones aimed at educating the public using such strategies as public policy

[4] See, e.g., Abrams 2020 and the "APA's commitment to addressing systemic racism" at www.apa.org/about/apa/addressing-racism.

[5] Such as the Women's Institute for Science, Equity and Race; see Francis and Opoku-Agyeman 2020 and the WISER website at www.wiserpolicy.org.

debates, history tutorials, and science marches. Those proposals, like the ones of the second set, are intended to loosen the hold on the sciences of a whole battery of values not frequently conducive to widespread flourishing – corporate interests, fundamentalist Christian values, right-wing political values, and anti-regulation, market fundamentalist values (those are the values that loomed large in the war reports), as well as the perverse incentives and nonincentives pervading contemporary science (those are the values of current scientific culture that lie behind the replication crisis, broken peer review system, and inconsequential busywork of much contemporary science). But such proposals do not at the same time strengthen the hold on the sciences of the legitimate social values that are to replace the others, or even help to make explicit what those legitimate social values are.

By contrast, distinguishing between research shaped by legitimate social values and research shaped by illegitimate ones, and distinguishing between the legitimate and illegitimate ways in which such shaping is to occur, are important projects in contemporary philosophy of science – are, in fact, the "new demarcation problem" many philosophers of science are now investigating (e.g., Holman and Wilholt 2022). And feminist as well as other philosophers of science have already made important contributions to the project (for recent contributions concerned with race, or both race and gender, see Fernandez Pinto 2018; Kourany 2020; Biddle 2020; Havstad 2021). At the same time, many other philosophers of science are now committed to dealing with a wide range of other socially important projects connected with this one, as shown by the workshops, publications, and other activities of groups such as the Consortium for Socially Relevant Philosophy of/in Science and Engineering, the Joint Caucus of Socially Engaged Philosophers and Historians of Science, and the Society for Philosophy of Science in Practice.[6] And, of course, all these philosophers of science are especially well equipped to deal with such projects. For normative questions, ethical/political as well as epistemic, and the arguments and counterarguments that go along with them, are emphasized in the training of philosophers of science, as in the training of all philosophers, which is just the kind of background that is helpful here.

So, strengthening the hold on the sciences of the legitimate social values that are now missing from science is a project to which we philosophers of science might very effectively contribute. Done successfully it will help to

[6] For more information about these groups and their activities, see their websites at https://srpoise.org, https://jointcaucus.philsci.org, and www.philosophy-science-practice.org.

prevent the three sets of problems previously described from continuing (hence, call it the *prevention* project). But it will not dispel the damage already done by those problems – the "contestable, unreliable, unusable, or flat-out wrong" (Sarewitz 2016) information that is now part of our accepted scientific knowledge as well as the crucial gaps in information and missed opportunities that are also there. Is some sort of rectification now called for, and if so, what sort of rectification and how might it be accomplished? This is a second project to which philosophers of science might contribute, and it is especially pressing with regard to the third set of science's problems previously discussed: the ones having to do with race. In order to see this, start with a thought experiment.

Imagine a race in which half the runners have been made to carry heavy weights on their shoulders, and imagine that midway through the race there is a desire to make the race a fair one. What might be done to achieve this goal? One possibility would be to stop the race, take the weights off the shoulders of the runners who are carrying them, and then resume the race. This would hardly do the trick, however, for the disadvantage of the weights for the first half of the race would not have been overcome. A second possibility would be to stop the race, transfer the weights from the one group of runners to the other, and then resume the race. This would equalize the disadvantage of the weights for the two groups and thereby yield a fair race, but at the cost of treating the previously unweighted runners in the same cruel way the first group had been treated. By contrast, a third possibility would avoid this problem while still producing a fair race. It would be to give the previously weighted runners a head start for the second half of the race, providing an advantage to compensate for the previous disadvantage without harming the other runners in any way.

This last possibility is the idea of affirmative action elaborated during the US civil rights era in Martin Luther King's 1964 book *Why We Can't Wait* and Lyndon Johnson's 1965 graduation address at Howard University. Both men used a race metaphor to make the justification of their idea clear. King framed it this way: "It is obvious that if a man is entered at the starting line of a race three hundred years after another man, the first would have to perform some impossible feat in order to catch up with his fellow runner." "Something special" needs to be done "for him now to balance the equation and equip him to compete on a just and equal basis" (1964: 165). Johnson framed the metaphor slightly differently: "You do not take a person who, for years, has been hobbled by chains and liberate him, bring him up to the starting line of a race and then say, 'you are free to compete with all the others,' and still justly believe that you have been completely fair. Thus it is not enough just to open the gates of opportunity. All our citizens must have the ability to walk through those gates" (1965). In other words, to make the race of our thought experiment fair the previously weighted runners have to be given "something special," some kind of head start after their weights are removed – enough of a head start so that they "all . . . have the ability" to win, that is, are all now as likely to win the race as the other runners.

The thought experiment given here helps us consider how we might deal with science's centuries-old treatment of Black people and other racialized groups. It suggests and at the same time offers an assessment of three possible responses. The first response amounts to removing all racist values from science (the weights on half the runners) and replacing them with egalitarian values (all runners free of weights, in other words treated equally). Such a response would dramatically increase the gathering of facts serving the interests of Black people and other racialized groups while still continuing the gathering of facts serving the interests of previously privileged groups. It would ensure that all future research would always generate information helpful to all – the prevention plan described previously. But like the first possible fix in our thought experiment (weights removed after half the race is over), it would do nothing to overcome the disadvantages of the past – the huge inventory of facts gathered over the centuries that continue to serve the interests of only some while they undermine the interests of all the rest. The situation portrayed in this first response, in other words, would exactly correspond to the man in King's metaphor who starts a race three centuries after his fellow runner, though the time difference in this case might be quite a bit longer than three centuries.

But what if the racist values of the past were replaced, now and for the next few centuries, not with egalitarian values but, instead, with values privileging the previously unprivileged, leading to research focused on the previously unprivileged. The facts gathered would then be about *their* needs and experiences, exploits and accomplishments, with methods and concepts and assumptions and questions supporting that aim. Like the second possible fix in our thought experiment (weights transferred from the one group to the other after half the race is over), this *would* overcome the disadvantages of the past for all of these individuals, for it would eventually yield equal inventories of facts serving the interests of all. But it would do this at the cost of treating the previously privileged in the same unconscionable way Black people and the other racialized groups had been treated in the past (and in fact are still treated now). Such an inegalitarian science, in short, would be as unacceptable as the present and past inegalitarian sciences.

This leaves the third possible response offered in our thought experiment, the affirmative action response, which seems to be the only acceptable response. It calls for an *epistemic* affirmative action program for science, one in which research serving the previously privileged would continue while research serving the others would be given extra advantages (like a head start for the previously weighted runners). The problem is that this leaves the nature of the extra advantages completely undefined. It also leaves undefined the conditions under which such an epistemic affirmative action program would be applicable – whether it would apply, for example, to the first (war on science) and second (perverse incentives and nonincentives) sets of problems confronting science as well as the third (relating to social biases such as racism). So, working all this out is a second project – a *rectification* project – to which philosophers of science might contribute.

At least one additional project might be pursued by philosophers of science: setting out, explaining the merits of, and applauding the many cases of science that *do* fulfill Bacon's promise, especially the heroic work currently being done regarding the most pressing global challenges now confronting us (the speed with which the COVID-19 vaccines were developed, their effectiveness, and the antiviral treatments for the disease now available are especially obvious examples). This additional project, this *celebration* project, would include, as well, an analysis of the political and social (including hiring and funding) conditions under which exemplary science has been enabled. Such a project would be important for a number of reasons. For one thing, it would help to give a concrete understanding of the goal that Bacon defined for science, including real, full-blooded illustrations in contrast to the abstract, utopian characterizations provided in Bacon's *New Atlantis* and other works. For another thing, it would anticipate and help to disarm the possible negative use by current science denialists of the prevention and rectification projects' critical work. For a third thing, it would help to balance the picture of science provided by philosophers of science, allowing science's strengths and successes to be fully appreciated as well as science's shortcomings.

In short, three projects – a prevention project, a rectification project, and a celebration project – would seem to be necessary if there is to be any hope of saving the specifically humanist science Bacon promised. And philosophers, happily enough, can have a central role to play in all three.

CHAPTER 10

Science and Justice

Beyond the New Orthodoxy of Value-Laden Science

David Ludwig

The New Orthodoxy of Value-Laden Science

In the face of climate change, the COVID-19 pandemic, and rising anti-science populism, an unlikely alliance of scholars has emerged to "regain some of the authority of science," as Bruno Latour puts it in an interview with *Science* (Vrieze 2017). Historians, philosophers, and sociologists of science, who have long operated in competing intellectual niches, find a common calling in highlighting the existential importance but also increasingly fragile position of science in society. The aim of this chapter is to trace the emergence of this new intellectual orthodoxy and its defense of science in times of global challenges. While the new orthodoxy conveys important insights about the interface between science and society, I argue that it also neglects the roles of science in enabling the exploitation of people and the destruction of ecosystems. I conclude that it is time to move beyond the new orthodoxy by addressing the intricate relationship between science and global justice.

In her TED talk "Why We Should Trust Scientists," historian of science Naomi Oreskes (2014; see also 2021) sets the stage with two salient issues: climate change and public health. Oreskes emphasizes that we need to trust scientists when it comes, for example, to a warming planet or vaccines. This is not because science is infallible, but rather because scientists collectively gather and evaluate evidence. Scientific consensus may be wrong, but it provides the best judgment that societies have when facing complex social-environmental challenges. In his essay "Science as craftwork with Integrity," sociologist of science Harry Collins (2021: 297) does not only recommend trust but even love for science: "We should love science other than that which is visibly corrupt, because basing political decisions upon it gives rise to the best decisions." Collins' love is qualified in ways similar to Oreskes' trust: Science is not characterized by its infallible objectivity but by its sophisticated craftwork. While science can be corrupted, noncorrupted science provides the best craftwork we have in addressing global challenges such as climate change and the COVID-19 pandemic.

Latour's authority, Oreskes' trust, and Collins' love for science provide a striking contrast with the legacy of the field of science and technology studies (STS). While philosophy of science became increasingly depoliticized in the postwar period (Reisch 2005), STS emerged as an interdisciplinary meeting ground of scholars who were often "involved in or influenced by counter-cultural and radical activities from the late 1960s, '70s and '80s" (Taylor and Patzke 2021) and challenged science as a social system that is intertwined with oppressive social realities of "racism, imperialism, heterosexism and class oppression" (Harding 1991). But the stakes are too high to focus exclusively on critique (Latour 2004b). Collins et al. (2020: 1) even go a step further in arguing that "STS erodes the cultural importance of scientific expertise and unwittingly supports the rise of populism." History, philosophy, and sociology of science needs to move beyond such a performance of critique, toward a serious understanding of scientific expertise and integrity. As philosopher of science Philip Kitcher (2020: 119) points out, recognition of scientific expertise has become a truly existential matter as failure to respond to climate change will leave us with "a world so bereft of resources, so buffeted by a different climate, that no voice within it could rise to mourn and accuse."

None of the scholars cited here want to return to an unquestioned authority of science. Science is not properly characterized in terms of value-free objectivity and "scientists invariably bring biases, values, and background assumptions into their work" (Oreskes 2021: 64). Science is not some kind of infallible "magic" but rather a very specific kind of "craftwork" (Collins 2021: 304) that can go wrong and can be corrupted. The answer to global crises is not an old-fashioned scientism that preaches from the pedestal of certainty and value freedom. On the contrary, we need "science with a human face" that is reflexive about its complex entanglement with society, honest about its own limitations, and still able to produce the most reliable knowledge about global challenges such as climate change, food production, loss of biodiversity, public health, social inequality, soil erosion, and sustainable energy production.

In an admittedly polemical simplification, I want to call this broad position the *New Orthodoxy of Value-Laden Science*. Talk of a New Orthodoxy is apt as the picture is promoted by many of our most prominent science scholars and synthesizes major insights from history, philosophy, and sociology of science. Talk of a New Orthodoxy is polemical as it glosses over the many substantial differences between the scholars that are thereby lumped together. Scholars such as Collins, Kitcher, Latour, and Oreskes engage with science through different intellectual traditions and styles of reasoning that have often created explicit disagreements (e.g., Collins and Yearley 2010 versus Callan and Latour 2010) and that remain reflected in overlapping but still distinct communities of research who identify with labels such as "sociology of science," "science and technology studies," or "history and philosophy of science."

However, the intellectual diversity of these scholars makes their converging voices all the more remarkable. From interviews in *Science* (Vrieze 2017) to TED talks (Oreskes 2014) to features in the *New York Times* (Kofman 2018), disagreements of academic scholarship vanish in the background of a publicly articulated vision of the role of science in society. Roughly, this common vision contains four elements. First, an existential *diagnosis of the fragility of science* in the face of a planetary crisis. Science is indispensable for addressing global challenges such as climate change and the COVID-19 pandemic but simultaneously threatened by rampant anti-intellectualism and anti-science populism. Second, an *opposition to the ideal of value-free science* that downplays the historical and social embedding of research in order to present science as an unquestionable authority of pure objectivity. Third, an *endorsement of "science with a human face"* that acknowledges the deep entanglement of science and values but stresses the epistemic integrity and success of value-laden science that is not epistemically corrupted. Fourth, an *emphasis on the public importance of science* that requires qualified authority, trust, and even love in the face of existential planetary crises.

The New Orthodoxy synthesizes insights from decades of historical, philosophical, and sociological debate about the nature of science and its relations with society while bracketing remaining scholarly disagreements. Methodologically, it reflects the waning of a simple dichotomy between realist defenders of science who highlight value-free objectivity and constructionist critics who highlight the historical and social contingency of science. While this dichotomy is familiar from debates about the so-called science wars of the 1990s, so is its rejection as a false dichotomy (Carrier et al. 2004). Yes, science is always embedded in values. Science is always shaped in sociocultural contexts and therefore does not lead to an absolute and subjectivity-free description of "the world as it is independent from our experience" (Williams 1985: 139). No, that does not mean that "anything goes" and it does not mean that reality somehow collapses into mere social constructions. It also does not mean that we have to give up on scientific objectivity or that scientists lack epistemic authority when rejecting the claims of climate change denialists or anti-vaxxers.

Politically, transcending "science war" dichotomies also suggests a realignment of the relations between science and society. According to the New Orthodoxy, the question is not anymore whether science needs to be defended against postmodern and poststructuralist obscurantists or criticized as relying on false claims of value freedom and universality. Instead, the question is how to develop a middle ground that aligns science and society in reasonable ways and takes their complex relations into account. Instead of being isolated from society, science needs to inform policy while cultivating reflexivity about its own social character.

Contradictions in Framing Science

The New Orthodoxy provides a reasonable and well-balanced compromise that has been forged through major intellectual controversies about the nature of science and its relations to society. It incorporates legitimate criticism of absolutist interpretations of the objectivity, universality, and value freedom of science while simultaneously articulating a positive vision of the epistemic authority of science that provides a robust response to anti-science populism. The arguments of the New Orthodoxy are well suited to addressing the problem of anti-science populism but their extrapolation into a generalized defense of science risks invisibilizing contradictions that characterize the institutional reality of the science system. The risk of structural blindness is especially pressing in the New Orthodoxy's lack of engagement with the role of science in society beyond Europe and North America. Programmatic statements in Oreskes' *Why Trust Science*, or Latour's *Down to Earth*, or Collins et al.'s *Experts and the Will of the People* depart from a rather uniform set of examples. Brexit and Donald Trump. Climate denialism and anti-vaxxers. Conspiracy theories and social media trolls. The Global South appears only if it conforms to this pattern, such as Jair Bolsonaro's attack on the Brazilian science system and evidence-based governance. Indeed, the Brazilian case illustrates that anti-science populism is not an issue exclusive to the Global North (Reyes-Galindo 2021). However, it is misleading to address global contestations of science exclusively through the problem of anti-science populism.

The New Orthodoxy does not explicitly deny that science has contradictory and sometimes exploitative roles on a global scale. In fact, most proponents of the New Orthodoxy would probably accept many of the arguments of this chapter. However, the New Orthodoxy de facto invisibilizes such issues by simply not talking about the complicity of science in global exploitation while presenting seemingly general defenses of science. This issue of epistemic silencing (Dotson 2011; Spivak 1988) becomes most salient when contrasting commentary from the New Orthodoxy with scholar activism that centers on questions of global justice. For example, Colombian post-development scholar Arturo Escobar challenges trust in science by arguing that "science has become the most central political technology of authoritarianism, irrationality, and oppression of peoples and nature" (2018: 89). According to Escobar, the science system is implicated in the production of global injustice in two ways. First, Escobar argues that science often constitutes a vehicle for "violent development" (2018: 89) in the Global South, where it contributes to neoliberal agendas of growth and modernization that deepen global economic inequality while redistributing the social and environmental burdens of biodiversity conservation, food production, and resource extraction onto the Global South. Second, Escobar argues that science functions as "a reason of state" that "even standardizes the formats of dissent" (2018: 89). Alternative visions of societies and environments are commonly articulated by social movements and scholars in the Global South who mobilize local philosophical resources such as *Buen Vivir* in Latin America (Varea and Zaragocin 2017), *Ubuntu* in Southern Africa (Simba 2021), and *Mātauranga Māori* in Aotearoa/New Zealand (Watene 2016). However, such alternatives remain invisible in mainstream development as they are not couched in academic vocabulary and therefore fail to adhere to formats of dissent that are defined by the science system. Despite notable exceptions in feminist scholarship (Harding 2010; Wylie 2015), they also remain invisible in mainstream philosophy of science that theorizes science almost exclusively through its expression in the Global North.

Escobar's perspective on science as promoting narrow agendas of growth and modernization is mirrored in contributions by scholar activists beyond Latin America, including the work of the Indian ecofeminist Vandana Shiva. Shiva's (1991) influential *The Violence of the Green Revolution* inverts the narrative of agricultural modernization in the second half of the twentieth century as the most shining success of humanitarian research that elevated much of the "Third World" out of hunger and poverty. Written in the wake of the Bhopal disaster and a decade-long armed conflict in Punjab, Shiva states that "two decades of the Green Revolution have left Punjab ravaged by violence and ecological scarcity. Instead of abundance, Punjab has been left with diseased soils, pest-infested crops, waterlogged deserts, and indebted and discontented farmers. Instead of peace, Punjab has inherited conflict and violence" (Shiva 1991: 11). According to Shiva, the web of economic, environmental, social, and religious conflicts in Punjab is not simply a failure of policy but was co-created by a science system that "offers technological fixes for social and political problems, but delinks itself from the new social and political problems it creates" (1991: 19). Shiva argues that the contradictions of the science system are obscured by a tendency to take credit for its societal benefits (e.g., climate change mitigation, poverty reduction, public health) while externalizing negative and destructive impacts as mere issues of misguided application and policy. "The tragic story of Punjab is a tale of the exaggerated sense of modern science's power to control nature and society, and the total absence of a sense of responsibility for creating natural and social situations which are totally out of control" (Shiva 1991: 21).

The perspectives of scholar activists such as Escobar and Shiva are also reflected in many social movements in the Global South such as the "Rhodes Must Fall" movement in South Africa. The Fallist movement emerged in 2015 at the University of Cape Town in protest against a statue commemorating the British colonialist and mining magnate Cecil Rhodes (1853–1902) but quickly turned into a broader protest movement against the colonial and apartheid legacy of the South African university system. The omnipresence of Rhodes in South African academia became challenged as representing a university system that served colonial oppression and often still remains inadequate – for example, in its student fees and admission procedures – for purposes of contemporary South African society. As most clearly expressed in a variation "Science Must Fall" (Harris 2021), a part of the movement pushed even further in locating the problem not merely in colonial symbols or administrative issues but also in the very structure of South African science – the problems that are prioritized by researchers, the questions that are asked, the methods that are employed, the theories that are taught, the interventions that are derived. In this sense, Harris (2021: 113) describes the Fallist movement as demanding a "path of accommodation and inclusion [that] leaves intact the integrity of scientific explanation while at the same time allowing for the possibility of tapping into African knowledge for a different type of edification."

Contradictions in the Science System

The examples of Escobar, Shiva, and Fallism exemplify framings that radically differ from the New Orthodoxy as expressed in Latour's authority, Oreskes' trust, and Collins' love for science. Of course, it may turn out that this is just an issue of framing that can be resolved through more nuanced analysis that highlights the qualified character of the New Orthodoxy's defense of science. Defending science as "craftwork with integrity" (Collins 2021), for example, is intimately linked to criticizing science that lacks integrity because it has been epistemically corrupted by corporate influence, political ideology, or other factors. The suggestion is not to trust every scientist but to trust science as a collective endeavor of evaluating evidence and establishing a consensus of experts (Oreskes 2021).

Highlighting this qualified case for trust may be seen as creating a middle ground for embracing some claims of scholar activists in the Global South, while rejecting others. Indeed, the influence of big corporations in areas such as agriculture and public health is worrying and justifies some of the concerns that Escobar and Shiva are articulating. The legacies of colonialism and apartheid did not magically vanish from the South African university system but require continued scrutiny as exemplified by Rhodes Must Fall. At the same time, science cannot be reduced to issues of corporate or colonial corruption as noncorrupted science remains the most reliable guide for addressing global challenges such as climate change or food security. In this sense, the New Orthodoxy may be seen as offering a compromise that acknowledges the need for critical reflexivity about epistemic corruption while dampening the sharp edges of activist criticism toward the science system as a whole.

Such a compromise fails, however, insofar as it frames all criticism of epistemically noncorrupted science as anti-science populism. For example, consider academic responses to Shiva's critique of genetic modification and mainstream agricultural development. When invited to speak at an event of Students for a Sustainable Stanford in 2019, for example, forty-five leading scientists from European and North American institutions signed an open letter condemning Shiva's alleged "use of anti-scientific rhetoric to support unethical positions" based on "preposterous," "ridiculous," and "nonsense" statements (Tabliabue et al. 2019). Positioning Shiva as an "anti-science populist" in analogy to climate change denialists or anti-vaxxers is also reflected in an article in the *New Yorker* with the title "Seeds of Doubt" (Specter 2014), in a variation of Oreskes and Conway's (2010) book *Merchants of Doubt*, which focuses on epistemic corruption of scientists by tobacco and oil industries rather than the contribution of agricultural sciences to the exploitation of people and planet.

There is plenty of room for criticism of Shiva's often relentlessly polemic engagement with mainstream agricultural sciences. Reducing her critique to anti-science populism, however, exposes a fundamental misunderstanding that risks being reinforced through the framing priorities of the New Orthodoxy. Contradictions at the interface of science and society are not merely the product of epistemic corruption. They do not only appear when academic integrity is seduced by corporate influence or political ideology. The case of agriculture highlights that the science system as a whole, and not just its epistemically corrupted fringes, is producing contradictions in the sense that scientific knowledge is indispensable for addressing social-environmental crises but is also often a driving force in creating them.

Much of this remains off the radar of public interventions of the New Orthodoxy that tend to focus on a narrow set of disciplines (often climatology and epidemiology) in an equally narrow set of societal contexts (often the UK and USA). In programmatic articulations of the New Orthodoxy, this narrow frame of reference finds a reliable expression in stage setting that involves trustworthy scientific actors such as the Intergovernmental Panel on Climate Change or the Centers for Disease Control and Prevention versus populist advisories from Trump to anti-vax Facebook groups. If the frame of reference is defined this way, many contradictions of the science system indeed become invisible, and the dominant concern becomes the defense of well-established but publicly contested scientific evidence.

The problem with this frame of reference, however, is that it invisibilizes large parts of the science system that affect social-environmental systems. Addressing this as an issue of reference frames allows an analogy with a familiar debate in the philosophy of science about the diversity of scientific practice (Radder 2012). Rather than assuming that a theory of the nature of science in general can be formulated through case studies from fundamental physics or evolutionary biology, philosophy of science has come to emphasize the diversity of disciplines from archaeology to biomedical sciences to engineering sciences to microbiology – not because fundamental physics or evolutionary biology do not matter but because the reality of scientific practice is too heterogeneous to be assessed through a narrow set of reference sciences. By analogy, engagement with the interface of science and society needs to look beyond a narrow set of examples from climatology or epidemiology – not because these fields do not matter but because the political structure of scientific practice is too heterogeneous to be assessed through a narrow set of reference sciences. The following sections, therefore, develop both critical and constructive diagnoses of social roles of science through a focus on disciplines and issues that are largely ignored by the New Orthodoxy.

The Case of Agricultural Production

Agriculture constitutes one of the most important junctions of science and society. The dramatic transformations of agricultural production shape the lives of billions of people around the world. Processes of "depeasantization" illustrate the scale and pace of these transformations: between 1991 and 2017, employment in agriculture fell from 58.01 percent to 36.55 percent in Nigeria, from 69.51 percent to 39.07 percent in Bangladesh, and from 55.31 percent to 17.51 percent in China (World Bank 2021a). However, focusing on depeasantization efforts and declining rates of agricultural employment only scratches the surface of the dramatic social effects of shifting agricultural production. As van der Ploeg (2018: 1) points out, "there are far more peasants in the world than ever before in human history. In absolute numbers, even the most conservative estimates suggest that there are between 500 and 560 million peasant farms in today's world, and this number is continually increasing." Peasant farming does not only continue to provide the livelihood basis for roughly two billion people, but depeasantization is also often intertwined with complex processes of repeasantization in the light of consequences such as urban poverty as well as declining profit margins for many farmers who compete on global commodity markets.

Transformations of agricultural production are worlds of contradictions. Scientific contributions to these transformations represent some of the brightest and darkest dimensions of the intersection of science and society. On the one hand, there is a positive narrative about a wide range of disciplines – for example, agronomy, chemistry, engineering, genetics, hydrology, plant breeding, and soil sciences – that have contributed to increasing yields and decreasing rates of hunger. Scientific contributions to pushing the boundaries of agricultural productivity have been so prominent in the challenge of "feeding the world" that they even produced a Nobel Peace Prize winner, Norman Borlaug, commonly described as the "father of the Green Revolution."

On the other hand, it has become widely recognized that generic appeals to decreasing rates of hunger only tell one part of a much more complex story. Food insecurity has actually been on the rise again since 2014 (von Grebmer et al. 2020) and has spiked since the COVID-19 pandemic in the light of reinforcing effects of "climate, conflict, zoonotic diseases and pests, as well as economic shocks" (World Bank 2021b). Scientific research has not only failed to mitigate this trend but has also contributed to deepening this crisis through cash crop monocultures that are vulnerable to economic and environmental disruption, and through unsustainable production systems that contribute to droughts, loss of biodiversity, soil erosion, and other environmental factors that drive food insecurity (La Via Campesina 2020).

Furthermore, rates of food insecurity are only one relevant factor that is not always positively correlated with other relevant factors such as rates of poverty (Gentilini and Webb 2008). Science-led increases in agricultural productivity often come in the form of "technological packages" of large-scale intensive agriculture that produce cheaper commodities through new seeds, fertilizers, pesticides, machines, seeding techniques, value chains, and so on. Even where these interventions have increased the availability of cheap food, they have often simultaneously driven land grabbing of peasant farms, rural unemployment, crumbling communities due to out-migration, and the explosion of urban underclasses (Sumberg, Thompson, and Woodhouse 2012). Societal contradictions are therefore deeply embedded in processes of agricultural modernization – for example, by rapidly increasing urban underclasses while simultaneously making food more cheaply available to them. In this way, agricultural modernization often creates and connects spaces of poverty (rural spaces for creating food commodities as cheaply as possible, urban spaces of expendable peasant labor) and spaces of richness (concentrated ownership across food value chains, affluent consumer markets) on a global scale (van der Ploeg 2018: 93).

While it is possible to highlight contradictions of agricultural production at a general level, it is often more informative to address specific cases of scientific knowledge production and the specific interventions they enable. For example, genetic modification (GM) constitutes a salient issue at the interface of science and society with many more specific case studies. GM has a lot of potential for agricultural productivity that is only further increased through the rapid development of novel gene-editing technologies that promise ease and precision in manipulating targeted genes (Shah, Ludwig, and Macnaghten 2021). Beyond abstract talk about potential, there is also plenty of real life evidence. Proponents of GM crops often focus on Bt cotton as the shining example of a "pro-poor" technology with straightforward benefits for farmers (Ali and Abdulai 2010). Containing a gene from the bacterium *Bacillus thuringiensis,* Bt cotton produces a toxin that kills bollworms. Growing Bt cotton can therefore reduce risk of crop failures, costs of inputs such as pesticides, and health risks associated with widespread pesticide application. Especially in India, the largest cotton producer in the world, the introduction of Bt cotton in 2002 became associated with narratives of "technological triumph" with adoption rates over 90 percent, increasing yield, and reduced pesticide application (Kranthi and Stone 2020).

The narrative of Bt cotton as a triumphal "pro-poor" technology is commonly contrasted with a counter-narrative, publicly most visible in Shiva's characterization of Bt cotton as "Seeds of Suicide" (Shiva et al. 2000) that create debt and dependency on global markets, allegedly causing an epidemic of farmer suicides in India. Almost thirty years after the approval of Bt cotton, it has become increasingly clear that these narratives of triumph and counter-narratives of failure capture parts of a complex and highly contradictory story (Kranthi and Stone 2020). Initially developed for large-scale farms in North America, Bt cotton did not turn out to be a universal "pro-poor" technology but had wildly different effects in different agrarian and ecological contexts (Glover 2010). Take the case of Burkina Faso, which approved Bt cotton in 2008. It was hailed as a "role model" for agricultural development in Africa with quickly rising adoption rates (2 percent in 2008, 70 percent in 2014) and sharply declining insecticide use (Pertry et al. 2016). In the midst of this developing story of technological triumph, the Burkinabè cotton sector announced that it would cease producing Bt cotton, ending GM crop production in Burkina Faso entirely. As Luna and Dowd-Uribe (2020) point out, a wide range of problems had accumulated. Most importantly, the shorter fiber length of Bt cotton compared to conventional Burkinabè varieties made the former less profitable on global markets and created substantial losses for Burkinabè cotton companies. Luna and Dowd-Uribe (2020) highlight the problem that the marginalization of Burkinabè stakeholders (local farmers, researchers, and companies) led to distorted external studies of the alleged success of Bt cotton that misrepresented local realities and culminated in an abrupt collapse of GM crops in Burkina Faso. The contradictory effects of the introduction of Bt cotton in Burkina Faso reflect the complex (economic, ecological, social) dynamics of GM-based agriculture in Africa, which have led to only three out of fifty-four countries in Africa commercializing any GM crops whatsoever (ISAAA 2019).

Cases such as Bt cotton in Burkina Faso provide an entry point for engaging with the complexity of the interface of science and society – both in its potential for improving local livelihoods and its reality of often failing to realize this potential. And indeed, historians, philosophers, and sociologists of science have produced excellent scholarship on issues of global agricultural production (Curry 2017; Hicks 2015; Lacey 2015; Millstein 2015; Motta 2014). However, this scholarship does not fit well into framings of the New Orthodoxy that contrast reliable scientific consensus with anti-science populism. Despite the contested role of large agrifood companies such as Monsanto, the majority of proponents of GM crops are clearly not "Merchants of Doubt" (Oreskes and Conway 2010) that trade epistemic integrity for corporate benefits; rather, they often include the most influential researchers in fields such as plant genetics at the most prestigious research institutions of the Global North. As a consequence, criticism of GM crops has often been rejected as "antiscience zealotry," as Norman Borlaug famously put it, or even as a "crime against humanity," as claimed in 2016 in an influential letter of 127 Nobel Prize laureates (Biddle 2018). History, philosophy, and social studies of science have the potential to highlight the need for a more substantial debate that acknowledges science as a key actor in addressing and producing global injustices in agricultural production. As much as research has the potential to improve agricultural production in ways that actually improve livelihoods, the reality of agricultural production often makes science central to the production of a wide range of injustices (e.g., environmental destruction, economic inequality and poverty, and health hazards).

Despite its undeniable virtues, the New Orthodoxy risks obscuring this complex and contradictory picture. Kitcher's (2011) discussion of GM crops in *Science in a Democratic Society* provides a striking example by developing a vision of "well-ordered science" in which citizens are tutored by scientists and eventually learn that there "is nothing special, or especially risky, about genetic modification of organisms" (2011: 567). Kitcher's discussion takes as its starting point a public ignorance of genetics (e.g., endorsements of the statement "GMOs [genetically modified organisms] contain genes, but ordinary organisms do not") and a "picture of genes as mysterious little agents of evil, inserted into healthy foods by the wicked minions of agribusiness" (2011: 567). Given such a framing, the contestation of GM crops indeed seems largely analogous to the contestation of vaccines by anti-vaxxers: While there is scientific consensus about the safety of many GM crops and vaccines, rampant ignorance about the actual science and diffuse concerns about "big business" regarding everything from Monsanto's seeds to Pfizer's vaccines leads to the rejection of technologies that are literally saving the lives of millions of people.

While Kitcher frames his discussion in terms of the knowledge deficit of citizens about genetics, he does not consider the knowledge deficit of scientists about the social-environmental context in which GM crops are implemented. Tutoring appears as a unidirectional process in which scientists already hold all the relevant expertise and other stakeholders are negatively characterized through their lack of expertise. However, the case of GM crops illustrates that it is crucial to recognize the diversity of situated knowledges (Haraway 1988) and that it is often the scientists who need tutoring about the social-environmental ramifications of scientific knowledge production. This lack of engagement with contested realities of agricultural production is also apparent in the way Kitcher's discussion characterizes GM opposition as "largely a European phenomenon" while "not much heard" among "many of the world's people, particularly in Africa and parts of Asia, [whose] current agriculture is unable to provide them [. . .] with ways of reliably growing the food they need" (2011: 318). The reality, however, is that GM adoption in the Global South has been hesitant at the policy level and publicly deeply contested. Burkina Faso is no exception. In 2018 (ISAAA 2019), GM crops covered 2.9 million hectares on the African continent – not even a quarter of Canada's 12.7 million hectares. In Asia, the largest producer is India with 11.6 million hectares, but only GM cotton and no other crops. Apart from Indian cotton, the whole of Africa and Asia combined cultivates less GM crops than Canada and less than 20 percent of the USA's 75 million hectares. Competing with the agricultural output of GM production in the Americas would risk the livelihoods of millions of farmers across Africa and Asia. Opposition is so strong that only three African countries (Eswatini, South Africa, and Sudan) commercialize any GM crops whatsoever.

While Africa and Asia illustrate hesitant GM adoption at the policy level, Latin America illustrates the public contestation of GM agriculture. For example, Brazil is the second biggest producer of GM crops in the world and GM varieties dominate the production of soy, maize, and cotton with an overall adoption rate over 90 percent (ISAAA 2019). The social contestation of GM crops in Brazil highlights the contradictions between visionary statements of biotechnological benefits "for the poor" and the economic reality of GM crops being part of technological packages that require land- and resource-intensive monocropping of cash crops for industrial use and export. GM agriculture is therefore often associated with a devaluation of traditional peasant production as underdeveloped and a push for agricultural industrialization that dispossess peasants and makes their labor expendable. It is therefore no surprise that peasants have been driving the resistance against GM crops in Brazil, most notably the Landless Workers' Movement (MST). The roughly 1.5 million members of the MST embody many of the contradictions of agricultural production and of modernist development projects such as the construction of the Itaipú hydroelectric dam in Paraná that resulted in the eviction of more than 10,000 mostly Indigenous or peasant families. In the MST case, opposition to GM crops is therefore not driven by affluent consumers, as imagined by Kitcher, but is part of a wider agrarian struggle for peasant livelihoods in rapidly globalizing agrifood commodity markets.

None of this is to suggest that GM crops only have negative effects in Brazil or the Global South more generally. But it is simply misleading to characterize its contestation as "a European phenomenon" that derives from the privilege of not having to worry about food security. Just as I was writing this chapter, the Court of Justice of Paraná, Brazil, confirmed the responsibility of the multinational biotech company Syngenta for the murder of the peasant farmer and activist Valmir Mota de Oliveira, who was killed on an experimental GM field by a corporately hired militia (Brasil247 2021). Syngenta is not some shady "merchant of doubt" who aims to undermine the established consensus of agricultural sciences. On the contrary, the position of Syngenta at the very heart of agricultural science is difficult to miss from my office at Wageningen University and Research. The president of my university, the "world's leading" agricultural university (WUR 2021), joined the nine-member board of directors of Syngenta in 2019 (Kleis et al. 2019). If only the contradictions of agricultural production could be modeled along the lines of familiar cases of climate change denialism or anti-vaxxers that demand a firm stance with the scientific mainstream against a vocal minority of "merchants of doubt." Unfortunately, such a model is deeply misleading in many cases. The contradictions of agricultural production are embedded in our best science at the very heart of the science system.

Science as a Site of Injustice

The case of agriculture is not a strange outlier but illustrates a more general discrepancy between the potentials and realities of scientific knowledge production in global contexts. Indeed, scientific knowledge production has enormous potential for addressing social-environmental challenges while mitigating inequality. Agricultural sciences are a shining example of this potential as they can contribute to making food more affordable, more nutritious, and more sustainable for current and future generations. The reality of the agricultural sciences, however, not only highlights this potential but also the point that science can become a site of injustice that actually deepens inequality and social-environmental crises.

There may be a possible world in which the science system is entirely aligned with the public good wherever it is efficiently defended against epistemic corruption. In the actual world, however, the science system is deeply entangled with economic and governance regimes that also turn it into a source of justice and injustice. Agriculture may be an especially salient example, but similarly obvious stories could clearly be told in other domains, such as the health sciences. The ethically and politically outrageous handling of intellectual property regimes during the COVID-19 pandemic, which often prioritized corporate profits in the Global North over vaccine access in the Global South (Krishtel and Malpani 2021), provides just one straightforward example of contradictions in the health domain of similar magnitude to those in the agricultural domain.

Contradictions also appear in domains such as biodiversity conservation, which typically have more pristine reputations for being directed toward the common good. While corporate influence in agrifood and health domains raise relatively straightforward concerns about science as a source of injustice, fields such as conservation biology may appear as uncontroversially positive cases: scientific contributions to conserving biodiversity are of existential importance for all of humanity and the planet as a whole. There is no question that scientific contributions to biodiversity conservation are urgently needed and involve research in a wide range of disciplines such as conservation biology, ecology, engineering, environmental sciences, economics, ethnobiology, geology, management studies, policy studies, soil sciences, and sustainability sciences. Again, however, one-sided stories about scientific contributions to saving biodiversity risk distorting a complex picture. As political ecologists have documented for decades (Bryant and Bailey 1997), not only the destruction but also the conservation of biodiversity is embedded in economic and governance structures that commonly deepen rather than address global inequality.

Indigenous peoples, peasants, and other marginalized communities are indeed often most directly threatened by the destruction of biodiversity through industrial agriculture, logging, mining, and other forms of resource extraction. However, this does not mean that they are always beneficiaries of biodiversity conservation. There are countless counterexamples. "Green grabbing" (Fairhead, Leach, and Scoones 2012), including the expulsion of Indigenous communities for the creation of conservation areas free of humans, provides an example. The criminalization of traditional and subsistence forms of resource extraction offers another case in point (Boelens, Guevara-Gil, and Panfichi 2009). Yet another example are human–wildlife conflicts that almost exclusively affect marginalized communities "when wildlife forage on crops, attack livestock, or otherwise threaten human security" (Treves et al. 2006: 383). As biodiversity has increasingly become a commodity for "green capitalism," familiar contradictions appear in global biodiversity governance: As in the case of food commodities, biodiversity is also most cheaply produced in spaces of poverty to be consumed from spaces of richness – from carbon offsetting markets to ecotourism (Büscher and Fletcher 2020). Opportunity costs for the production of biodiversity are simply the lowest in spaces of poverty. Biodiversity regimes often contribute to stabilizing or actively creating those spaces by making other forms of economic activity illegal and concentrating economic benefits in the hands of large producers of biodiversity, such as owners of large carbon offsetting plantations or wildlife parks. "Science-led" or "evidence-based" approaches to biodiversity conservation are by no means a guarantee of resolving or even mitigating these tensions. On the contrary, the transformation of biodiversity into a form of capital (e.g., in ecotourism) and into a commodity (e.g., in carbon offsetting) are shaped by the mainstream producers of scientific knowledge.

Of course, it would be disingenuous to blame the science system for all injustices in domains such as agriculture, biodiversity, and health. However, it would be equally disingenuous to hail the science system for its potential to "feed the world," "save biodiversity," or "achieve global health" without addressing the reality of the science system with its wide range of both positive and negative effects in these domains. This does not mean ignoring the potential of the science system but rather not conflating potential with reality. A sober assessment of the current state of relations between science and society is crucial for developing normative visions of relations that actually harness the positive potential of the science system. The following section moves toward such a positive vision by emphasizing the role of three justice dimensions – distribution, recognition, and representation – for outlining an account of just science.

Science as a Site of Justice: Distribution, Recognition, Representation

My labeling of a "New Orthodoxy of Value-Laden Science" highlights the formative role of debates about values in creating an intellectual middle ground that transcends the dichotomies of the "science wars": Values permeate scientific practice from the choice of research questions to methods to theories to dissemination. At the same time, appropriately situated values do not undermine the epistemic authority of science and create entry points for substantial conversations about socially engaged and democratically legitimized science. While there is indeed a lot to be learned from debates about "science and values" (Brown 2020; Douglas 2009; Elliott 2017), they are no substitutes for debates about "science and justice." First, much of the "science and values" debate has been focused on making a general case for the legitimacy of values rather than trying to identify *just* values in science (e.g., Ludwig 2016). As such, the debate is helpful for navigating theoretical issues such as expertise, objectivity, or relativism but often provides much less guidance for engagement with the politics of scientific practices in contested social-environmental settings. Second, the focus on values can encourage a methodological individualism that focuses on the values of individual scientists in making certain choices (e.g., about conceptual framings, inductive risks, and theory choices) rather than the economic and governance structures in which these choices are embedded.

Rather than limiting the analysis to values in scientific practice, this section therefore outlines a broader, justice-oriented perspective. Political philosophy provides a wide range of frameworks for debates about justice (Kolm 2002) that also suggest different angles for debates about just science. For example, procedural accounts of justice will highlight stakeholder participation in science, while substantive accounts of justice will directly address the impact of science on livelihoods and well-being. Although it may be philosophically interesting to aim for one fundamentally unified account of justice, engagement with the messy reality of scientific practice suggests a multidimensional framework that can facilitate discussion of heterogeneous dimensions of scientific practice that relate to the production of heterogeneous (in)justices. Fraser's (2009) account of global justice provides such a framework by highlighting two substantive dimensions (distribution and recognition), and one procedural dimension (representation), that are of immediate relevance to a positive vision of just science.

Distribution: One angle for thinking about just science is provided by debates about *distributive* justice. Scientific research shapes a wide range of practices with direct effects on the global distribution of benefits and burdens across and within societies. Some effects are of a direct, economic nature – for example, research facilitates novel technologies that lead to commercialized innovations with often varied effects on different societies and on different members within a society. At the same time, scientific research is also central to a wide range of further issues of distributive justice, such as exposure to environmental hazards, access to health services, and access to educational resources.

The food system illustrates the broad and differential effects of science on distributive justice. As argued in the previous section, research in fields such as agronomy, engineering, genetics, organic chemistry, plant breeding, and soil science has contributed to a radical transformation of food systems with differential impact on stakeholders. For many stakeholders, agricultural modernization has made food more accessible, as reflected in declining long-term rates of undernutrition. As previous sections have highlighted, however, the reality is much more complex. Not only have global rates of undernutrition been on the rise recently, but an exclusive focus on rates of undernutrition obscures the social and environmental price of agricultural modernization in many areas of the world. The reduction of production costs of food has often come with dispossession of land and loss of labor for peasant populations, creating novel spaces of poverty of enormous scales. Distributive concerns also extend beyond food itself toward issues such as exposure to environmental hazards such as synthetic fertilizers and pesticides. In all of these cases, scientific contributions are complex and multidimensional: For example, new seed varieties can reduce the need for synthetic fertilizers and pesticides and thereby reduce exposure to environmental hazards. At the same time, synthetic fertilizers and pesticides are themselves a product of scientific research and dependency on such chemical inputs is a mark of science-led industrialization of agriculture.

Distributive justice provides a lens for substantial engagement with such complex causal effects of agricultural research on the distribution of material benefits and burdens. Indeed, increasing the productivity of agriculture has the potential to contribute to distributive justice. Scientific research that contributes to agricultural sustainability is indispensable for addressing a wide range of distributive justice issues. At the same time, potential impact is not the same thing as actual impact, and the food system illustrates how deeply the current state of agricultural research is implicated in the production of distributive injustices. From the perspective of distributive justice, a focus on just science therefore highlights the importance of transforming the role of science in society for redistributing its diverse benefits and burdens, such as income, stable access to food, food safety, nutritional diversity, health hazards, or environmental degradation.

Recognition: While distribution is at the center of many justice concerns, it has been widely argued that justice is not exhausted by matters of distribution but also raises complex questions of recognition (Young 1990). As Fraser and Honneth (2003) put it: "Whether the issue is indigenous land claims or women's carework, homosexual marriage or Muslim headscarves, moral philosophers increasingly use the term 'recognition' to unpack the normative bases of political claims." Issues of recognition are closely entangled with issues of distribution, but the former often do not reduce to the latter. A woman who is sexually harassed in the workplace may be negatively affected in her career and income but clearly experiences injustices beyond such distributive effects. An Indigenous community that loses its land loses much more than simply the distributive benefits of control over natural resources. Thus, Fraser (2009: 377) stresses "the demand for recognition of people's standing as full partners in social interaction, able to participate as peers with others in social life. That aspiration is fundamental to justice and cannot be satisfied by the politics of redistribution alone."

In the case of the global food system, concerns about recognition are most clearly reflected in the expansion of political activism from food security to food sovereignty (Noll and Murdock 2020). While the concept of food security is typically operationalized in distributive terms through stable access to nutritious and safe food, the influential Declaration of Nyéléni defines food sovereignty as "the right of peoples to healthy and culturally appropriate food produced through ecologically sound and sustainable methods, and their right to define their own food and agriculture systems" (Forum for Food Sovereignty 2007). Food sovereignty expands the scope of food security along two dimensions. First, the recognition of cultural (e.g., culinary, farming, fishing, hunting) practices and values that are crucial to the identities and self-determination of peoples. Even when agricultural intensification provides secure access to food, it may still constitute misrecognition of Indigenous peoples or peasants whose community structures, food practices, and ways of relating to environments are dismantled in the process. Second, the idea of food sovereignty highlights how recognition often turns out to be a condition for distributive justice. As Iris Marion Young (1990: 22) already argued, an exclusive focus on distributive indicators often "ignores and tends to obscure the institutional context within which those distributions take place, and which is often at least partly the cause of patterns of distribution." The institutional context of agricultural modernization in the Global South is often based on misrecognition of local communities and food systems that contributes to unjust patterns of distribution – for example, through dominance of exogenous cash crops that replace Indigenous food crops but are vulnerable to crop failures or market fluctuations.

Expanding the scope of concern from distribution to recognition provides important and challenging lessons for an account of just science. While distributive concerns are of crucial importance, they need to be complemented through serious intercultural dialogue about the structure of the science system and a recognition of global epistemic diversity including the knowledge of Indigenous communities (Chilisa 2019; Rivera Cusicanqui 2010; Solano and Speed 2008; Vijayan et al. 2022). Modern science and technology are deeply disruptive in peoples' lives, and the food system provides some of the most dramatic illustrations of this, having fundamentally transformed rural spaces through dynamics of depeasantization and repeasantization, as described in previous sections. Not all forms of disruptive change are negative, but they are fraught with contradictions that can (and will) be evaluated in radically different ways from different, culturally situated standpoints. There is no "view from nowhere" in evaluating the global ramifications of science through a neutral set of distributive indicators. This lesson is especially challenging for scientists in the Global North who may be inclined to think of just science through well-intended distributive indicators rather than serious intercultural dialogue that recognizes heterogenous aspirations, needs, practices, and values.

Representation: Nancy Fraser (2009) identifies distribution and recognition as "first-order questions of substance." In the domain of agriculture, they include: How do transformations of agricultural productivity affect profits and wages, and whose? How do they affect patterns of land ownership and issues such as land grabbing? What are the effects on local community structures, from capital accumulation to division of labor to migration patterns? What are the effects on culinary cultures and diets? Who is exposed to what kinds of environmental and health hazards? What are the effects on local agrobiodiversity? How do they interact with processes of deforestation and soil erosion? What are the effects on community resilience in the face of disruptive events such as climate change and economic shocks? What are the effects on local relations with ecosystems such as leisure activities and spiritual connections?

Second-order questions of representation address the ways in which these first-order questions are negotiated. In the agricultural context, representation is crucial for two reasons. First, due to the entanglement of various issues of distribution and recognition that make evaluations of first-order questions deeply contested: How to weigh cheaper access to food against increased exposure to environmental hazards? How to weigh benefits for one group of stakeholders (say, the urban poor) against burdens for another group (say, the rural poor)? What is the weight of recognizing cultural dimensions of food sovereignty compared to more straightforward distributive aspects of food security? Second, issues of global justice often involve deep procedural inequality in negotiating these first-order questions. Agricultural development constitutes a prime example as it usually involves a dramatic discrepancy between dominant actors (e.g., corporations, donor countries, nongovernmental organizations [NGOs], scientists) and those who are most profoundly affected by interventions (e.g., Indigenous communities, peasants, urban underclasses). Second-order injustices of representation therefore often feed back into first-order injustices of distribution and recognition, since the former are often shaped by the interests of dominant actors. And even interventions that focus on benefits for marginalized communities can deepen injustices if they are grounded in a paternalistic second-order mode that evaluates first-order issues *for* rather than *with* these communities. For example, an NGO and a local community may have very different priorities in evaluating the complex ramifications of introducing a new cash crop for issues of distribution and recognition.

Expanding the scope of this discussion from first- to second-order questions of justice has important implications for a positive perspective of just science, as it highlights procedural aspects of the interface of science and society. Indeed, these procedural concerns have become increasingly prominent in science governance, reflecting broad shifts toward "transdisciplinary research methods," "participatory action research," and "public engagement" (Ludwig and Boogaard 2021). Especially in development contexts, a wide range of debates about "inclusive development" reflects a reckoning with the paternalistic legacy of the science system that highlights epistemic diversity and the need to codevelop interventions *with* (rather than merely *for*) marginalized groups (Ludwig et al. 2021). Second-order questions of representational justice thus have substantial implications for a positive vision of just science. It is not sufficient to incorporate first-order questions of distribution and recognition into research projects. The science system needs to become more inclusive and responsible in shaping practices together with affected stakeholders (Wittrock et al. 2021).

Fraser's distinction between distribution, recognition, and representation provides a helpful heuristic for engaging with questions of just science.

On the one hand, it provides an angle for critical engagement with contradictions of the science system that often remain invisible in debates about climate change denialism, anti-vaxxers, and other forms of epistemically corrupt anti-science sentiment. While these debates clearly matter, epistemic integrity does not guarantee just science. Beyond this critical attitude, however, an account of just science also provides an entry point for positive visions of the science system that aim to address the contradictions it produces. Scientific research can contribute to a more just distribution of resources, just as it can be shaped by an intercultural recognition of diverse standpoints and create spaces for their representation in scientific practice.

Lovable Science

Polemics aside, there are many important insights in the literature that I have lumped together as the "New Orthodoxy." Yes, the world is facing social-environmental crises that require urgent action. Indeed, science is indispensable for addressing these crises. And yes, this requires challenging anti-intellectualism and anti-science populism as exemplified by climate change denialism and anti-vaxxers. Furthermore, much of the literature of the New Orthodoxy reflects an understandable frustration with the legacy of critique in STS (Latour 2004b), which has often focused on a negative program of challenging scientific authority rather than a positive program of aligning science and society. Against this backdrop, Collins' (2021) plea for loving science can be situated in a wider humanist tradition that recognizes that "the application of the fruits of scientific investigation by reason is crucial to shaping a better, collective future" (see the Introduction to this volume).

There are many reasons to highlight this humanist tradition in the light of global challenges, and it finds a beautiful expression in Collins' call for loving science. Loving science, however, should motivate us to strive for lovable science. And epistemic integrity is not enough. Large parts of the science system are epistemically successful and still play deeply contradictory roles in both addressing and producing social-environmental crises. Science that is deserving of our love demands not only epistemic but also political integrity in confronting its impact on the world. Or, to put it as a slogan, *lovable science is just science*.

Engaging with science through first-order questions of distribution and recognition as well as second-order questions of representation opens spaces for a positive vision of both epistemic and political integrity in science. Realizing a humanist perspective on lovable science therefore demands an equally critical and constructive attitude. Engagement with the contested and sometimes fragile position of science in society is most convincing when showing that a more just science system is possible – that there can be science that is deserving of our love. Historians, philosophers, and sociologists of science have a lot to contribute to developing such positive and disruptive perspectives on the position of science in society. Indeed, such perspectives are a crucial part of the legacy of political philosophy of science from Otto Neurath to W. E. B. du Bois to Paul Feyerabend to Sandra Harding to Paulo Freire. Rather than simply accepting that "critique has run out of steam" (Latour 2004b), however, this requires a constructive reading of critique that diagnoses current contradictions in order to open new directions for a more just interface of science and society.

CHAPTER 11

The Human Sciences and the "Theory of Women"

Catherine Wilson

A number of influential commentators embrace the view that the human sciences can deliver knowledge relevant to morality, the design of institutions, the framing of laws, and political life. For Daniel Dennett (1996: 268), "ethics must somehow be based on an appreciation of human nature – on a sense of what human nature is or might be like and what a human being might want to have or be." Steven Pinker, in *The Blank Slate: The Modern Denial of Human Nature,* maintains that

> The new sciences of human nature can help lead the way to a realistic, biologically informed humanism. . . . They promise a naturalness in human relationships, encouraging us to treat people in terms of how they do feel rather than how some theory says they ought to feel. They offer a touchstone by which we can identify suffering and oppression wherever they occur . . . They renew our appreciation for the achievements of democracy and of the rule of law. And they enhance the insights of artists and philosophers who have reflected on the human condition for millennia. (2002: xi)

The principle that "ought" can't be derived from "is" is technically correct. Nevertheless, according to these writers, what "is" can provide guidance for what "ought" to be, assuming agreement in underlying values, such as relief of suffering and oppression. The empirical perspective invites us to look beyond interchangeable Kantian noumenal egos with abstract rights and obligations and to consider people's endowments and desires, and their fit or lack of fit with the social conditions they live in. The more we can learn about human nature, it seems, the more humane and the less wasteful our institutions and practices can be made to be. Frustration results when needs are not satisfied, when people are required to behave in ways that are unnatural for them, and when talents and interests have too little room for development. The recommendation to adopt a scientific, rather than a purely philosophical, approach to designs for living is accordingly sound to the extent that the sciences can shed light on human needs, capabilities, and sources of satisfaction and dissatisfaction.

Nevertheless, any suggestion that we can discover, not only some, but all of people's needs and abilities merely by examining their choices and their successes and failures is naive. In a society that limits choices and that restricts opportunities for the development and display of talents, these will not be revealed.[1] Nor does empirical observation distinguish between morally acceptable needs and the so-called perversions or the hunger for power and domination. The human sciences – anthropology, biology, and comparative ethology – are supposed to take us under the surface to show us what is really going on: what human nature, undistorted by culture, really looks like. But anthropology began as a study of cultural, ethnic, and sexual difference, not simply as a study of what makes humans human. And the long history of "scientific racism" and "scientific sexism" can undermine confidence in the ability of anthropology and biology to promote democracy and respect for other people's feelings.

Where "scientific racism" was a comparatively modern development, women and their parts have been observed, anatomized, weighed, and measured – and thereby found wanting – since ancient times. More recently, the theory of evolution by natural selection has inspired countless writers, beginning with Charles Darwin, to consider its applications to social and political life. Yet the prescriptions and policies claimed to be rooted in Darwinian science, first by "sociobiologists" and later by "evolutionary psychologists," range from the disappointing to the disturbing. They have appealed to concepts of inheritance, innateness, and evolutionary significance in order to parade values – or at least to sigh over inevitabilities – that have an authoritarian and archaic cast to them. In the latest version of biology-is-destiny, we were told that it is the biological goal of all living things to leave a lineage, and that insofar as males and females are differently specialized to maximize their personal chances of passing on their genes, we can expect the socio-economic-political world to be permanently constructed on the basis of difference, unless misguided ideological fanatics succeed in forcing social equality, and with it misery on us. Claims about male–female sameness, including that "the mind has no sex," belong, it is implied, to a philosophical fantasy world.

This chapter is written to dispel the nagging suspicion – or the frank accusation – that the real world, not the world of noumenal selves and their posited equality, contains forces and constraints, rooted in biology

[1] See Wilson 2004: 99–117 for further discussion.

and human ecology, that make the equality of men and women neither fully achievable nor really desirable. While I agree that there is sexual specialization as a result of natural selection and that it functions importantly in the historical explanation of the division of labor, I believe three long-standing myths have been overturned by research in the human sciences: the myth of female cognitive inferiority; the myth of female domesticity; and the myth of female natural monogamy.[2] To that end, I focus on research in psychology, anthropology, and primatology that has upended the sociobiological "theory of women," with its echoes of ancient theorizing, that began to appear in the literature in the late 1940s and that has persisted in widely cited articles and popular books. The still-to-be-digested revisions of these sciences are not products of specifically feminist research; they belong to ordinary science. But they awaited the political and social changes that brought women, who asked new questions and noticed new phenomena, into the natural and social sciences.

The "theory of women" comprising the three myths was enough to explain women's exclusion from important offices and activities and to justify behavioral restrictions on women and liberties for men. It was held to explain an important and uncontroversial set of observations. Until quite recently, in what vocational roles did the ordinary person most often find women? Certainly not in the top ranks of commerce, politics, scientific research, literary criticism, public architecture, or the arts – these roles were believed accordingly to lie outside their competence. As Charles Murray asked in 2005, "Where are the female Einsteins?" (Murray 2005b).

According to Richard Lynn (2017: 9–42), men beat women not only at tennis, golf, and footraces but also at thinking. Human males over the age of sixteen, he reported (2017: 145–156), are better than women at Raven's Progressive Matrices, a test of nonverbal reasoning through pattern analysis that does not depend on cultural knowledge, and better at winning top prizes in science, chess, Scrabble, and bridge. As well as not being found among the decision makers, winners, and influencers, women were not observed seeking and collecting packs of sexual servants. Mostly, women were to be found at home, taking care of things, and as active in the world in helping others and providing for their needs. The normative image emerging from observation was that of a human lacking a first-class

[2] Recent research has also forced a revision of the traditional view that "female" is a scientifically precise term and the traditional view that there are only two sexes. See Fausto-Sterling 2012. This chapter does not go into these interesting issues; I refer here to "males" and "females" and "men" and "women" in the rough, vernacular sense of most social science.

intellect, but possessing propriety, altruism, and charm. There were obvious deviations from the norm, but in fiction and in real life, they were liable to be ridiculed, feared, diminished, obstructed, or tragedified and punished.

I now turn to the three myths in order.

The Myth of Cognitive Inferiority

The notion that high levels of intelligence are a sex-linked trait shaped by natural selection goes back to Darwin, who was convinced of not only the intellectual inferiority of women but also their lesser creativity, coordination, and even sensory acuity. Darwin supposed that the inheritance of acquired characteristics ("habit"), as well as natural selection on the male sex, had produced this abundance of excellence.

> The chief distinction in the intellectual powers of the two sexes is shewn by man's attaining to a higher eminence, in whatever he takes up, than can woman – whether requiring deep thought, reason, or imagination, or merely the use of the senses and hands . . . The greater intellectual vigour and power of invention in man is probably due to natural selection, combined with the inherited effects of habit, for the most able men will have succeeded best in defending and providing for themselves and for their wives and offspring. (Darwin 1981/1871: 327)

Modern humans have large brains,[3] and our proliferation across the globe and the extinction of our hominin cousins is often ascribed to this anatomical feature. Such large, heavy brains are metabolically costly. Though the brain accounts for only 2 percent of the weight of the human body, it uses 20 percent of the organism's energy budget just to maintain and regulate basic bodily functions. Why did such an expensive-to-feed organ evolve? What benefits did it bring its possessors?

Following Darwin's hypothesis, on one view popular in the 1960s, the modern human brain evolved to enable male humans to solve complex survival problems in the environment of early adaptation; men are thus implicitly the developers and owners of these large brains. In the famous Lee and DeVore anthology, *Man the Hunter*, William Laughlin declared that "hunting is the master behavior pattern of the human species"

[3] For comparison, the brain of *Australopithecus australensis*, living about three million years ago, was about 450 ccs, comparable to that of a modern chimpanzee; that of *homo habilis*, 2 million years ago, was about 600 ccs. Modern humans, who have existed for at least 250,000 years, are far larger in overall size and have average cranial capacity of 1,250–1,400 ccs.

(Laughlin in Lee and DeVore 1968: 307). Bipedalism was suggested to have evolved to free the hands for spears, boomerangs, and bows and arrows, and to have led to the evolution of language and intelligence to support the coordinated activity of early hunters. Laughlin detailed the skills required in the hunter: knowledge of animal types and habits; scanning, stalking, and immobilizing; and the retrieval and transportation of the carcass. The growth of the neocortex allegedly allowed for abstract thought, imagination, impulse control, and other well-developed and typically human capacities, and was driven by this hunting adaptation.

On another view, that of Geoffrey Miller (2000: 97), brain growth was driven by sexual selection on male humans. According to Miller, developed mentality was a male display feature, analogous to the peacock's tail, with clever, artistic men preferred by early women as mates. Miller went on to argue that the struggle for wealth, position, and deference through cultural production is the specifically human form of antler-locking, head-butting, or biting and chasing that determine male "access" to fertile females in many other mammalian species. Men are more driven than women are, he thought, to create objects and structures that can make them famous and celebrated, or at least esteemed and respected. Female variants who allocated too much time to the pursuit of status and mating opportunities, on this theory, were out-reproduced by more specialized maternal competitors, and male variants who allocated too much time to direct paternal care were out-reproduced by more specialized promiscuous status-seekers. Implicit in both accounts is that the feeding role and baby care do not require much raw inventive, strategic, and computational brain power, so nature could skimp on this endowment for women and compensate with extra emotional responsiveness.

The notion that male hunting drove the increase in brain size because men but not women needed to be intelligent is no longer regarded with much favor. Animals with much smaller brains can locate, track, slay, and transport prey, and they can strategize and coordinate their kills with others. And where sexual selection is concerned, although humans use their large, evolved, and culturally developed brains for social, cultural, and intellectual purposes, and even to attract mates – for modern men and women both place "intelligence" high on their list of desiderata – a top-level mind is unlikely to be a male display feature females lack. In the peacock case, it is disadvantageous for a female to have a fancy tail. Like many female birds with relatively exposed nests, for her own safety, she must remain "drab" and inconspicuous. In birds that build nests relatively inaccessible to predators, both sexes are brightly colored.[4] It is hard to see of what protective advantage it could be for women to be less intelligent and less artistic than males.

In any case, the Miller account of the evolution of a sex-linked trait is unsustainable, simply because decades of testing have shown that men and women vary little or not at all with respect to intelligence, currently understood as including memory, learning, problem solving, flexibility in behavior, language fluency, creativity, speed of understanding, and ability to function in social settings (Colom 2000: 57–68).[5] The higher cortical functions needed for planning, designing, and governing, considered in the abstract, are the same in men and women. They have essentially the same ability to recognize patterns and perform inferences. Humor, aesthetic appreciation, and language use are not very different. And apart from some small number of tasks concerned with spatial orientation, nimbleness, perception, and fluency, favoring one sex or the other, men and women differ little in problem-solving ability.

An important datum is nevertheless the "two tails" phenomenon. There are more men at the very top and the very bottom of the IQ scales: men are more variable than women in this regard and in other regards. This has an important biasing effect in social judgment, as I explain later. For now, it is sufficient to remark that in the view of the most recent researchers sex differences in variability do not account for sex differences in high-level achievement. Neither male-favoring cognitive differences nor the number of males in the upper regions of the IQ distribution are large enough to explain the predominance of men in science and engineering (Brush 1991; Halpern et al. 2007).

What do we actually know about brain size, intelligence, and its evolution? We think of our brains as mainly used for planning, strategizing, inventing, and reasoning, but this is an error. Most of the volume of the brain is devoted to sensory perception, the regulation of movement, and physiological homeostasis and periodicity: the regulation of bodily processes, especially the release and suppression of hormones and neurotransmitters. Lynn (1994: 257–271) ascribed male success in general to men's larger brains,[6] but with the exception of a few traits, such as those related to anger and empathy, and, potentially, spatial abilities, male/female differences in behavior, interests, and mental health are not reliably

[4] In kingfishers, woodpeckers, toucans, parrots, and turacos, the females are as brilliantly colored and conspicuous as the males (Wallace 1889: 273).

[5] See also Halpern and LaMay 2000: 229–246.

[6] See also Lynn and Irwing 2004: 481–498.

correlated to brain anatomy (Eliot 2020: 63–82).[7] Larger organisms in any case require more neurons to regulate and control their bodies (Sowell et al. 2007: 1550–1560). According to Lisa Eliot: "Most male/female brain differences are attributable to body size; thus, all brain structures are 10 percent larger in males, but after accounting for individuals' total brain volume, sex/gender explains only ~1 percent of the variance in structural volumes at both the cortical and subcortical level." Women have on average thicker cerebral cortices, associated with greater intelligence, than men (Ritchie et al. 2018: 2959–2975).

Evolutionary theory recognizes a number of coexisting conditions for increasing brain size. These include concentrated nutrition on account of its high caloric requirements; a system for cooling the brain, which cannot sweat inside the skull; and parturition of relatively underdeveloped infants. These features would have had to evolve in step with the trend toward larger brains. The following main contenders to the man-the-hunter theory and the peacock's tail theory regard selection pressures as operating on both sexes and are based on the concepts of longevity, sociality, and expertise.

Kristen Hawkes' "Grandmother Hypothesis" (Hawkes et al. 1998: 1336–1339) proposes that human evolution involved the coevolution of three features: long life, protracted infancy and childhood, and large brains. Large brains are found only in animals with relatively long gestation, long juvenile periods, delayed reproduction, and long life – animals such as whales, elephants, and humans – committed to the "K-strategy" of reproduction, by contrast with the "r-strategy" of trying to maximize the number of offspring in a short lifetime,[8] in the hope that some manage to survive. According to Figueredo et al. (2006: 246): "Traits associated with a high K-strategy in humans are long-term thinking and planning, commitment to long-term relationships, extensive parental investment, existence of social support structures, adherence to social rules (e.g., altruism and cooperation), and careful consideration of risks."

Like large brains, long childhood is a prima facie evolutionary puzzle. A protracted period of nutritionally dependent, nonreproductive childhood is expensive in biological terms; an organism that can shave a few months off its period of dependency on others and begin to reproduce would seem to have an evident advantage. Hawkes' hypothesis links the K-

[7] See also Halpern and Wai 2020: 317–345 and Blinkhorn 2005: 31–32.

[8] For the origin of these terms (in "*Kapazitaetsgrenze*" and "rate") see https://en.wikipedia.org/wiki/R/K_selection_theory.

strategy to brain evolution, at the same time explaining the cultural importance awarded to grandmothers, especially maternal grandmothers. Hawkes proposed that long childhood and larger brains coevolved with a lengthening human lifespan, including a lengthy post-reproductive phase. While chimpanzee females retain their fertility until they die – the lifespan of a chimpanzee in the wild is about forty-five – the fertility of human females peaks in the early twenties and begins to decline by the early thirties, coming to a decisive end sometime in their early fifties, when hormonal cycling falters and then comes to a halt. Yet a hunter-gatherer female who lives into her mid-forties can expect to live another twenty or so years. Hunter-gatherer males as well tend to outlive their reproductive span when they are not victims of homicide.

Hawkes argued that selection for cessation of reproduction and for a period of post-reproductive vitality in women enabled them to shift their effort from the care and feeding of extra children up to the point of their own demise to the care and feeding of grandchildren. This shift allowed for a prolonged period of nutritional dependency in childhood and the slow maturation of a large brain, a process which in turn gave an even greater selective advantage to hardworking grandmothers (Davison and Gurven 2022: 1–12). Other theorists have argued that the elderly are essential to the human way of life because of their stored knowledge – for example, of unusual foods to be eaten in times of famine (Shipton 1990: 369).

Longevity in turn requires brain redundancy. "The brain," observes Nick Humphrey (1999: 2),

> is . . . a fragile organ, which is vulnerable not only to external knocks, and internal hemorrhages and tumors, but also to intrinsic processes of cell death and decay. By the time we reach middle age, the brains of every one of us will almost certainly have suffered significant structural damage. Yet the fact is that the majority of us will not yet have suffered any obvious intellectual loss.

The reason is that we have more than enough brain to make up for it. The chief purpose of the overly large brain may be to furnish back up power for essential physiological and psychological operations, keeping senescence and senility at bay.

Robin Dunbar's (2003: 163–181) socialization hypothesis proposes that the brain coevolved with larger tribe sizes and greater interdependence of their members. This created a need for language and for "social intelligence," for keeping track of one's relations with a multitude of others, for understanding, predicting, and directing their behavior and adjusting one's own to it, for political outwitting and outmaneuvering, and for recognizing and punishing wrongdoers. In a related vein, Dean Falk and Sarah Hrdy have proposed that infants and children, together with their minders, prompted the evolution of the human mind, jointly evolving a propensity for shared attention and playfulness. Hrdy (2009) suggests that human babies evolved into charming, demanding, interactive beings – perhaps even into linguistic beings – in order to engage attention from not only their hormonally saturated mothers but also from others who were less thoroughly primed but whose nervous systems were vulnerable to this type of stimulation. Such babies might have grown up into sociable, mind-reading adults who were in turn better able to soothe and care for active and curious young babies. Falk (2004: 498–501) points out that human babies, unlike ape infants, cannot cling to the fur of their mothers. They need to be set down – parked – so that their mothers can do other things. Mothers needed to warn, control, and reassure when at a distance, and language and empathy permit this.

A third current theory is expertise: Hillard Kaplan and his coauthors (2000: 149–186) have argued that not only expert tracking and hunting but also food preparation practices coevolved with a longer human lifespan, and a larger brain that permitted learning, practice, and mental storage. John Skoyles (1999: 1–14) has suggested that large brains coevolved not with general intelligence, which is measured on tests by the speed with which one can spot patterns or complete inferences, and which is measurably normal even in brains of only 750 cubic centimeters (ccs), as compared with the normal brain of around 1,300 ccs, but with the capacity to learn through practice and refinement.

Nonhuman animals, as Descartes noted, are expert but in limited domains, such as nest building and prey snatching, and need little practice to perform activities necessary to survival. Humans, by contrast, can master a variety of skills but only by observing, submitting to instruction, and engaging in solo effort involving much trial and error. They are motivated to learn new arts and to stick with learning even when frustrated. Hunting and foraging animals, as noted, do not need human-level intelligence to find food, but the human way of life requires considerable expertise. The usual diet of the environment of early adaptation was apparently rather chewy and sour. Many vegetable foods require boiling, steeping, or pounding if they are to be made tender, sweet, or nonpoisonous, and considerable technological inventiveness was required to make containers for transporting, processing, and storing. The remarkable tools and techniques of hunter-gatherer women for building shelters, making garments and ornaments, taxidermy, tattooing, and pottery making, as well as all aspects of food preservation, has been well documented. As one social historian comments (Mason 1929: 158), the modern lucrative employments "originated with her."

While Lynn's claims for 3–5 points higher male IQ might seem puzzling, since IQ test are currently formulated so that the average IQ of men and women is the same,[9] the more interesting aspects of his argument concern the so-called right tail of the IQ distribution and mathematical ability.[10] When tests are scaled as described, men are, as noted, more frequently found than women at the high and low ends of the intelligence spectrum. The ability to perform spectacular feats of mental calculation and the hyperfocus on abstract objects associated with tech workers is more common in men. Although the size of the right tail varies from culture to culture and has decreased in recent years, it is still there. Jonathan Wai and colleagues, in their longitudinal study (2010: 412–423), found that in the top 0.01 percent of mathematics SAT students, the male advantage declined from 13.5:1 to 3.8:1 over the decade between 1980 and 1990. The reasons for this dramatic improvement in women's test scores have not been fully explored, but the most attractive hypotheses must focus on the motivational changes in women and their parents and teachers, brought about by the feminist movement, that encourage the cultivation and display of competence and competitiveness.

Two-thirds of the population falls within the IQ range of 85–115, and most academics are found in the 120–140 regions, comprising about 11 percent of the population. At the very top of the scale, where IQs of 160 and above and dazzling mathematical abilities are found, often along with striking personal qualities,[11] the ratio of men to women is about 6:1. We are talking here about very few people – about 0.03 percent – of the population, or 3 persons per 10,000; approximately 1 in every 20,000 of those is a woman. Yet the cultural worship of the male "genius" – usually associated with mathematical and musical accomplishments known to the general public – operates to the disadvantage of the entire female sex.

Reilly, Neumann, and Andrews (2022) found that men estimate their own intelligence as higher than it is when measured, and women estimate

[9] What is the purpose of this stipulation? One might wonder. There are many purported tests of intelligence, some of which favor women, some men. To bypass arguments about what intelligence *really* is and which test *really* tests for it, a test on which the sexes score the same on average offers a pragmatic solution.

[10] Lynn's choice of Raven's Progressive Matrices as a proxy for a general intelligence test arguably reflected a bias in favor of a specific ability. For criticism of the choice, see Gignac 2015: 71–79.

[11] For discussion see Persson 2007: 19–34.

theirs as lower than it is when measured. Married men tend to believe they are more intelligent than their wives (Kidder, Fagan, and Cohn 1981: 239–255). A number of men I have queried in academic life have admitted to the belief that they are intellectually superior to every woman, or nearly every woman, they have ever met. The actual shape of the IQ distribution shows how mistaken this belief is. At any given level, running into someone of the opposite sex with more IQ points than you varies, but given the statistics above, it's bound to be a routine occurrence in any professional milieu. If we installed people in university posts and paid them according to their IQ points alone, the composition and emoluments of our academic faculties would look very different than they do.

Yet the more frequent surfacing of male "geniuses" at such tasks as mathematics, architecture, musical composition, and military conquest in the record of civilization redounds to the credit of the male sex as a whole in our minds. While we would not want to have Bobby Fischer or Isaac Newton on the Supreme Court, it is thanks to the Bobbys and the Isaacs, as well as the Alberts, the Ludwigs, the Alexanders, and the Napoleons, that all men benefit in reputation.

Why do well-educated men tend to believe they are cleverer than all the women they meet? I suspect it is that they feel that they (and not the women they meet) belong to the club of Einstein, Newton, Beethoven, and the top men in their fields, and they know intuitively that they do *not* belong to the club of all the men on the left-hand tail. All women are seen as, by comparison, pretty much average.

The Myth of Female Domesticity

"There is no society," David Barash told us in his 1979 book *The Whisperings Within*,

> historically, or in recent times, in which women have not borne the primary responsibility for childcare. ... In all societies, men do men things and women are left holding the babies. ... If male mammals are generally less involved than their mates in caring for offspring, what do they do? Males tend to achieve fitness by making themselves as attractive as possible to females, then rely largely on the females to take it from there. Often, they compete with other males, either for direct access to mates or for access to resources which help them acquire mates. (1979: 126–128)

In this passage, Barash springs from historical time and its written record – beginning about 5,000 years ago – back to the early days of mammalian evolution. He fails to take into account what can be inferred from the study of contemporary hunter-gatherers, whose forms of social organization are theorized as corresponding to those of the anatomically and neurologically modern humans of the environment of early adaptation and the Paleolithic period. Although the few remaining hunter-gatherers are not living fossils, and although even they have been altered by contact with explorers and occupiers, governments, and trade, their mode of life offers compelling clues as to "human nature" as it whispered within before being shushed and shouted over by the development of civilization through the invention of agriculture and metalworking.

Barash's claim that women have always devoted more time than men to direct childcare is true, but his statement that women are "left holding the babies" is misleading. The second sentence should have read: There are only a few hunter-gather societies in which women do not bear the primary responsibility for finding and furnishing the majority of food for the community. Only under civilization did some women become dependent on male effort and earnings to sustain themselves and their children, and even under civilization, poor women have continued to be the major providers for the household.

Hunting is a nutritionally and culturally significant activity, but the Lee and DeVore volume – as its own editors pointed out in the introduction – might as well have been titled less sensationally "Man the Hunter-Gatherer," or even "Man the Gatherer," for repeatedly the point was made that most of the calories consumed by savannah, desert, and forest-dwelling people are vegetable foods, roots, shoots, fruits, and berries, and that this food was gathered mainly by women. Meat is highly valued by hunter-gatherers for its taste and for its fat, protein, and mineral content. But only in the far northern regions colonized late by humans, where the growing season was short and grasses predominated, has meat been found to compose the bulk of the human diet. For southern populations, it would be more accurate to say that hunting is 25 percent of the food-getting behavior pattern of the human species, and foraging, scavenging, and gathering, 75 percent (Lee in Lee and DeVore 1968: 30–48).

Most hunter-gatherer societies observe a division of labor. This is not because women are not strong enough to kill animals, or because they have limited spatial abilities and cannot find their way around, or because they are too burdened with children. In the Arctic, it has been reported, women will occasionally hunt seal or caribou on their own, or bring down deer with sticks, ropes, dogs, or nets. The Agta women in the Philippines hunt deer and pig in the forest, alone or with the help of dogs, and communal hunting or solitary hunting of small game is an activity performed by both sexes (Estioko-Griffin and Griffin 1981: 121–154). Latter-day female gatherers range, in groups or very often alone, as much as five to seven miles from camp with their sacks and their digging sticks, while, in many cases, carrying an infant or very young child or in the middle to late stages of pregnancy. Children over the age of two or three are not taken into the bush where they are a nuisance to working mothers: They are too heavy to carry, they cannot walk fast, and they get thirsty and tired; they remain behind with other caregivers, such as grandmothers. While both men and women forage, men do not tend to collect and transport plant materials using nets, containers, or other means, or to share plant foods they have gathered with others as women do.

A point that emerged in *Man the Hunter* is that while gathering is obligatory for human societies, hunting is more or less optional. Many men in hunter-gatherer and indeed in hunter societies are reluctant to hunt, do not enjoy it, or are not very good at it. In the far north, where there is almost nothing else to eat, women must nag men to get them to hunt. Among the Hadza of the Serengeti plateau in East Africa, one of the last hunter-gatherer societies surviving to be studied, only men in their late teens, twenties, and thirties were successful hunters, and about half had failed to kill even one animal the entire year (Woodburn in Lee and DeVore 1968: 54) Older men in their fifties and sixties did not hunt but worked alongside women and sometimes alongside children. "Only about 60 percent of the population of Bushmen in the Kalahari Desert appeared to be working at all" (Lee and DeVore 1968: 36). What did people do the rest of the time? They visited, chatted, gambled, manufactured or repaired things, or rested. This, we can reasonably suppose, was the "master behavior pattern" of our species.

Nothing, accordingly, could be further from the human template than the housewife-at-home-with-the-children. Adult women in hunter-gatherer societies spend time in the company of other adult women in physically demanding and essential economic activity outside the home; children are taken care of by older children and by elderly relatives who are no longer as mobile. Women have not evolved to be round-the-clock hearth-hugging nurturers. They need and gain psychological satisfaction from moving around in the world, exploring their terrain, facing uncertainties and dangers, and bringing back food and necessities to their dependents, both men and children.

The frequently heard argument that it is best for infants and young children to be cared for by the mother alone has no basis in anthropology or psychology and is certainly not a consequence of our evolutionary history. If women are hormonally and neurologically primed to devote themselves full time to infants and toddlers and to gain satisfaction from doing so, why do somewhere between 10 and 25 percent of women – the higher figure pertains to the lower socioeconomic brackets – develop a tragic array of symptoms, neglecting their babies and young children, physically abusing them, and developing a sense of apathy toward their own lives? Hormonal explanations for post-partum depression have been discredited; and evolutionary explanations have lately come to the fore. Studies from across the globe suggest that depression may be the effect of being unable to cope alone and a signal to others, a cry for more help (Hagen 1999: 325–359).

An important feature of the human way of life is what Hrdy terms "alloparenting." Human infants, as noted, cannot cling; they must be carried on the hip or in slings or backpacks. As such they are shareable; they can be handed around to female relatives, to fathers, and to older children. Hrdy (1999) proposes that early humans were "co-operative breeders," that human mothers relied on a spectrum of "allomothers," not only grandmothers but also fathers, siblings, and other boys and girls for infant and childcare. Where a chimpanzee mother carries her offspring and jealously guards it until it is able to nourish itself, a human mother is willing to relinquish her baby into the arms of others immediately after birth, and other humans are willing to provide it with nutrition and care, freeing the mother for economic tasks and social interaction, and enabling her to become pregnant again, given the short window of peak fertility she has available to her. Cooperative breeding arises in a number of species; it is evidently consistent with selfish genes, though it cannot entirely be explained by kin selection, insofar as distantly or unrelated individuals, including foster children, are observed in babysitting roles.[12]

Barash's comment that "men do men things" while women mind the baby proved to be not only anthropologically but also historically uninformed. Throughout history, women have always worked outside the home. In ancient Mesopotamia, Wright (1996: 89), following Kang (1973), states, women worked in "harvesting and irrigating fields, carrying and winnowing barley, hauling barley into granaries," and also in "milling, weaving, and loading goods into boats." Medieval texts and illustrations cite and depict women in a variety of skilled occupations, as butchers,

[12] The advantage to the individual cooperator may come in the form of expectations of reciprocity, practice, or social stabilization. See Page et al. 2019: 115–116 and Valentine et al. 2020: 1037–1055.

ironmongers, hatters, shoemakers, bookbinders, painters, goldsmiths, innkeepers, veterinarians, and farmers.[13] But the adaptation of women to work outside the home and outdoors and their drive to feed their families had a consequence. For human men, paternal investment is optional, where not enforced by law or rigid custom, and where men have not been forced into provider roles, they can demand status or high wages if they are to work, while most women, except in the historically rare cases of middle-class prosperity, needed to perform manual work of low status outside the home. Even in poor societies, men have more leisure and more discretion over their time, their expenditures, and their choice of activities than women (Dasgupta 1995: 308ff.). In wealthy societies, this extra freedom and leisure enabled them to monopolize the high-status tasks of culture and civilization.

The Myth of the Naturally Monogamous Female

The notion that men are naturally polygamous and women naturally monogamous is a favored topic for journalists and popularizers of evolutionary psychology. While its alleged ramifications are extensive, the credulity attached to it comes as something of a surprise. For in philosophy, theology, and fiction, strict fidelity to a partner was considered the ideal, but real women were known to be alluring, fickle, deceitful, liable to illegitimate pregnancy, and accordingly dangerous.

This myth is pernicious in many ways. It interprets the battle of the sexes as a conflict between a male need for freedom and gratification and a female demand for food, shelter, and money. It confuses female choosiness with a lack of sexual drive. Its proposal that men select sexual partners on the basis of face and figure, whereas women select sexual partners on the basis of status and income, creates a flattering halo around the predatory behavior of older men toward younger women. It tells women not only what they may and may not do in order to be natural and normal but also what they ought to feel rather than what they do feel.

How did this myth get into the books? It was supposed to follow from gamete size and number. We often read that female fertility is a scarce resource while male fertility is nearly boundless. A man, it is supposed, can impregnate hundreds of women in a single year. He can allegedly father healthy children even in advanced old age, while a woman can produce at

[13] Working women who were not "heads of households" were left out of administrative occupational records. See Swanson 1989 and Sharar 1983.

most about twenty children over her entire reproductive lifespan, and she is liable to give birth to genetically abnormal infants toward the end of that period in her forties. Women invest more physiological and behavioral effort into gestation and lactation; men must, it is supposed, invest more physiological and behavioral effort on attempted impregnation. According to A. R. Bateman in a 1948 article cited over 4,000 times, independent of the particular mating system, "it is a general law that the male is eager for any female without discrimination, whereas the female chooses the male" (Bateman 1948: 352). "It pays males to be aggressive, hasty, fickle and undiscriminating," declared E. O. Wilson (1978: 124) some years later. If a man were given "total freedom to act," he maintained, he could produce "thousands" of descendants. Matt Ridley (1994: 172–173) remarks that: "In … human terms, men can father another child just about every time they copulate with a different woman, whereas women can bear the child of only one man at a time."

A man's best reproductive strategy, on this view, is allegedly to pursue "mating opportunities" with as many of the "scarce resource" young females as possible; rape is the unfortunate backup option where seduction fails or takes too long. According to David Buss and David Schmitt writing in 1993, in an article cited over 5,000 times, the constraints on male reproductive success involved problems of identifying "accessible" and fertile women and "minimizing commitment and investment" (Buss and Schmitt 1993: 206). (They forgot to mention what is in real life the number one constraint on male reproductive success: not being much liked and trusted by women.) A woman's best corresponding strategy a priori was to identify the "best genes" and to try to maximize their possessor's commitment and investment. Having found the best provider her face and figure could attract, there would be no need to look further. His promiscuity should be of no concern to her unless it threatens her food supply or protection. Her infidelity would be, by contrast, unacceptable to him, as his provisioning and protecting efforts would then be directed to the offspring of his biological competitors. It has even been maintained that girlfriends and wives do not object to their partners' "physical" infidelity but only to "emotional" infidelity that might lead to the abdication of the provider.

What do we actually know about sexual strategies, choice, and refusal? Given the variance in male quality, along with greater female direct investment, it pays females to be choosy and males to be competitive. But the average number of offspring generated per male in a community is the same as the average number generated per female, though some individuals will do better or worse than average (Einon 1998: 413–426). It is inadvisable to try to derive social consequences from comparative gamete size, which holds across the taxa with their many and varied solutions to the problems of reproduction.

In any case, a female preference for lifelong sexual exclusivity is found in very few animals and is not a characteristic of our nearest primate relatives, the common chimpanzees and bonobos, whose lineages branched off from our extinct common ancestor 3–5 million years ago. Like human beings, chimpanzees have preferences and aversions: Females prefer males who remain near them, groom them, and offer them food; males prefer older females to younger, which is unsurprising given that females remain fertile all their lives and that maternal experience is correlated with survival of the offspring. Chimpanzees have three main patterns of sexual association: consortships, in which a female and a male sequester themselves from the rest of the group, remaining together as a sexually exclusive pair for as long as a month; possessive relationships, in which a dominant male tries to monopolize a female in estrus within a group setting by remaining close to her and fighting off other males; and opportunistic mating, in which several males take turns with a single female in estrus (McGrew in Dahlberg 1981: 35–74). Because a female chimpanzee, like a human female in a hunter-gatherer society, is likely to bear only four to five live young during her lifespan, it is evident that most sexual behavior will not result in conception. Sex serves other roles: release of tension, practice, research, making friends. The smaller bonobo has recently drawn attention for its hypersexuality, including female–female, male–infant, and male–male as well as male–female interactions (de Waal 1990: 378–393). Nonreproductive mating cements the social group and reduces hostilities and tensions. Female bonobo anatomy, as well as newer discoveries regarding human anatomy, puts paid to the notion that the clitoris was never more than a residual organ serving no function in motivating mating.[14]

Chimpanzees are distinguished from human beings by a number of important features. One is the recurrence of estrus, which in chimpanzees is fully apparent to males and highly motivating to both males and females. Humans seem to retain something of this feature, but in a dampened form: women at mid-cycle when they are most fertile become more aware of attractive male scents and vice versa. Another feature is the absence of

[14] And now even the snakes; see Folwell, Sanders, and Crowe-Riddell 2022; O'Connell, Sanjeevan, and Hutson 2005: 1189–1195. This research contradicts the claims and arguments of Lloyd 2006.

paternal involvement: Male chimpanzees do not know who their offspring are, and it does not concern them; human beings by contrast attach importance to social fatherhood even in conditions where biological fatherhood is not known, or where it is less significant than the paternal or avuncular role played by a man who has a relationship with the child's mother.

The same three chimpanzee patterns of consortship, aggressive possessiveness, and casual frolicking, as well as bonobo-type homosexual and "pedophilic" activity, appear in human relationships. Humans experience romantic attachments which involve an emotional focus on a single individual, and these can occur at all stages of the life cycle, including its nonreproductive phases, from childhood to old age. Fights over women are common in most societies and a major cause of homicide, as jealousy is a frequent cause of femicide (Taylor and Jasinski 2011: 341–362). And there is one-off casual sex, willingly entered into by both parties in the absence of romance or possessiveness. Women who do not need an unrelated provider or the status and security conferred by marriage and who are outside the control of their elders are motivated by the same drives as men: curiosity, practice, and the thrill of seduction, and they compete for attention. Female–female rivalry, though ignored by male writers of evolutionary psychology, drives the plots of many operas and soap operas. The notion that men and women experience jealousy differently has also been effectively punctured (Harris 2004: 62–71). Both sexes stalk and obsess. Men, being larger, stronger, more irritable, and with more access to weapons, are more prone to express lethal violence.

Casual talk of "mate choice" with regard to humans is misleading in not distinguishing between different sorts of "mates." What makes human society strikingly different from primate society is the social institution of marriage, a form of behavior that is clearly related to human interdependency, long childhood, a need to minimize social frictions between and within groups, and a need to avoid inbreeding. "Mate choice" in this regard has little to do with the preferences exhibited by speed daters. From a comparative ethnological perspective, contemporary Western courtship and marriage practices where young people do their own choosing among people they already know are unusual.

For most of human prehistory, as well as human history, people did not select their own marriage partners; they were selected for them by their parents or close kin, and this system prevails in many parts of the world today. According to R. S. Walker and colleagues (2011), "it is probably safe to conclude that an important selective pressure on the evolution of human mate choice, certainly more than any other species, has been the direct, deliberate, and conscious intervention of parents and other close kin on the sexual lives of their descendants." In a 2007 comparative study of 190 hunter-gatherer societies, Menelaos Apostolou (2007) found that arrangement, or required approval of marriage partners by parents or close kin, was the primary mode of marriage in 96 percent of his sample. Walker and his coauthors established in turn through genetic analysis that around 85 percent of offspring in these groups were indeed offspring of the married couple. "Reproductive skew" in men – the ability of some to beat the averages while other men fall short – in the earliest human societies appears to have been minimal, with little variance among men (Anderson 2006: 513–520). By contrast, Hrdy (1999: 83) has noted a greater than expected variance in the number of children born to them among women.

Male preferences for particular waist-to-hip ratios or large busts in candidates for marriage, as revealed in answers to questionnaires, are unlikely to have driven the evolution of the female form (Singh 1993: 293–307).[15] These are not necessarily the attributes parents looking for wives for their sons put at the top of their lists. Although reported preferences may involve what people think they ought to prefer and may be different from the preferences of 100,000 or 30,000 years ago, the traits preferred in pre-industrial societies were similar for both sons in law and daughters in law: Surveyed parents cited emotional stability, dependable character, good health, desire for home and children, and pleasing disposition (Apostolou 2010: 695–704). At least as far as revealed preferences are concerned, parents wanted someone *nice* for their children. While the 15 percent or so of nonmarital children would afford more opportunity for reproductive skew, and for criteria such as the waist-to-hip ratio to come into play, we can cautiously conclude that in the human case:

(1) For most men, generating children with a partner picked out by the parents was the best way to maximize fitness, since most children were born from such unions.
(2) Competition by men with other men to impregnate the most likely future mothers extramaritally, and the recruitment of extramarital

[15] Though purely aesthetic preferences are likely to play a role in evolution, the hourglass shape appears to represent optimal fat storage for more reliable lactation. See Low, Alexander, and Noonan 1987: 249–257.

sexual partners by women, nevertheless had significant reproductive advantages for both sexes.[16]

(3) Most differences between human males and human females, regarding sexual motivations and their results, are smaller than commonly supposed (Andersen, Cyranowski, and Aarestad 2000: 385–389).[17]

(4) Natural selection likely favored nice individuals of both sexes, on the grounds that persons with unpleasant personalities were less likely to be awarded marital partners.

The natural reproductive window for men and women is not as different as is often assumed. Human males continue to produce sperm throughout their lives, but researchers (Rossman 1978: 71) have noted that "[t]here is no functional parameter of aging that falls off more steeply than sexual performance." In the absence of modern pills and potions, male sexual energy declines earlier than in females, and few hunter-gatherer males father children after the age of 50 (Buller 2005: 220). More important, where maternal age is decisively correlated only with the risk of Down syndrome, paternal age is a risk factor for psychiatric disorders, including schizophrenia, autism, bipolar disorder, and mental disability.[18] The fiction of healthy lifelong male sexuality and fertility that is supposed to incline women toward older, successful male partners, needs to be discarded. Young women's alliances with very old high-status men may be to their financial or status advantage, or reflect the latter's gratitude, kindness, and understanding, but as future fathers for their offspring, they were never ideal candidates from the biological point of view.

Psychobiosocial Explanation: The Role of Comparative Advantage

One argument formerly heard that has since disappeared is that academic standards for research and teaching would decline with the entry of women into the higher ranks of the academy. In fact, the opposite has happened, with much sharper competition and the generation of so much new

[16] Hrdy's research in *Mother Nature* (1999: 235–265) suggested that women benefit from uncertainty of paternity. See also Hrdy 2006: 131–160 and Hoquet 2020: 223–231.

[17] Baumeister claimed that men's sexuality was more fixed, women's more influenced by the mores of the time.

[18] The incidence of autism is alleged to double with each decade of paternal age over forty. See Reichenberg et al. 2006: 1026–1032. See also de Kluiver et al. 2017: 202–213. Respect for neurodiversity as one encounters it is one thing; being indifferent as to whether these conditions might arise in one's own future children is another altogether.

knowledge. This should not be surprising since we now know that women are as cognitively well endowed as men. We know from Cole and Zuckerman's 1987 study that women with children as well as without evolved to be productive workers outside the home, and that, despite their greater choosiness, their appetites and behavior over the course of the life cycle, when not constrained by socioeconomic pressures, are more like those of men than earlier scientific writing and recent journalism proposed.

This leaves us with an explanatory puzzle: namely, how to explain the frequency with which women, until recently, have been found in domestic roles and not in the lucrative and visible professions and in positions of political and economic authority. The hypothesis of a deeply rooted but inexplicable misogyny is not helpful. While we are now aware of centuries of learned discourse on the incompetence and moral undependability of women, this discourse has to be understood as the effect of the observed frequencies as well as their reinforcing cause. And here we must point to certain average differences between men and women that have nothing to do with cognition or the taste for exploration, social participation and economic contribution, and freedom, but that have long worked to the disadvantage of women.

First women are smaller and weaker than men. Women are 90 percent of men's height and 80 percent of their weight. They are less muscular, with 30 percent lesser upper-body strength. Having a smaller vocal apparatus, they speak in higher and softer, accordingly more childlike, voices. Further, as noted by Pinker: "Women experience basic emotions more intensely, except perhaps anger. Women have more intimate social relationships, are more concerned about them, and feel more empathy toward their friends . . . Men have a higher tolerance for pain . . . Women are more attentive to their infants' everyday cries" (2002: 345). Such qualities can be enhanced or diminished by social learning; men can certainly be taught to weep, to cultivate intimate friendships, to shrink from contact sports and dangerous occupations, and to find gratification in soothing and playful contact with infants and young children. Nevertheless, the path of least resistance does not lead that way; it is not the path to respect and social rewards for men. For women, such qualities – especially daintiness, sensitivity, and caring behavior – were valued and inculcated, especially by other women.

Two important drivers of the division of human labor that relegated most women to maintenance tasks were first, the employment of the principle of comparative advantage, and second, the discovery of the advantages of domestication: the cultivation of other living things for their utility. The principle of comparative advantage says that efficiencies are generated when a group doesn't try to manufacture every socially desired product itself but concentrates on what it can do quickly and well and trades its surplus for things others can make quicker and better. Rather than trying to make soap and baskets, it's best just to specialize in soap and trade with the neighbors for their baskets.

The ultimately pernicious division of labor began with a simple and comparatively innocuous division of labor between gathering for subsistence and hunting for meat, from the greater allocation of aggregate female effort to maintenance and handicrafts and male effort to recreation and warfare. Even looking aside from the responsibilities of care for infants, a small advantage in size, strength, and insensitivity to pain on one side, and in dexterity and visual memory on the other, generates efficiencies. In the exit from the state of nature to herding, farming, and city dwelling, the domestication of animals was followed by the institution of human slavery; ancient civilizations were uniformly slave civilizations. The disadvantages of urban overpopulation and the problems of sexual competition, jealousy, and violence suggested a neat and feasible solution: Lock up the unenslaved women and assign them the backup maintenance tasks. Women's greater vulnerability to coercion and intimidation followed from their smaller size and economic dependency.[19] In accord with the "belief in a just world," people in subordinate positions are assumed to be there because of lacks and failings on their part (Lerner 1980).

Under civilization, as the need for the management of large populations became critical, and as occupations multiplied, institutions such as schools, courts, armies, kitchens, laundries, and workshops appeared, which, once dominated by members of one sex, became inaccessible to members of the other. As the tendency toward formal education and credentialing moved from the crafts into the professions in the early modern period, these habits of exclusion were retained. Men came to occupy specialized roles that were innovative and sometimes dangerous; they became mariners, explorers, and construction workers, later scientists and financiers. The feedback process entrenched role divisions further.[20]

[19] Aristotle took women's size to be proof of their imperfection (Touraille and Gouyon 2008). Dominance in the nonhuman animal world and to a surprising extent in the human world is a matter of size. See Rowell 1974: 131–154.

[20] As E. O. Wilson remarks in *On Human Nature* (1978: 11) in another context: "A small evolutionary change in the behavior pattern of individuals can be amplified into a major social effect by the expanding upward distribution of the effect into multiple facets of social life." See for a full discussion Boulding 1992.

There was accordingly a logic behind the system of different spheres that made it difficult to question. What enabled a rethink was eighteenth-century anthropology. European philosophers speculating on the state of nature with the help of travelers' reports of "found peoples" began to understand how the classless and relatively egalitarian small societies of prehistory had given way to the hierarchically organized slave societies of antiquity and to the tyrannies of their own times. There had always been sporadic uprisings and rebellions of slaves and peasants, and complaints from women, but these had not been guided by a historical theory of the formation of castes and classes, and by forceful challenges to colonial and aristocratic domination. Now, for the first time, it could be seriously questioned whether monarchy, slavery, and patriarchy were just and efficient.

Conclusions

Robert Trivers and Irven DeVore maintain that, because there are biological, genetic, and natural components to our behavior, "we should start setting up a physical and social world which matches [our] ... tendencies."[21] Charles Murray (2005a: 13) cautions in turn that "specific [social] policies based on premises that conflict with scientific truths about human beings tend not to work."

What is usually meant by such recommendations, as many of my quotations show, is that we need to turn the clock back. E. O. Wilson in *On Human Nature* (1978: 128) maintained that the world that matches our tendencies involves a division of labor along the traditional lines. Murray and Wilson maintain that "social engineering" and top-down directives such as quotas and affirmative action policies are harmful in forcing people into environments where they do not feel or perform well.

Accordingly, the human sciences have been accused, with good reason, of presenting a "theory of women" that offers to explain and rationalize their subordination. But when political philosophers fret too much over scientism and essentialism or turn their backs on science as too ideologically corrupted to trust, they do a disservice to the many investigators who have observed and experimented carefully. These researchers too took women as objects of study, and in many – though not all – cases they were women whose experiences and interests enabled them to notice different phenomena and to pose different questions. Further, by resisting

[21] Filmscript for *Doing What Comes Naturally*, quoted in Caplan 1978: 321.

science, philosophers fail to address the powerful charge that human beings are not "blank slates." We should not fear this accusation, I have argued, because the ongoing investigation of human nature turns out to underwrite the shakeup in the professions and family life of the recent era, with its undeniable gains for women individually and for the wider society.

To be sure, we often read about the problems of women in formerly male-dominated occupations – about things that "tend not to work" – ranging from low pay and lack of recognition, to workplace harassment and unwelcome solicitation and bargaining for advantage, to the second shift of maintenance and caring responsibilities. The problems did not arise in the old system of separate spheres, and perforce they would not arise in the dystopian world some would like to "set up." But ultimately the old system proved not to work by our own improved standards of efficiency, fairness, and personal fulfillment. Its failure forces us to address these new problems separately and in their own terms.

Pinker refers to the "insights of artists" that the human sciences can validate. Marriage, in turn, is one of the top preoccupations of dramatists, novelists, and filmmakers, and while fiction is a mixture of idealization, demonization, and real-world knowledge, the insight gained from art and science, alas, may be that there is no solution, including locking up the women, for human partnering that solves all problems. Monogamy is hard, because the world is full of temptations to which both sexes are liable, and because people change their minds about their current partners in light of experience. Polyandry and polygamy are hard because, while some people appear to be free of jealousy, most know what it is like to suffer its torments, and these arrangements make household economics complicated. Single motherhood is hard; children need more than a single caregiver and benefit from the care and teaching of fathers. Fathers in turn want an appropriate share of parenting. A more humane and scientifically aware society would not take lifelong sexual exclusivity with no lapses from either men or women for granted. It would lay out less romantically but more objectively an account of the advantages of long-term faithful cooperation, and at the same time it would offer us models of fairness and amicability that can operate when partnerships are disrupted.

CHAPTER 12

Toward More Inclusive Science

New Challenges and Responsibilities for Scientists, Philosophers, and Citizens

Stéphanie Ruphy

Introduction

Once upon a time, science was widely held to be and advocated as a key source of progress in most if not all dimensions of our lives. In order to vindicate sustaining massive public investment in science after the close of World War II, Vannevar Bush, chief scientific advisor to President Franklin D. Roosevelt, who played a key role in American science policy at that time, made it very clear in his seminal science policy treatise "Science – The Endless Frontier" that "[s]cientific progress is one essential key to our security as a nation, to our better health, to more jobs, to a higher standard of living, and to our cultural progress" (Bush 1945: 2). This centrality of science and innovation has only intensified since Bush's time and cannot be overstated today. More than ever, science retains its role as the main engine of economic growth and a key contributor to most other areas of activity in societies. For instance, behind recent, massive governmental support for the development of quantum engineering lie major challenges of national strategic independence. And, of course, the COVID-19 crisis has vividly reminded us of our direct dependence on science not only for preventing or curing diseases but also because nowadays many political decisions directly shaping our daily lives are based on scientific expertise.

In short, scientific development impacts our lives, directly or indirectly, to unprecedented degrees. Admittedly, while the works of Newton or Darwin radically altered the physics and biology of their times, they had much less impact on the lives of their lay contemporaries. Today, however, the widely acknowledged centrality of science is associated with more differentiated attitudes toward the impact of science on society. Surveys of public opinion about science suggest that over the past fifty years or so, trust in researchers has remained, globally, very high compared to other professional categories, but a strong, unconditional deference to science has progressively given way to more conditional support: The idea that science brings benefits to humanity is no longer taken for granted over the whole range of scientific disciplines. More people now hold that "science does as much harm as good" rather than "science does more good than harm" (Boy and Rouban 2019).[1] Interestingly, some surveys suggest that these different attitudes toward science go hand in hand with an increasing demand for the involvement of lay citizens in the choices and decisions shaping scientific development.[2] This should come as no surprise. When many dimensions of one's daily life are impacted by scientific developments that are not necessarily deemed beneficial, one may indeed want to have a say in these choices.

In light of these changes, my general aim in this chapter is to investigate the prospects of a more inclusive science to better fulfill humanist expectations. In other words, to what extent and under which conditions would involving lay citizens in the scientific enterprise increase the relevance and benefits of its outputs to society? My take on the notion of humanist expectations toward science is rather straightforward: Expecting science to bring progress and human flourishing mainly means expecting that the outputs of research and innovation are well aligned with the various needs and interests of the citizens of a society at a given time in its history.

Public engagement with science comes in many shades, depending on the nature of the engagement and the phase of scientific inquiry at which it occurs. Central to the purpose of this chapter is a discussion of the phase of choice of research questions and priorities, since reducing the gap between what science delivers and what society needs depends directly on the way the agenda of research and innovation is set. I thus start with a brief description of how research priorities are defined in most "research-intensive" countries and explain why it is hardly surprising that this gap exists between the outputs of scientific inquiries and society's needs. To set the stage for the discussion of the prospects of a more inclusive science to reduce this gap, I present the many faces of citizens' involvement with science as well as relevant background features of our "participative societies." The bulk of the chapter examines, for various types of public engagement, the potential benefits of a more inclusive science, and also epistemological, cultural tensions and sticking points potentially thwarting

[1] The situation may vary from one country to another but the general trend toward a more differentiated attitude is shared among many European countries.

[2] The use of the term "citizen" in the context of a discussion of citizen science may raise exclusionary concerns. For the fact is that not all members of society affected by scientific developments have citizenship. My use of the term "citizen" in this chapter includes these members.

its humanist prospects. I discuss in particular new responsibilities and challenges for scientists, including new expectations regarding professional training and the ethics of research.

Setting the Research Agenda: Current Systems of Governance of Science and Their Limits

Who are the main actors today involved in the setting of research agendas? The answer may of course vary to some extent from one country to another, but sociological studies of science organization identify common, dominant features (e.g., Gläser and Velarde 2018). There exist in most "research-intensive" countries national agencies directly involved in the shaping of the research agenda or coordinating strategic committees. Just to name a few, Japan and the United Kingdom each have a "Council for Science and Technology Policy," the United States has its "National Science and Technology Council," and Switzerland its Conseil suisse de la recherche (Swiss Council for Research). In France, the Conseil stratégique de la recherche (Strategic Research Council) is explicitly in charge of "identifying and proposing a limited number of big research and technological priorities to prepare and construct the future of France." Who, you may ask as a citizen eager to find out who decides the public research priorities of your country, serves on this council? Not surprisingly, the majority comprises very distinguished French scientists (mostly from the natural sciences), a few representatives of big French companies, and three elected representatives.[3] The composition of the French Strategic Research Council illustrates the dominant players in the field in most countries: scientists, representatives of private sector interests (the market, in short), and politicians. Looking into further details would reveal a complex interplay between these actors. But what matters for our purposes is assessing to what extent those actors are the right ones to fulfill the humanist expectation of a better alignment between what society needs and what scientific research delivers.

Two preliminary qualifications are in order here. The first spells out a key background philosophical commitment of the rest of the chapter; the second is essentially conceptual and terminological.

First, my take on the notion of humanist expectations toward science is nonobjectivist, that is, the very notions of "human flourishing" or

[3] Conseil stratégique de la recherche, Wikipedia, https://fr.wikipedia.org/wiki/Conseil_strat%C3%A9gique_de_la_recherche.

"common good," etc., that a humanist science would help to promote should be approached in a nonobjectivist way. In other words, I am committed to the idea that the outputs of a humanist science, in the context of our democratic societies, should contribute to meet the needs and interests of their citizens, *as identified and expressed by them.*[4] This nonobjectivist approach can be contrasted with an objectivist, substantialist approach, according to which the citizens' needs and interests to which science should respond can be defined independently (or partly independently) of what citizens themselves would identify and express as being their needs and interests. Later discussion (in the third section) of our increasingly participative societies buttresses this commitment.

The second qualification concerns the nature of the problems addressed by science: A distinction will be made between "endogenous" problems and "exogenous" problems (Bedessem and Ruphy 2019: 2). An "endogenous" problem is encountered and defined internally by scientists within the course of a scientific inquiry, and its relevance and interest are judged solely according to epistemic or practical considerations internal to scientific communities. By contrast, an "exogenous" problem is identified outside (or partly outside) a scientific field and evaluating its relevance and interest incorporates interests and needs of other components of society (and not only of scientific communities). "Grand societal challenges" such as developing "secure, clean and efficient energy" or "inclusive, innovative and reflective societies" are typical exogenous (encompassing) problems,[5] whereas the search for the Higgs boson in particle physics is a rather newsworthy example of an endogenous problem. With these two qualifications in hand, let us now return to the question of who sets, or should set, scientific research agendas.

The Scientists (Epistemic Elitism)

Let us start with the prospects of "epistemic elitism," as Philip Kitcher puts it, to refer to the idea that "the wise experts can be expected to know what's objectively in human interests" (2001: 138). Are scientists today in the best position to define research priorities fulfilling humanist expectations? There are several reasons to seriously doubt it. Daniel Sarewitz (2016) for

[4] I follow here, for instance, Kitcher's nonobjectivism when he elaborates his ideal of well-ordered science (Kitcher 2001). By contrast, Kourany's plea for research guided by "sound social values" partakes of an objectivist approach (Kourany 2010).

[5] These two examples are drawn from the Horizon 2020 program put forward by the European Commission.

instance points out that the current functioning and internal reward systems of scientific communities do not spontaneously favor the orientation of scientific agendas toward the resolution of exogenous problems. Career-enhancing drives (publishing papers in highly ranked journals, Nobel Prizes, and the like) in particular may even pull in the other direction, producing more esoteric knowledge, valued first and foremost by your peers, without much consideration of direct usefulness for society. As Sarewitz puts it, not mincing his words: "Advancing according to its own logic, much of science has lost sight of the better world it is supposed to help create. Shielded from accountability to anything outside itself, the 'free play of free intellects' begins to seem like little more than a cover for indifference and irresponsibility" (2016: 40). Independently of this lack of an internal propensity to address exogenous problems, epistemic elitism can be challenged on the more fundamental and simple grounds that epistemic expertise in a particular field of research does not guarantee relevant epistemic expertise when it comes to grasping which exogenous problems should be addressed first and foremost to fulfill the needs and expectations of a society as it exists at a certain point in its history. When they aim at finding out what people think or need, the human and social sciences might, admittedly, help to provide this kind of expertise, but the fact is that they are currently only very marginally involved in the setting of big research priorities.

The Market

On the face of it, the prospects of relying on the private sector might seem a bit better. After all, in societies with market-driven economies, doesn't a market-driven science respond to *some* needs and interests of the citizens of these societies? Answering this question would take us back to more general political considerations. In particular, the extent to which a market economy can meet the needs and interests of society is notoriously controversial, with assessments varying according to political and other commitments. In any case, it seems safe to contend that if solely shaped by economic interests (be it directly through private sector actors or indirectly through public–private agreements), the research agenda would not be responsive to the *whole* range of needs and interests of society but only to a limited (albeit central in our capitalist societies) subset of it. What would evidently not be addressed are public interests that do not intersect with those of the private sector, as rightly emphasized by a large critical literature on the "commercialization" or "commodification" of science (e.g., Radder 2019).

Elected Representatives

Here is where, one might hope, our elected representatives could step in to make sure that public interests are sufficiently served as well, or even solely served (depending on your political inclinations), by publicly funded research. After all, aren't elected representatives supposed to act on the *whole* range of interests and needs of their constituents? Well, their capacity to do so is notoriously questioned in our contemporary democratic societies. Later I outline general considerations that shed light on the diminishing appreciation by citizens of representative forms of democracy. Let us just note for the moment that biases toward short-term, practical goals, collusion with private sector actors, etc., are often mentioned as grounds for resisting a direct shaping of the research agenda by politicians.

Responsible Research

In light of the previous remarks, the existence of a gap between what science actually delivers and citizens' needs and interests should come as no surprise. A couple of years ago, an editorial in the influential scientific journal *Nature* (2017), entitled "Researchers Should Reach Beyond the Science Bubble," made it very clear: "the needs of millions of people in the United States (and billions of people around the world) are not well enough served by the agenda and interests that drive much of modern science." The Human Genome Project is taken as an example of a successful scientific story but with mixed impacts on society. In addition to new insights in genomics, it did create firms and jobs, but "rather than trickling down through society, these benefits of discovery science arguably deepen the pool of wealth and privilege already in place – creating expensive new drugs that most people cannot afford." And the editorial concluded with a plea for more social responsibility: "science organizations – universities, funders, supporters and the rest – should look harder at social problems and opportunities and seek ways for science to help."

This piece in *Nature* is one example among many of the expression of a growing demand for more accountability and social responsibility from research actors. On the institutional side, this demand is reflected, for instance, in the notion of "Responsible Research and Innovation" (RRI) put forward by the European Commission, aiming at fostering "the design of inclusive and sustainable research and innovation." But how should this social responsibility be exercised when epistemic elitism is no longer, at least from a normative point of view, a live option? Direct public participation has become the favorite answer of a growing number of scientific institutions and governing bodies. Through its appeal to RRI, the European Commission promotes it explicitly: "societal actors (researchers, *citizens*, policy makers, business, third sector organizations, etc.) work together during the whole research and innovation process in order to better align both the process and its outcomes with the values, needs and expectations of society" (ETHNA 2025, emphasis added).

Before assessing its prospects, let me put public participation in science in the broader perspective of an increasing demand for more direct participation by citizens in various areas of public and political life, starting with a few examples.

Participative Societies

In election campaigns, for example, citizens are sometimes directly consulted by a party to build up its political priorities. Some mayors reserve parts of municipal budgets to be spent according to priorities defined by public consultation. More sophisticated and deliberative forms of citizen consultation are set up to feed into the elaboration of national plans by governments or assemblies. A noticeable recent example is the Citizens Convention for Climate set up in France by President Emmanuel Macron.[6] Such participative forms of democracy are often presented by democracy theorists as a means to redressing the weakening of traditional representative forms of democracy, at both national and local levels.

More broadly, direct participation of citizens may be considered an appropriate response to the following six changes in contemporary democratic societies (Blondiaux 2008: 24–28). (1) *Increasingly complex societies.* Our societies are more and more divided into specialized "subsystems" calling for the existence of distinct spaces of negotiation and governance; direct participation of citizens in these governance processes may serve to meet democratic expectations. (2) *Increasingly divided societies.* Here, the focus is more philosophical than sociological. Our pluralist democratic societies are characterized by divergent views on what is good or bad, without the ability to directly overcome these differences by referring to common values or principles. Hence the necessity to implement spaces for deliberation where citizens can justify their disagreements and work on reaching consensus. (3) *Increasingly reflexive societies.* Overall levels of

6 Citizens Convention for Climate, Wikipedia, https://en.wikipedia.org/wiki/Citizens_Convention_for_Climate.

knowledge and proficiency of lay citizens have increased. At the individual level, deference to experts is not unconditional and lay or experiential knowledge can be put forward as a counterpoint or as an addition to certified knowledge provided by scientific institutions. Standpoints of lay citizens can then be expected to be taken into account in decision processes. (4) *Increasingly disobedient societies.* In response to individual or local acts of insubordination, often linked to health or environmental issues, citizens' consultations appear as a means to prevent or diffuse such resistance, sometimes labeled in a somewhat derogatory way as the NIMBY (not in my backyard) syndrome. (5) *Increasingly defiant societies.* A decline in confidence in institutions and between citizens has been extensively described and discussed by sociologists. Direct participation of citizens may be promoted, especially at local scales, as a means to recreate social ties. (6) *Increasingly ungovernable societies.* The preceding five changes feed into a final one: In many liberal democracies, states and political decision makers appear more and more powerless to impose decisions from the top downward.

Blondiaux's six propositions, built on various seminal works by sociologists and philosophers such as John Rawls, Jürgen Habermas, Ulrich Beck, and Niklas Luhmann, allow us to make sense of the significant development of participatory devices in many areas of public and political life: In order to cope with this crisis of governability, governing bodies see the development of various mechanisms for citizen participation as a means to increase their political power of action. And science is, or should be, no exception to this general trend toward more direct involvement of citizens, given its centrality in our societies and the multiple levels of imbrication between science, public life, and politics. This partly explains my earlier commitment to nonobjectivism: In more participative societies, when it comes to defining their needs and interests in terms of research outputs, citizens should be directly involved.

Let me now briefly describe the various forms that public engagement may take in science.

The Many Faces of Citizens' Engagement in Science

Nonparticipative Forms of Engagement

A minimal, traditional form of involvement with science is exemplified in the public understanding of science. The associated notion of "science literacy" has become a multifaceted notion, reflecting various, growing demands of mastering developments in scientific knowledge. Given the centrality of science in our daily lives, science literacy is commonly promoted as essential to "help people live interesting, responsible, and productive lives" (American Association for the Advancement of Science 1994: xi). In this traditional approach, citizens remain passive receptors of scientific knowledge or, in more recent takes on the notion of scientific literacy, passive receptors of knowledge about science as a social enterprise (Slater, Huxster, and Bresticker: 2019), without any direct participation in the process of knowledge production itself.

At the other end of the spectrum lies another long-standing and multifaceted form of engagement with science, to wit, public contestations of science.[7] In that case too, lay citizens remain outside the process of knowledge production.

Participative Forms of Engagement

The current diversity of participative forms of involvement with science, where nonprofessional inquirers are involved in the very process of knowledge production, has given rise to a variety of classifications. Following the commonly used classification proposed by Bonney et al. (2009), my discussion distinguishes between "contributory," "collaborative," and "co-created" science.

In the first kind of participatory practice, *contributory science*, involvement of nonprofessionals is limited to the phase of data collection: Citizens act as passive or active data collectors and are not involved in the phase of defining the problems to be solved or in the phase of interpreting and producing the results. Such crowdsourcing programs, in which any interested citizen can participate, constitute the most widespread type of participatory practices and have a long history in fields such as astronomy and environmental sciences. *Collaborative science* corresponds to a stronger form of engagement of specific populations identified by scientists as sharing expertise or skills. In agronomic research, for instance, programs in plant breeding take advantage of the practical knowledge of farmers to improve productivity. In biomedicine, the experiential knowledge of groups of patients is now commonly considered a key ingredient in the success of the development of a treatment. Participation thus goes well

[7] An often cited historical example is the nineteenth-century protest by the Luddites in England against textile machinery and, more broadly, against the impacts of scientific and technological developments on the quality of human lives.

beyond data collection: Nonprofessionals can also be involved in the design of methods and the interpretation of results. In *co-created science*, by contrast with contributory and collaborative science, the initial formulation of the problem to be solved is not made by scientists but by citizens, who in this case are better described as stakeholders.[8] This corresponds to a stronger form of participation: To resolve problems that stakeholders have themselves identified, scientists collaborate with them at every stage of the scientific process, from the co-construction of the initial problem as a research question to the collection and interpretation of data and the production and diffusion of results. "Community-based research" is another common label for this strongest form of engagement, reflecting the *local* character of the problems to be solved when, for instance, a group of people faces an environmental risk such as the pollution of a lake, or is affected by a rare genetic disease.

Admittedly, even taken together, these three participative forms of scientific inquiry still represent today only a very small fraction of global scientific knowledge production. However, in several research fields with direct societal impact, such as the environmental sciences and biomedical sciences, they occupy a more central stage and are increasingly supported by research institutions.

Participation in the Setting of Global Research Priorities

The last kind of citizens' involvement I consider here is the participation of lay citizens in decision processes concerning *global* research priorities, that is, research priorities affecting *all* citizens. In contrast with the previous forms of citizens' involvement, this form of involvement remains largely programmatic. As briefly described in the second section, current systems of governance of science do not include mechanisms for citizen participation – or when they do the actual participation of citizens remains anecdotal. Sure enough, various types of participatory mechanisms have been set up to consult citizens on specific issues in the domain of science and technology (e.g., nanotechnology), such as the pioneering "consensus conferences" organized by the Danish Board of Technology in the late 1980s. However, existing participatory mechanisms are rarely designed to address the broader issue of what the big priorities of science in response to

[8] Following the literature on public deliberation (Kahane and Lopston 2013), "stakeholders" refers here to a group of people who are directly affected by a problem or by the various ways it may be resolved.

society's needs and interests should be. That is where philosophers might step in, proposing ideals of democratization of the research agenda. For example, the ideal of "well-ordered science" developed by Philip Kitcher (2001) has been widely discussed in the philosophy of science. In a nutshell, in well-ordered science, the problems addressed by scientists are those selected by a group of deliberators, tutored by scientific experts, who dedicate themselves to revising their preferences in light of the preferences of others (Kitcher 2001: chapter 10).

Assessment of the Humanist Prospects of Public Engagement in Science

To assess the prospects of a more inclusive science as regards the reduction of the gap between science's outputs and society's needs, after some quick comments on nonparticipative forms of engagement, I then discuss forms of participation that do not impact scientific life globally, and turn to the assessment of participation in the setting of global research priorities in the next section.

Public Understanding of Science and Contributory Science

The humanist prospects of nonparticipative forms of citizen involvement, such as the public understanding of science, have been well identified for a long time. Having some cognitive access to our most important scientific insights into the world is consensually held to contribute toward having a meaningful life for at least three reasons (Shen 1975): "Practical science literacy" helps people to make individual decisions in their everyday lives, "cultural science literacy" helps people to appreciate scientific achievements, and "civic science literacy" allows people to reach considered decisions about public issues that have scientific components.

When participation is limited to the collection of data, as in the case of contributory science, the humanist prospects of citizens' involvement are in the same vein. For one can reasonably expect increased science literacy in the three previously mentioned dimensions from citizens involved in scientific inquiry as data collectors. But what about the prospects of greater science literacy when one adopts the deflationary approach to the notion of humanist expectations advocated earlier? Otherwise put, to what extent may increased science literacy help to reduce the gap between what science delivers and what society expects and needs from science? By itself, greater science literacy won't help to bridge the gap as long as the decision processes establishing global science policies are not open to lay citizens. Nonetheless, it seems reasonable to think that science literacy should at least raise general awareness of the centrality of science in our societies and, consequently, of the necessity to democratize the setting of its research agenda. Lacking conclusive empirical studies of such correlations, let me move to the prospects of the second type of participation, to wit, collaborative science.

Collaborative Science

In the case of collaborative research, the epistemic benefits brought about by involving a lay population with specific skills or experiential knowledge in scientific inquiry are better known (e.g., Bedessem and Ruphy 2020). A paradigmatic and well-documented case of successful contributions of lay expertise is the contribution by AIDS patients to research aiming at understanding and curing the disease (Epstein 1995; Godlee 2016). Here, the benefits went beyond epistemic gains: It also brought about more *actionable* scientific findings, that is, scientific findings more easily translatable into therapeutic care well adapted to the specificities of living with this new disease, as documented by the AIDS patients themselves. Collaborative research programs in agronomy also illustrate this benefit of more actionable findings: Involving farmers having experiential knowledge of a particular local context allows for the production of knowledge and recommendations well adapted to that context, and is hence more useful to the population concerned. By allowing the production of more actionable findings, a more inclusive science in the form of collaborative science thus allows for more directly relevant and useful outputs, thereby contributing to the reduction of the gap between what science delivers and what people need. Conditions of success in fulfilling humanist expectations toward science thus correspond to conditions of success in collaborative science.[9] Let me just mention here that a key factor of successful collaborative science is the ability of professional researchers to communicate and interact with nonprofessionals. This is certainly still a cultural and professional challenge for scientific communities since these kinds of interactive skills are rarely part of the regular training of future scientists (remember that "among peers" has been the rule for a long time in science, with peer evaluation in particular playing a central role in many phases of scientific endeavor).

[9] Those conditions are discussed more extensively in Bedessem and Ruphy 2020.

Co-created Science (Community-Based Research)

As regards the question of reducing the gap between the outputs of scientific inquiry and the needs and interests of citizens, the answer is even more straightforward for community-based research. In this case, since the problems to be addressed are identified by the stakeholders themselves, the issue is moot: Research programs are conceived from the beginning to contribute directly to respond to the needs and interests of concerned groups of citizens. However, opening the very process of the production of knowledge to stakeholders gives rise to various epistemological and political challenges.

Let us consider first an epistemological risk (discussed in more detail in Bedessem and Ruphy 2020). When research programs are developed mainly by local communities to contribute toward solving specific problems they are facing (hence exogenous problems for scientific communities), this may lead to a fragmentation of the research agenda overall into a juxtaposition of unrelated research questions needing to be resolved in isolation. From a purely epistemological point of view, such fragmentation may be deemed problematic for the overall dynamics of the research fields concerned. The reason is, in short, the following: When exogenous problems are chosen in light of their urgency from a political or practical point of view, rather than in light of their potential epistemic interest for the development of a research field, the resolutions of these problems are unlikely to open new lines of inquiry that will increase fundamental knowledge in the research fields concerned.[10] Moreover, the kinds of research questions addressed in co-created research science may not be cutting-edge questions, and therefore may not be very attractive for professional scientists. In any case, the key normative question is whether epistemological considerations should prevail when it comes to valorizing one type of research over another. I suggest that it should not. Defending a utilitarian view of science today – as Vannevar Bush did eighty years ago – requires that we equally valorize the work of scientists engaging in community-based research. Sure enough, it is up to researchers to decide to engage in co-created research or in blue sky, basic research (or in both for that matter); at the end of the day, it is a matter of personal, political, and ethical choice. However, as briefly mentioned in the second section,

[10] This line of argument is only valid for exogenous problems in the specific context of co-created science. In other contexts, see Bedessem and Ruphy 2019 on the epistemologically positive impact on the dynamics of a research field of addressing exogenous problems.

the current internal reward system of scientific communities does not really encourage scientists to work alongside communities and stakeholders to contribute toward solving practical problems defined by the latter. And as in the case of contributory science, changes in the training of scientists (or a subset of them) is also called for to facilitate interactivity with nonprofessionals. Overall, beyond financial support, more incentives to engage in inclusive research are needed from scientific institutions and scientific communities.

Another challenge results from a prima facie tension between the inclusion of stakeholders in scientific research and traditional expectations of objectivity and impartiality, since in co-created research, the very questions being asked are chosen in relation to the stakeholders' interests. Two levels of concern should be distinguished here. First, one may worry that when inquirers have stakes in the output of the inquiry, they might be tempted to take some liberties with the usual standards of good practice which guarantee the reliability of the results, in order to channel them toward what they consider desirable conclusions. The concern is understandable but calls for more empirical study. Departures from standards of research integrity are already notoriously difficult to document within traditional scientific communities. More work needs to be done to find out whether this concern is more serious in the case of community-based research.[11]

Meanwhile, let us discuss the second level of concern, which takes us to the political issue of unbalanced processes of production of scientific expertise (Sarewitz 2004; van der Vegt 2018).[12] Consider the production of expertise on a multifaceted issue such as, for example, an environmental or health security issue, for which various co-created research programs are developed, each aiming at addressing a limited dimension of the issue, in relation to the interests of the stakeholders involved. Depending on the play of power between stakeholders, you might end up with biased scientific expertise on the issue *overall* (even if the expertise developed in each individual program is not biased at all), because some aspects of the issue may remain understudied. Justin Biddle (2018) offers a detailed analysis of this phenomenon in the case of genetically modified organisms

[11] As discussed in Bedessem and Ruphy 2020: 641, interestingly, some studies (e.g., Yamamoto 2012) suggest that as stakeholders, participants may pay more attention to the existence of potential conflicts of interest of professional scientists, thereby perhaps attenuating the risk of diminished objectivity and impartiality. In any case, it is not (yet) unnecessary to remind ourselves that an awareness of the domain of research integrity is needed in any type of research.

[12] This point is discussed in more detail in Bedessem and Ruphy 2020: 642.

(GMOs). To sum up the basic idea, the food industry favors the production of expertise on yield increase, whereas anti-GMO NGOs favor the study of environmental impacts. In light of these considerations, what can be expected from decision makers, scientific institutions, and also individual scientists? Decision makers, together with scientific institutions, especially public ones, should make sure that no aspect of the issue is understudied so that they can act on the basis of unbiased expertise (overall). This requires that public scientific institutions and funding agencies in particular should favor research on topics that tend to be understudied, in order to compensate for the effects of unbalanced power. For what matters for a functioning democracy is that when decisions have to be made based on scientific expertise, there are no blind spots in the expertise available. Regarding individual scientists, it seems reasonable to expect that they should show their hands, by being transparent about the roles they choose to play when producing (reliable) knowledge of a limited aspect of a phenomenon, in relation to their own values and interests. In other words, being an "Issue Advocate," to follow Roger Pielke's (2007) terminology,[13] is perfectly acceptable, both epistemologically and politically.

The discussion so far has focused on assessing the prospects of opening the process of producing knowledge and expertise to better respond to *local* needs and interests. Let us turn now to a more overarching, global perspective on the setting of the research agenda.

Assessment of Participatory Devices in the Setting of *Global* Research Priorities

Directly involving citizens in the setting of *global* research priorities is, admittedly, at least on paper (and if we opt for nonobjectivism), the best way to reduce the gap between the actual needs and interests of all citizens and the needs and interests that are currently shaping research agendas. Let us now investigate the various possible impacts that such direct participation would have on scientific life and the consequent, new responsibilities for researchers and scientific institutions, leaving aside the multifaceted and thorny issue of how a direct shaping of the global

[13] Pielke (2007) proposes a typology of four idealized roles for scientists engaging in decision making: the "Pure Scientist," the "Science Arbiter," the "Issue Advocate," and the "Honest Broker of Policy Alternative." When acting as an Issue Advocate, a scientist "focuses on the implications of research for a particular political agenda. Unlike the Pure Scientist, the Issue Advocate aligns him/herself with a group (a faction) seeking to advance its interests through policy and politics" (2007: 15).

research agenda by citizens could be implemented concretely.[14] This discussion is structured around the identification of three tensions or sticking points, starting with issues of the legitimacy of the very demand for social responsibility that underlies humanist expectations toward science.

Legitimacy of the Demand for Accountability

I emphasized earlier a growing demand for social responsibility and accountability in the sense of being directly useful to society. This demand could be rejected on the simple grounds that direct social utility is just not a legitimate demand on science, contra currently predominant, institutional science policy discourses and philosophical views (e.g., Kitcher 2001; 2011; Kourany 2010; Radder 2019). This rejection of a demand for direct social utility is still endorsed by some influential practicing scientists, usually as part of a plea for more money for blue sky research. In 2014, Sir John Cadogan, a well-known British chemist, and forty-one other Fellows of the Royal Society, expressed very clearly their reluctance to address societal challenges:

> The nature of all politics and politicians means it is easier for our paymasters to feel comfortable about the proclaiming of programmes relating to Energy, Health, Materials, Climate Change, the Hydrogen Economy and so on, rather than to announce, let alone trumpet, that money is available for scientists to follow their curiosity in their own disciplines. (Cadogan 2014)

This resistance to direct social utility is hardly something new. In 1955, the famous physicist Richard Feynman expressed similar concerns with the shaping of the research agenda to fulfill societal needs, but on the slightly different grounds that scientists are just not good at solving societal problems:

> From time to time, people suggest to me that scientists ought to give more consideration to social problems – especially that they should be more responsible in considering the impact of science upon society... And it seems to be generally believed that if scientists would only look at these very difficult social problems and not spend so much time fooling with less vital

[14] In other words, let us set aside the (in principle) multiple shortcomings of and difficulties encountered by participatory processes at global scales. Recall (from the second section) that effective participatory processes in the setting of global research priorities have not yet been implemented in real life.

> scientific ones, great success would come of it. It seems to me that we do think about these problems from time to time, but we don't put full-time effort on them – the reason being that we know we don't have any magic formula for solving problems, that social problems are very much harder than scientific ones, and that we usually don't get anywhere when we do think about them. (1955: 13)

Cadogan's and Feynman's standpoints sum up two views on the very nature of scientific research that are culturally still entrenched in scientific communities and beyond. First, curiosity and the urge to discover the secrets of Nature are widely held as the most central motivation for engaging in scientific inquiry. Therefore, scientists should be left free to follow their curiosity when inquiring about the world (rather than being expected to solve societal problems), all the more because they are more successful when doing so. And this takes us to the second view, which is about comparative success in solving problems, depending on whether the problem is defined internally by scientists – endogenous problems in our terminology – or in light of considerations external (or at least partly external) to the inner dynamics of a scientific field (i.e., exogenous problems).

The bottom line of a Feynman-type reluctance to accept the idea of socially responsible science is that scientists are more successful when addressing endogenous problems than when addressing exogenous ones. Kuhn's (1962) defense of the social irrelevance of research problems on resolution efficiency grounds is in the same vein. From an epistemological point of view, it would be hard to deny that addressing exogenous problems raises the additional challenge of translating social issues into tractable research problems and may very well diminish efficiency and success of scientific inquiry. But, again, should epistemological considerations prevail when it comes to the shaping of the research agenda? This question can be addressed as part of the broader, fundamental question of who should decide what the very aims of science should be.

It is now commonly acknowledged that the pursuit of exogenous problems has become more prevalent in the past few decades. Seminal contributions from science and technology studies (STS) have extensively studied this trend, describing in particular the evolution of modes of research funding and the setting of research priorities. For instance, Henry Etzkowitz (2003) proposed the concept of a triple helix of entrepreneurial science to describe the intertwining of government, industry, and academia. The much discussed contrast between "mode-1" and "mode-2" proposed by Michael Gibbons et al. (1994) emphasized a shift from a traditional academic, discipline-based mode of production of knowledge toward a more interdisciplinary, application-oriented one.

In another paper (Ruphy 2019), I proposed that we reformulate our understanding of these changes in terms of a shift toward more pressing and targeted expectations. When Vannevar Bush advocated massive public support of science on utilitarian grounds, he advocated at the same time complete scientific freedom as regards the setting of research agendas: "Scientific progress on a broad front results from the free interplay of free intellects, working on subjects of their own choice, in the manner dictated by their curiosity for exploration of the unknown. Freedom of inquiry must be preserved under any plan for government support of science" (1945: 12). This suggested connection between utilitarian expectations toward science and freedom of research topics followed from what is often called the "cascade" model of the relationship between science and society (e.g., Guston 2000b). In this model, society, via its governing bodies, gives "blind delegation" (Wilholt and Glimell 2011) to research communities to conduct their business. In particular, policies of research oversight and funding are limited, in order to inject money into scientific communities without setting any thematic priorities. According to this model, often considered a lost paradise by many scientists, the main aim of researchers is to fill a reservoir of knowledge, following their curiosity, and from this reservoir of knowledge will eventually emerge, in short or long terms – who knows, research being unpredictable – all kinds of things beneficial to society, especially technological innovations. Researchers in public institutions know all too well that we have significantly departed from this cascade model.[15] How should we make sense of this transformation?

A possible reading of the decline of the cascade model is, I suggest, properly understood in terms of an evolution of our expectations of science. We no longer expect more knowledge and more innovation *tout court*, but more knowledge and more innovation in specific priority domains, corresponding to specific needs, and sometimes urgently so, in light of challenges encountered by our societies (climate change, an aging population, you name it). In earlier work (Ruphy 2019), I proposed that we consider this shift toward more pressing and more targeted expectations as the other side of the coin of the very success of science and innovation in

[15] This departure is well documented, for instance, in Guston 2000b. It is also emphasized in the *Nature* editorial mentioned earlier: "Just telling the same old stories won't cut it. The most seductive of these stories – and certainly the one that scientists like to tell themselves and each other – is the simple narrative that investment in research feeds innovation and promotes economic growth" (Nature 2017).

our knowledge societies. As soon as science becomes a key element of so many aspects of the development of our societies, it is understandable that expectations from other components of society, including of course public science funders, should become increasingly pressing and specific. Otherwise put, there is a shift from an "offer mode" toward a "demand mode." In the former, scientific inquiries are mainly oriented by endogenous problems and produce new knowledge that, in turn, may lead to very useful exogenous developments. The development of the now ubiquitous laser is a paradigmatic success story of this view of science as filling a reservoir of knowledge for later applications. By contrast, in the "demand" mode, scientific inquiries are mainly oriented by exogenous problems – say, the demand for a cure for a new virus-borne illness, or the need for strategic independence in cryptography.

The key, normative philosophical question is then the following: Is this shift toward more targeted and pressing expectations legitimate and desirable, or should it be resisted, and if so on what grounds? This question takes us back to the question with which I began: Who should decide what the very aims of science are or should be in our societies?

Philosophers of science are traditionally very good at discussing what the *epistemic* goals of science are or should be: discovering the laws of nature, providing objective explanations, for instance by making use of causal patterns (Potochnik 2017), etc. But should these epistemic aims be ends in themselves or just instrumental to practical ends? The traditional contrast here is between (in short) a primarily epistemic view and a primarily utilitarian view of the aims of science. Which one is the right view? I contend that the answer to this fundamental question should be *political.* In a democratic society, where research is (at least in part) funded by public money and plays such a central role in so many aspects of life, it should not be up to scientists (or for that matter philosophers) to decide what the aims and value of science are or should be. We should thus avoid any essentialist approach to thinking about these aims, and prefer instead a thoroughly political one. In other words, the question of which of the two traditional views of the aims and value of science should prevail is an open, political question; it should not be decided by invoking some putative essence of what science is about. Acknowledging this is certainly in tension with well-entrenched cultural views of science, widespread both in scientific communities and in the rest of society. But it is a necessary, preliminary step toward addressing the question of the legitimacy of the demands for accountability and social relevance. As I have just stated that invoking some putative essence of science is not an option to decide what the aims and value of science are, it is also not an option for rejecting such demands. So let us now question two other sources of resistance to accountability.

Tension between Accountability and Unpredictability

A second interesting source of resistance to more accountability in the sense of direct social utility invokes a tension between accountability and a central feature of scientific inquiry, namely its unpredictability. To put it very simply: How can one expect science to be socially responsible by delivering what society needs and values when one cannot predict what science will deliver? And even if one could predict the outputs and consequences of scientific inquiry, one may not be able to anticipate their acceptance by society. Moreover, one cannot always predict what society needs to know, sometimes urgently, as the COVID-19 crisis has reminded us vividly.[16]

One needs first to distinguish between two kinds of unpredictability in science (Bedessem and Ruphy 2019). "Unpredictability" may sometimes refer to unforeseen practical applications of fundamental knowledge. The laser is a paradigmatic case of this first type of unpredictability: The development of this technological device in the early 1960s (Maiman 1960) was evidently not foreseen as an application of the theoretical developments of quantum mechanics that took place decades before. A second type of scientific unpredictability concerns the occurrence of unexpected results or observations in the course of scientific inquiry, leading to the opening of new lines of research and discoveries. A paradigmatic case of this kind of unpredictability is the famous accidental observation by Alexander Fleming of the blocking effect of a fungus on the proliferation of bacterial colonies (Fleming 1929) that led to the development of antibiotics.

My point here is about the first type of scientific unpredictability: Should we value the prospects of unforeseen applications as paramount when facing specific, pressing, urgent, or otherwise important social or societal issues, the resolution of which could be facilitated by science? Taking seriously humanist expectations of science invites, I suggest, a negative answer. Once the shift toward more pressing and more targeted expectations is deemed legitimate – and recall that this is, I contend, a political issue – research oversight policies should favor research programs

[16] There are many examples beyond the COVID-19 case. Consider for instance the pressing need for knowledge about radicalization processes when a country faces terrorist attacks.

mainly oriented by exogenous problems, aiming at responding to identified needs. If we already know that we urgently need better energy storage devices (etc.), why should we still place so much value on the hypothetical development of the next laser decades down the line? Laser-type unpredictable outputs may remain preeminently valuable so long as long-term contributions to economic growth and competitiveness are viewed as the central expectation for science, that is, when one mainly expects from science breakthrough innovations that open new markets. But a properly functioning democracy may (hopefully) broaden and diversify its expectations for science and opt for, if needed at a certain time, more targeted and short-term expectations (e.g., focusing on health, environmental, and strategic independence issues), making laser-type unpredictability a less valuable feature of science. As the British scientist and political activist J. D. Bernal put it some time ago: "Although it is true that we do not know what we may find, we must, in the first place, know where to look" (1939).

It thus turns out that the humanist aim of reducing the gap between what society needs and the outputs of scientific inquiry requires us to downplay the value of scientific unpredictability (as unforeseen applications). Here again, this calls for a significant cultural change for both many practicing scientists and much of the rest of society.

Loss of Autonomy

A third interesting and common reason to resist a growing demand for accountability in the sense of direct social utility is to invoke some putative negative epistemological effects of a loss of scientific autonomy when it comes to the choice of research questions.[17] In a nutshell, the argument put forward by proponents of autonomy is that the shaping of the research agenda by exogenous issues hampers the epistemic fecundity of science. In other words, or so the "unpredictability argument" goes, research whose agenda is set according to external considerations is less hospitable to the flourishing of the unexpected in inquiry, and hence less fecund, than research whose agenda is freely set internally by scientists following their curiosity and favoring the resolution of endogenous problems. A well-known and somewhat lyrical formulation of the unpredictability argument

[17] I do not comment here the very human reluctance to give up or share power as grounds for resisting more accountability, since this is not specific to decision makers regarding *science*.

is given by Michael Polanyi in his classic essay "The Republic of Science" (1962: 62):

> Any attempt at guiding research towards a purpose other than its own is an attempt to deflect it from the advancement of science ... you can kill or mutilate the advance of science, you cannot shape it. For it can advance only by essentially unpredictable steps pursuing problems of its own and the practical benefits of these advances will be incidental and hence doubly unpredictable

I will not come back here to the issue of the desirability of unpredictable applications but instead focus on the second type of unpredictability: occurrences of the unexpected in the course of scientific inquiry.

I have argued elsewhere that the unpredictability argument has many weaknesses. It is hardly convincing as a defense of the autonomy of science and the pressure of exogenous problems may actually favor the occurrence of the unexpected (Bedessem and Ruphy 2019). Leaving these contentions aside here, however, even if the unpredictability argument were to hold as an argument supporting the choice of endogenous problems, the question would arise again: Should epistemological considerations prevail over all others? After all, one might very well choose to prioritize the resolution of urgent or pressing (social or societal) problems at the possible cost of some (temporary) loss of epistemic fecundity.[18] And again, this should be a matter of *political* choice.

Conclusion

Humanist commitments regarding science in terms of relevance and benefits for society operate at two different levels, local and global, each raising specific challenges. In this chapter, I first discussed various ways in which lay citizens may engage in the process of producing knowledge and expertise, alongside professional scientists, and spelled out how public engagement at local scales may allow us to reduce the gap between science's outputs and society's needs. Three main, interrelated challenges were identified: (1) the need for more incentives from scientific institutions and communities to engage in citizen science programs; (2) the need for an evolution of the professional training of scientists and of cultural views on what kinds of science are worth pursuing; and (3) the need for an increase in individual awareness of the existence of political and ethical choices to

[18] Note that such political choices have been made in the past. Just think about the Manhattan project channeling research efforts toward well-defined, practical ends.

be made as regards the type of research one is willing to engage in as an individual researcher.

When tantamount to supporting stakeholders, the humanist commitment may appear on the face of it rather modest. However, it turns out to be very demanding in our inegalitarian democracies. For a humanist commitment regarding science requires us to ensure that *all* citizens and groups of citizens are afforded the chance to become epistemically well-equipped stakeholders and to assert their interests in the political arena. It is, admittedly, not solely the responsibility of scientists and science decision makers to ensure that the voices of all citizens are heard in a democracy. However, heightening vigilance within science so that the epistemic needs of underrepresented groups don't remain below the radar of scientific research because of unbalanced distributions of power in society at large is certainly called for.

At the more global scale of setting big research priorities, we have seen that calling for more relevance and benefits for *all* members of society impacts scientific life in several fundamental ways. It raises first the question of the legitimacy and desirability of a shift toward more targeted and pressing expectations concerning scientific research. Here the contribution of philosophy is to assess the very nature of the question and to argue (in my case) that it should be considered, in contemporary democracies, a *political* question. It also challenges the valuation of culturally well-entrenched features of science such as the valuation of unpredictability (as unforeseen applications). A complementary task is then to explore further the epistemological consequences of this shift for the dynamics of research fields, to identify epistemologically acceptable forms of limitation of scientific autonomy, and possibly to debunk other unfounded sources of resistance.

Another major philosophical task is to continue to explore the practical forms that a democratization of the setting of research agendas may take. It is difficult today to argue against the idea that citizens should have a say in the matter, but how exactly should that be accomplished? How should we articulate, for instance, the requirements of direct participation and indirect participation (via elected representatives)? To what extent is the implementation of participatory strategies at national scales compatible with the internationalization of science? These are undoubtedly crucial challenges to be met on the way to a more humanist science.

References

Abrams, Z. 2020: "A Time for Reckoning and Healing: Psychologists Have a Role to Play in Addressing Inequities and Achieving True Systemic Change," American Psychological Association. www.apa.org/news/apa/2020/reckoning-healing.

Adams, W. 1991: "Aesthetics: Liberating the Senses," in *The Cambridge Companion to Marx*, Terrell Carver (ed.), 246–274. Cambridge: Cambridge University Press.

Addams, J. 1930: *The Spirit of Youth and the City Streets*. New York: Macmillan.

2002/1902: *Democracy and Social Ethics*, C. H. Seigfried (ed.). Champaign, IL: University of Illinois Press.

AEA Executive Committee 2020: "Statement from the AEA Executive Committee," American Economic Association. www.aeaweb.org/news/member-announcements-june-5-2020.

Alberts, B., M. W. Kirschner, S. Tilghman, and H. Varmus 2014: "Rescuing US Biomedical Research from Its Systemic Flaws," *Proceedings of the National Academy of Sciences* 111: 5773–5777.

Ali, A. and A. Abdulai 2010: "The Adoption of Genetically Modified Cotton and Poverty Reduction in Pakistan," *Journal of Agricultural Economics* 61: 175–192.

Altham, J. E. J. 1986: "The Legacy of Emotivism," in *Fact, Science and Morality: Essays on A. J. Ayer's Language, Truth and Logic*, G. Macdonald and C. Wright (eds.), 275–288. Oxford: Basil Blackwell.

Althusser, L. 1977: *Lenin and Philosophy and Other Essays*, B. Brewster (trans.). London: New Left Books.

1998: "For Marx," in *Continental Philosophy: An Anthology*, W. McNeill and K. Feldman (eds.), 271–278. Oxford: Blackwell.

Altman, L. 2006: "For Science's Gatekeepers, a Credibility Gap," *New York Times*. www.nytimes.com/2006/05/02/health/02docs.html?pagewanted=1.

Altman, L. and W. J. Broad 2005: "Global Trend: More Science, More Fraud," *New York Times*. www.nytimes.com/2005/12/20/science/global-trend-more-science-more-fraud.html.

American Association for the Advancement of Science 1994: *Benchmarks for Science Literacy*. Oxford: Oxford University Press.

American Humanist Association 2003: "Humanist Manifesto III." https://americanhumanist.org/what-is-humanism/manifesto3/.

2023: "Our Mission." https://americanhumanist.org/about/our-mission/#.

Andersen, B. L., J. M. Cyranowski, and S. Aarestad 2000: "Beyond Artificial, Sex-Linked Distinctions to Conceptualize Female Sexuality: Comment on Baumeister," *Psychological Bulletin* 126: 385–389.

Andersen, H. K. and S. D. Mitchell 2023: *The Pragmatist Challenge: Pragmatist Metaphysics for Philosophy of Science*. Oxford: Oxford University Press.

Anderson, E. 2004: "Uses of Value Judgments in Science: A General Argument, with Lessons from a Case Study of Feminist Research on Divorce," *Hypatia* 19: 1–24.

Anderson, K. G. 2006: "How Well Does Paternity Confidence Match Actual Paternity?," *Current Anthropology* 47: 513–520.

Anonymous 2013a: "How Science Goes Wrong," *The Economist* 409: 13.

2013b: "Trouble at the Lab," *The Economist* 409: 26–30.

Apostolou, M. 2007: "Sexual Selection under Parental Choice: The Role of Parents in the Evolution of Human Mating," *Evolution and Human Behavior* 28: 403–409.

2010: "Parental Choice: What Parents Want in a Son-in-Law and a Daughter-in-Law across 67 Pre-industrial Societies," *British Journal of Psychology* 101: 695–704.

Ayer, A. J. 1936: *Language, Truth, and Logic*. London: Dover.

(ed.) 1959: *Logical Positivism*. New York: Free Press.

Babbitt, I. 1930: "Humanism: An Essay at Definition," in *Humanism and America: Essays on the Outlook of Modern Civilization*, N. Foerster (ed.), 25–51. New York: Farrar and Rinehart.

1955/1919: *Rousseau and Romanticism*. Boston: Houghton Mifflin.

Bacon, F. 1960/1620: *The New Organon and Related Writings*, F. H. Anderson (ed.) and J. Spedding, R. L. Ellis, and D. D. Heath (trans.). Indianapolis: Bobbs-Merrill.

1964/1603: "The Masculine Birth of Time," in *The Philosophy of Francis Bacon: An Essay on Its Development from 1603 to 1609, with New Translations of Fundamental Texts*, B. Farrington (ed. and trans.), 59–72. Liverpool: Liverpool University Press.

2000/1620: *The New Organon*, L. Jardine and M. Silverthorne (eds.). Cambridge: Cambridge University Press.

2008/1627: *The New Atlantis*. Project Gutenberg EBook #2434. www.gutenberg.org/files/2434/2434-h/2434-h.htm.

Badger, P. 2010: "What's Wrong with the Enlightenment?," *Philosophy Now: A Magazine of Ideas* 79 (June/July). https://philosophynow.org/issues/79/Whats_Wrong_With_The_Enlightenment.

Baker, M. 2016: "1,500 Scientists Lift the Lid on Reproducibility: Survey Sheds Light on the 'Crisis' Rocking Research," *Nature* 533: 452–454.

Baldwin, T. 2010: "Comment on Akeel Bilgrami's *Self-Knowledge and Resentment*," *Philosophy and Phenomenological Research* 81: 3–782.

Balietti, S. 2016: "Science Is Suffering Because of Peer Review's Big Problems: How to Reform the Journal Publication Process," *The New Republic*. newrepublic.com/article/135921/science-suffering-peer-reviews-big-problems.

Barash, D. 1979: *The Whisperings Within: Evolution and the Origin of Human Nature*. Middlesex: Penguin.

Bateman, A. J. 1948: "Intra-Sexual Selection in *Drosophila*," *Heredity* 2: 349–368.

BBC Horizon 2006: "A War on Science (Intelligent Design)." www.dailymotion.com/video/x226cec.

Bedessem, B. and S. Ruphy 2019: "The Unpredictability of Scientific Inquiry: The Unexpected Might Not Be Where You Would Expect," *Studies in History and Philosophy of Science* 73: 1–7.

2020: "Citizen Science and Scientific Objectivity: Mapping Out Epistemic Risks and Benefits," *Perspectives on Science* 28: 630–654.

Begley, C. G. and Lee M. Ellis 2012: "Drug Development: Raise Standards for Preclinical Cancer Research," *Nature* 483: 531–533.

Belluz, J., B. Plumer, and B. Resnick 2016: "The 7 Biggest Problems Facing Science, according to 270 Scientists," *Vox*. www.vox.com/2016/7/14/12016710/science-challeges-research-funding-peer-review-process.

Bennett, J. 2010: *Vibrant Matter*. Durham, NC: Duke University Press.

2018: "On the Call from Outside," *The Immanent Frame*. blogs.ssrc.org/tif/2010/08/18/on-the-call-from-outside/.

Bernal, J. D. 1939: *The Social Function of Science*. London: Faber and Faber.

Bernhard, P. 2021: "Sie Diskutieren Sehr Gern, Aber Sehr Dilettantisch: Carnaps Vorträge am Dessauer Bauhaus," in *Logischer Empirismus, Lebensreform und die deutsche Jugendbewegung*, C. Damböck, G. Sandner, and M. Werner (eds.), 302–327. Cham: Springer.

Betz, G. 2017: "Why the Argument from Risk Doesn't Justify Incorporating Non-Epistemic Values in Scientific Reasoning," in *Current Controversies in Value and Science*, K. C. Elliott and D. Steel (eds.), 94–110. Abingdon: Routledge.

Biddle, J. 2018: "Antiscience Zealotry? Values, Epistemic Risk, and the GMO Debate," *Philosophy of Science* 85: 360–379.

2020: "On Predicting Recidivism: Epistemic Risk, Tradeoffs, and Values in Machine Learning," *Canadian Journal of Philosophy* 52, 1–21.

Bilgrami, A. 2006: *Self-Knowledge and Resentment*. Cambridge, MA: Harvard University Press.

2010: "Replies to Baldwin and Normore," *Philosophy and Phenomenological Research* 81: 783–808.

2014: *Secularism, Identity, and Enchantment*. Cambridge, MA: Harvard University Press.

Bird, A. 2018: "Thomas Kuhn," in *Stanford Encyclopedia of Philosophy*, E. Zalta (ed.). plato.stanford.edu/archives/spr2022/entries/thomas-kuhn/.

Blackburn, S. 2013. "Disentangling Disentangling," in *Thick Concepts*, S. Kirchin (ed.), 121–135. Oxford: Oxford University Press.

Blanchard, O., B. Bernanke, and J. Yellen 2019: "A Message from the AEA Leadership on the Professional Climate in Economics," *American Economic Association*. www.aeaweb.org/news/member-announcements-mar-18-2019.

Blinkhorn, S. 2005: "Intelligence: A Gender Bender," *Nature* 438: 31–32.

Blondiaux, L. 2008: *Le Nouvel Esprit de la Démocratie*. Paris: Seuil.

Blum, P. R. 2007: "The Immortality of the Soul," in *The Cambridge Companion to Renaissance Philosophy*, J. Hankins (ed.), 211–233. Cambridge: Cambridge University Press.

Blumenberg, H. 1983: *The Legitimacy of the Modern Age*, R. M. Wallace (trans.). Cambridge, MA: MIT Press.

Boelens, R., A. Guevara-Gil, and A. Panfichi 2009: "Indigenous Water Rights in the Andes: Struggles over Resources and Legitimacy," *Journal of Water Law* 20: 268–277.

Boisvert, R. D. 1998: *John Dewey: Rethinking our Time*. Albany: SUNY Press.

Bonney, R., L. Hedi, L. Ballard, C. R. C. Jordan, E. McCallie, T. Phillips, J. L. Shirk, and C. C. Wilderman. 2009: *Public Participation in Scientific Research*. Washington, DC: Center for Advancement of Informal Science Education (CAISE).

Boudry, M. and M. Pigliucci 2017: *Science Unlimited?: The Challenges of Scientism*. Chicago: University of Chicago Press.

Boulding, E. 1992: *The Underside of History*. Menlo Park, CA: Sage.

Boy, D. and L. Rouban 2019: "La Science au Défi de l'Opinion Publique," *Revue Politique et Parlementaire* 121: 43–58.

Boyle, R. 2018: "Help Young Scientists," *Scientific American* 319: 62–64.

Brasil247 2021: "Justiça do Paraná Reafirma Responsabilidade da Syngenta Seeds por Assassinato de Agricultor do MST," *Brasil247*. www.brasil247.com/meioambiente/justica-do-parana-reafirma-responsabilidade-da-syngenta-seeds-por-assassinato-de-agricultor-do-mst.

Bright, L. K. 2017: "Logical Empiricists on Race," *Studies in History and Philosophy of the Biological and Biomedical Sciences* 65: 9–18.

Brinkmann, S. 2013: *John Dewey: Science for a Changing World*. New Brunswick, NJ: Transaction.

Bronowski, J. 1956: *Science and Human Values*. New York: Julian Messner.

1968: "Science as a Humanistic Discipline," *The Bulletin of Atomic Scientists* 24: 33–38.

1973: *The Ascent of Man*. London: Macdonald Futura.

Brooke, J. H. 1991: *Science and Religion: Some Historical Perspectives*. Cambridge: Cambridge University Press.

Brown, J. R. 2008: "The Community of Science," in *The Challenge of the Social and the Pressure of Practice: Science and Values Revisited*, M. Carrier, D. Howard, and J. Kourany (eds.), 189–216. Pittsburgh: University of Pittsburgh Press.

Brown, K. 2020: "The Big Secret in the Academy Is That Most Research Is Secret," *Academe* 106, Spring. www.aaup.org/academe/issues/106-1/big-secret-academy-most-research-secret.

Brown, M. J. 2010: "Genuine Problems and the Significance of Science." *Contemporary Pragmatism* 7: 131–153.

2013: "Values in Science beyond Underdetermination and Inductive Risk," *Philosophy of Science* 80: 829–839.

2015: "John Dewey's Pragmatist Alternative to the Belief–Acceptance Dichotomy," *Studies in History and Philosophy of Science* 53: 62–70.

2020: *Science and Moral Imagination: A New Ideal for Values in Science.* Pittsburgh: University of Pittsburgh Press.

Brush, S. G. 1991: "Women in Science and Engineering," *American Scientist* 79: 404–419.

Bryant, R. L., and S. Bailey 1997: *Third World Political Ecology.* Hove: Psychology Press.

Buller, D. 2005: *Adapting Minds: Evolutionary Psychology and the Persistent Quest for Human Nature.* Cambridge, MA: MIT Press.

Burge, T. 1986: "Intellectual Norms and Foundations of Mind," *Journal of Philosophy* 83: 697–720.

Buscher, B. and R. Fletcher 2020: *The Conservation Revolution: Radical Ideas for Saving Nature beyond the Anthropocene.* Miamisburg: Verso.

Bush, V. 1945: *Science – The Endless Frontier: A Report to the President on a Program for Postwar Scientific Research.* Washington, DC: National Science Foundation.

Buss, D. M. and D. P. Schmitt 1993: "Sexual Strategies Theory: An Evolutionary Perspective on Human Mating," *Psychological Review* 100: 204–232.

Cadogan, J. 2014: *Curiosity-Driven Blue Sky Research: A Threatened Vital Activity?* Cardiff: Learned Sociey of Wales.

Callan, M., and B. Latour 2010: "Don't Throw the Baby Out with the Bath School! A Reply to Collins and Yearley," in *Science as Practice and Culture*, A. Pickering (ed.), 343–368. Chicago: University of Chicago Press.

Cameron, C. 2023: "W. E. B. Du Bois and African American Humanism," in *Forging Freedom in W. E. B. Du Bois's Twilight Years: No Deed but Memory*, P. L. Sinitiere (ed.), 131–146. Jackson: University Press of Mississippi.

Caplan, A. L. (ed.) 1978: *The Sociobiology Debate.* New York: Harper and Row.

Carnap, R. 1935: *Philosophy and Logical Syntax.* London: Kegan Paul.

1937/1934: *Logische Syntax der Sprache.* Vienna: Springer. Trans.: *Logical Syntax of Language.* London: Routledge and Kegan Paul.

1950a: "Empiricism, Semantics and Ontology." *Revue Internationale de Philosophie* 4: 20–40.

1950b: *Logical Foundations of Probability.* Chicago: University of Chicago Press.

1959/1932: "Überwindung der Metaphysik durch logische Analyse der Sprache," *Erkenntnis* 2: 219–241. Trans.: "The Elimination of Metaphysics through Logical Analysis of Language," in *Logical Positivism*, A. J. Ayer (ed.), 60–81. New York: Free Press.

1963a: "Carnap's Intellectual Autobiography," in *The Philosophy of Rudolf Carnap*, P. A. Schilpp (ed.), 1–84. Chicago: Open Court.

1963b: "Intellectual Autobiography" and "Comments and Replies," in *The Philosophy of Rudolf Carnap*, P. A. Schilpp (ed.), 3–85 and 859–1016. LaSalle: Open Court.

1967/1928: *Der Logische Aufbau der Welt*. Berlin: Weltkreis-Verlag. Trans.: *The Logical Structure of the World*. Berkeley: University of California Press.

2013/1934: "Theoretische Fragen und Praktische Entscheidungen," *Natur und Geist* 2: 257–260.

2017: "Value Concepts (1958)," *Synthese* 194: 185–194.

2022: *Tagebücher. Band 2: 1920–1935*, C. Damböck (ed.). Hamburg: Meiner.

2022/1918: "Deutschlands Niederlage. Sinnloses Schicksal oder Schuld?," in *Logischer Empirismus, Lebensreform und die deutsche Jugendbewegung*, C. Damböck, G. Sandner, and M. Werner (eds.), 317–337. Cham: Springer.

Carrier, M., D. Dragoman, J. Roggenhofer, G. Küppers, and P. Blanchard (eds.) 2004: *Knowledge and the World: Challenges beyond the Science Wars*. Berlin: Springer.

Cartwright, N., J. Cat, L. Fleck, and T. Uebel 1996: *Otto Neurath: Philosophy between Science and Politics*. Cambridge: Cambridge University Press.

Carus, A. W. 2007: *Carnap and Twentieth-Century Thought*. Cambridge: Cambridge University Press.

Caspary, W. R. 2000: *Dewey on Democracy*. Ithaca, NY: Cornell University Press.

Cassirer, E. 1951/1932: *The Philosophy of the Enlightenment*. Princeton, NJ: Princeton University Press.

Cat, J. 2019: "Otto Neurath," in *Stanford Encyclopedia of Philosophy*, E. Zalta (ed). plato.stanford.edu/entries/neurath/.

CBS News 2020: "The War on Science." www.cbsnews.com/video/the-war-on-science/.

Chakravartty, A. 2017: "Scientific Realism," in *Stanford Encyclopedia of Philosophy*, E. Zalta (ed.). plato.stanford.edu/entries/scientific-realism/.

2018: "Truth and the Sciences," in *The Oxford Handbook of Truth*, M. Glanzberg (ed.), 602–624. Oxford: Oxford University Press.

Chakravartty, A. and B. C. van Fraassen 2018: "What Is Scientific Realism?," *Spontaneous Generations: A Journal for the History and Philosophy of Science* 9: 12–25.

Chang, H. 2022: *Realism for Realistic People: A New Pragmatist Philosophy of Science*. Cambridge: Cambridge University Press.

Chilisa, B. 2019: *Indigenous Research Methodologies*. Thousand Oaks, CA: Sage.

Chung, E. 2014: "Foreign Scientists Call on Stephen Harper to Restore Science Funding, Freedom," *CBC News*. www.cbc.ca/news/technology/foreign-scientists-call-on-stephen-harper-to-restore-science-funding-freedom-1.2806571.

Churchland, P. M. 1981: "Eliminative Materialism and Propositional Attitudes," *Journal of Philosophy* 78: 67–90.

1985: "The Ontological Status of Observables: In Praise of the Superempirical Virtues," in *Images of Science*, P. M. Churchland and C. A. Hooker (eds.), 35–47. Chicago: University of Chicago Press.

Churchland, P. S. 2011: *Braintrust: What Neuroscience Tells Us about Morality*. Princeton: Princeton University Press.

Cohen, G. A. 2000: *Karl Marx's Theory of History: A Defence*. Oxford: Clarendon.

Cokley, K. 2020: "Why Black Psychology Matters: Validating the Lived Experience of Black People," *Psychology Today*. www.psychologytoday.com/us/blog/black-psychology-matters/202007/why-black-psychology-matters.

Cole, J. R. and H. Zuckerman 1987: "Marriage, Motherhood and Research Performance in Science," *Scientific American* 256: 119–125.

Coleman, A. R. 2020: "The Near-Certainty of a Black Depression," *Vox*. www.vox.com/2020/5/6/21248330/coronavirus-black-americans-depression.

Collins, H. 2021: "Science as Craftwork with Integrity," in *Global Epistemologies and Philosophies of Science*, in D. Ludwig, I. Koskinen, Z. Mncube, L. Poliseli, and L. Reyes-Galindo (eds.), 296–307. Milton Park: Routledge.

Collins, H. M., R. Evans, D. Durant, and M. Weinel 2020: *Experts and the Will of the People*. Cham: Palgrave Macmillan.

Collins, H. M. and S. Yearley 2010: "Epistemological Chicken," in *Science as Practice and Culture*, A. Pickering (ed.), 301–326. Chicago: University of Chicago Press.

Colom, R., M. Juan-Espinosa, F. Abad, and L. F. García 2000: "Negligible Sex Differences in General Intelligence," *Intelligence* 28: 57–68.

Condorcet, J.-A.-N. de C. 1955/1795: *Sketch for a Historical Picture of the Progress of the Human Mind*. London: Weidenfeld and Nicolson.

Cooper, D. E. 1999: "Humanism and the Scientific Worldview," *Theoria: A Journal of Social and Political Theory* 93: 1–17.

2002: *The Measure of Things: Humanism, Humility, and Mystery*. Oxford: Clarendon.

Copenhaver, B. 2007: "How to Do Magic, and Why: Philosophical Prescriptions," in *The Cambridge Companion to Renaissance Philosophy*, J. Hankins (ed.), 137–169. Cambridge: Cambridge University Press.

Corneanu, S. 2011: *Regimens of the Mind: Boyle, Locke, and the Early Modern Cultura Animi Tradition*. Chicago: University of Chicago Press.

Cottingham, J. 2009: "What Is Humane Philosophy and Why Is It at Risk?," *Royal Institute of Philosophy Supplement* 65: 233–255.

Creager, A., E. Lunbeck, and L. Schiebinger (eds.) 2001: *Feminism in Twentieth-Century Science, Technology, and Medicine*. Chicago: University of Chicago Press.

Curry, H. A. 2017: "From Working Collections to the World Germplasm Project: Agricultural Modernization and Genetic Conservation at the Rockefeller Foundation," *History and Philosophy of the Life Sciences* 39: 5.

Dahlberg, F. 1981: *Woman the Gatherer*. New Haven: Yale University Press.

Damasio, A. 2004: *Looking for Spinoza*. New York: Vintage.

Damböck, C. 2022: "Einleitung zu Rudolf Carnap, 'Deutschlands Niederlage: Sinnloses Schcksal oder Schuld?'," in *Logischer Empirismus, Lebensreform und die deutsche Jugendbewegung*, C. Damböck, G. Sandner, and M. Werner (eds.), 317–337. Cham: Springer.

Damböck, C., G. Sandner, and M. Werner (eds.) 2022: *Logischer Empirismus, Lebensreform und die deutsche Jugendbewegung*. Cham: Springer.

Darwin, C. 1981/1871: *The Descent of Man*. London: John Murray.

Dasgupta, P. 1995: *An Inquiry into Well-Being and Destitution*. Oxford: Oxford University Press.

Davidson, D. 1978: "What Metaphors Mean," *Critical Inquiry* 5: 31–47.

Davies, T. 2006: *Humanism*. London: Routledge.

Davison, R. and M. Gurven 2022: "The Importance of Elders: Extending Hamilton's Force of Selection to Include Intergenerational Transfers," *Proceedings of the National Academy of Sciences* 119: 1–12.

De Kluiver, H., J. E. Buizer-Voskamp, C. V. Dolan, and D. I. Boomsma 2017: "Paternal Age and Psychiatric Disorders: A Review," *American Journal of Medical Genetics Part B: Neuropsychiatric Genetics* 174: 202–213.

De Melo-Martin, I. and K. Intemann 2018: *The Fight against Doubt: How to Bridge the Gap between Scientists and the Public*. New York: Oxford University Press.

De Regt, H. W. 2017: *Understanding Scientific Understanding*. New York: Oxford University Press.

De Ridder, J. 2018: "Kinds of Knowledge, Limits of Science," in *Scientism: Prospects and Problems*, J. de Ridder, R. Peels, and R. van Woudenberg (eds.), 190–219. Oxford: Oxford University Press.

De Ridder, J., R. Peels, and R. van Woudenberg (eds.) 2018: *Scientism: Prospects and Problems*. Oxford: Oxford University Press.

De Waal, F. B. M. 1990: "Sociosexual Behavior Used for Tension Regulation in All Age and Sex Combinations among Bonobos," in *Pedophilia*, J. R. Feierman (ed.), 378–393. New York: Springer.

Deagan, M. J. 1988a: "W. E. B. Du Bois and the Women of Hull-House, 1895–1899," *American Sociologist* 19: 301–311.

1988b: *Jane Addams and the Men of the Chicago School, 1892–1918*. New Brunswick, NJ: Transaction.

Dennett, D. 1996: *Darwin's Dangerous Idea*. New York: Touchstone.

Descartes, R. 1998/1677: *The World and Other Writings*, S. Gaukroger (ed.). Cambridge: Cambridge University Press.

Dewey, J. 1910: *How We Think*. Boston: D. C. Heath.

1915: "The Logic of Judgments of Practice," *The Journal of Philosophy, Psychology and Scientific Methods* 12: 505–523.

1922: *Human Nature and Conduct*. New York: Henry Holt.

1923/1916: *Democracy and Education*. New York: Macmillan.

1985/1931: "Science and Society," in *The Later Works of John Dewey, Volume 6, 1925–1953: 1931–1932, Essays, Reviews, Miscellany*, J. A. Boydston (ed.), 53–63. Carbondale: Southern Illinois University Press.

1987/1934: *Art as Experience*, in *The Later Works of John Dewey, Volume 10, 1925–1953: 1934, Art as Experience*, J. Boydston (ed.), 1–368. Carbondale: Southern Illinois University Press.

1988/1925: *The Later Works of John Dewey, Volume 1, 1925–1953: 1925, Experience and Nature*, J. A. Boydston (ed.). Carbondale: Southern Illinois University Press.

1989/1934: *A Common Faith*, in *The Later Works of John Dewey, Volume 9, 1925–1953: 1933–1934, Essays, Reviews, Miscellany, and A Common Faith*, J. Boydston (ed.), 1–58. Carbondale: Southern Illinois University Press.

2021/1930: "What Humanism Means to Me," in *America's Public Philosopher: Essays on Social Justice, Economics, Education, and the Future of Democracy*, E. T. Webber (ed.), 314–317. New York: Columbia University Press.

Dewey, J. and J. Tufts 1932/1908: *Ethics*, revised edition. New York: Henry Holt.

Dewulf, F. 2021: "The Institutional Stabilization of Philosophy of Science and Its Withdrawal from Social Concerns after the Second World War," *British Journal for the History of Philosophy* 29: 935–953.

Dotson, K. 2011: "Tracking Epistemic Violence, Tracking Practices of Silencing," *Hypatia*, 26: 236–257.

Douglas, H. 2009: *Science, Policy, and the Value-Free Ideal*. Pittsburgh: University of Pittsburgh Press.

2013: "The Value of Cognitive Values," *Philosophy of Science* 80: 796–806.

2014: "Pure Science and the Problem of Progress," *Studies in History and Philosophy of Science* 46: 55–63.

2017. "Why Inductive Risk Requires Values in Science," in *Current Controversies in Value and Science*, K. C. Elliott and D. Steel (eds.), 81–93. Abingdon: Routledge.

Dumit, J. 2012: *Drugs for Life: How Pharmaceutical Companies Define Our Health*. Durham, NC: Duke University Press.

Dunbar, R. 2003: "The Social Brain: Mind, Language, and Society in Evolutionary Perspective," *Annual Review of Anthropology* 32: 163–181.

Dupré, J. 1993: *The Disorder of Things: Metaphysical Foundations of the Disunity of Science*. Cambridge, MA: Harvard University Press.

2001: *Human Nature and the Limits of Science*. Oxford: Oxford University Press.

2002: "The Lure of the Simplistic," *Philosophy of Science* 69: S284–S293.

2003: *Darwin's Legacy: What Evolution Means Today*. Oxford: Oxford University Press.

Dupré, J. and B. Barnes 2008: *Genomes and What to Make of Them*. Chicago: University of Chicago Press.

Dutilh, N. C. 2020: "Carnapian Explication and Ameliorative Analysis: A Comparison," *Synthese* 197: 1011–1034.

Dutton, D. 2009: *The Art Instinct: Beauty, Pleasure and Human Evolution*. London: Bloomsbury.

Early, G. 2006: "The Quest for Black Humanism," *Daedalus* 135: 91–104.

Eco, U. 1999: *Kant and the Platypus: Essays on Language and Cognition*, A. McEwan (trans.). London: Seeker & C. Warburg.

Eddington, A. 1928: *The Nature of the Physical World*. Cambridge: Cambridge University Press.
Ehrenfeld, D. 1981/1978: *The Arrogance of Humanism*. New York: Oxford University Press.
Einon, D. 1998: "How Many Children Can One Man Have?," *Evolution and Human Behavior* 19: 413–426.
Eliot, L. 2020: "Sex/Gender Differences in the Brain and Their Relationship to Behavior," in *The Cambridge Handbook of the International Psychology of Women*, F. M. Cheung and D. F. Halpern (eds.), 63–80. Cambridge: Cambridge University Press.
Elliott, K. C. 2017: *A Tapestry of Values: An Introduction to Values in Science*. Oxford: Oxford University Press.
Elliott, K. C. and D. Steel (eds.) 2017: *Current Controversies in Value and Science*. Abingdon: Routledge.
Elster, J. 1996: "Rationality and Emotions," *Economic Journal* 106: 1386–1397.
Engber, D. 2016: "Cancer Research Is Broken: There's a Replication Crisis in Biomedicine – and No One Even Knows How Deep It Runs," *Slate*. slate.com/technology/2016/04/biomedicine-facing-a-worse-replication-crisis-than-the-one-plaguing-psychology.html.
Epstein, S. 1995: "The Construction of Lay Expertise: AIDS Activism and the Forging of Credibility in the Reform of Clinical Trials," *Science, Technology and Human Values* 20: 408–437.
Escobar, A. 2018: *Designs for the Pluriverse*. Durham, NC: Duke University Press.
Estioko-Griffin, A. and P. B. Griffin 1981: "Woman the Hunter, the Agta," in *Woman the Gatherer*, F. Dahlberg (ed.), 121–152. New Haven: Yale University Press.
ETHNA 2025: "About RRI." https://ethnasystem.eu/about-rri/.
Etzkowitz, H. 2003: "Innovation in Innovation: The Triple Helix of University–Industry–Government Relation," *Social Science Information* 42: 293–337.
Evans, G. 1982: *Varieties of Reference*. Oxford: Oxford University Press.
2018: "The Unwelcome Revival of 'Race Science'," *The Guardian*. www.theguardian.com/news/2018/mar/02/the-unwelcome-revival-of-race-science.
Fairhead, J., M. Leach, and I. Scoones 2012: "Green Grabbing: A New Appropriation of Nature?," *Journal of Peasant Studies*, 39: 237–261.
Falk, D. 2004: "Prelinguistic Evolution in Early Hominins: Whence Motherese?," *Behavioral and Brain Sciences* 27: 491–503.
Farr, J. 1991: "Science: Realism, History, Critique," in *The Cambridge Companion to Marx*, T. Carver (ed.), 106–123. Cambridge: Cambridge University Press.
Fausto-Sterling, A. 2012: *Sex/Gender: Biology in a Social World*. London: Routledge.
2020: *Sexing the Body: Gender Politics and the Construction of Sexuality*, updated edition. New York: Basic Books.

Feigl, H. 1949: "Naturalism and Humanism," *American Quarterly* 1: 135–148.
1981/1949: *Inquiries and Provocations: Selected Writings, 1927–1974*, R. S. Cohen (ed.). Dordrecht: Reidel.
Fernandez Pinto, M. 2018: "Democratizing Strategies for Industry-Funded Medical Research: A Cautionary Tale," *Philosophy of Science* 85: 882–894.
Fesmire, S. A. 1995: "Dramatic Rehearsal and the Moral Artist: A Deweyan Theory of Moral Understanding," *Transactions of the Charles Sanders Peirce Society* 31: 568–597.
2003: *John Dewey and Moral Imagination: Pragmatism in Ethics*. Bloomington: Indiana University Press.
Feyerabend, P. K. 1975: "How to Defend Society against Science," *Radical Philosophy* 11: 3–8.
1987: *Farewell to Reason*. London: Verso.
1993. *Against Method*, 3rd edition. New York: Verso.
1999: *Conquest of Abundance: A Tale of Abstraction versus the Richness of Being*, B. Terpstra (ed.). Chicago: University of Chicago Press.
Feynman, R. 1955: "The Value of Science," *Engineering and Science* 19: 13–15.
Figueredo, A. J., G. Váquez, B. H. Brumach, S. M. R. Schneider, J. A. Sefcek, I. R. Tal, D. Hill, C. J. Wenner, and W. J. Jacobs 2006: "Consilience and Life History Theory: From Genes to Brain to Reproductive Strategy," *Developmental Review* 26: 243–275.
Fine, A. 1998: "The Viewpoint of No-one in Particular," *Proceedings and Addresses of The American Philosophical Association* 72: 9–20.
Fleming, A. 1929: "On the Antibacterial Action of Cultures of a Penicillium, with Special Reference to Their Use in the Isolation of B. Influenza," *British Journal of Experimental Pathology* 10: 226–236.
Flynn, T. 2005: "Foucault's Mapping of History," in *The Cambridge Companion to Foucault*, G. Gutting (ed.), 29–48. Cambridge: Cambridge University Press.
Foerster, N. (ed.) 1930: *Humanism and America: Essays on the Outlook of Modern Civilization*. New York: Farrar and Rinehart.
Folwell, M., K. Sanders, and J. Crowe-Riddell 2022: "The Squamate Clitoris: A Review and Directions for Future Research," *Integrative and Comparative Biology* 62: 559–568.
Ford, T. N., S. Reber, and R. V. Reeves 2020: "Race Gaps in COVID-19 Deaths Are Even Bigger than They Appear," Brookings Institution. www.brookings.edu/blog/up-front/2020/06/16/race-gaps-in-covid-19-deaths-are-even-bigger-than-they-appear/.
Forum for Food Sovereignty 2007: "Declaration of Nyéléni." www2.world-governance.org/IMG/pdf_0072_Declaration_of_Nyeleni_-_ENG-2.pdf.
Foucault, M. 1970: *The Order of Things: An Archaeology of the Human Sciences*. London: Tavistock.
1977: *Counter-Practice, Language, Memory: Selected Interviews and Essays*, D. F. Bouchard (ed.). Oxford: Blackwell.

1978: *The History of Sexuality, Volume 1: An Introduction*, A. Sheridan (trans.). New York: Vintage.

1981: "Is It Useless to Revolt?," J. Bernauer (trans.), *Philosophy and Social Criticism* 8: 1–9.

Francis, D. and A. G. Opoku-Agyeman 2020: "Economists' Silence on Racism Is 100 Years in the Making," *Newsweek*. www.newsweek.com/economists-silence-racism-100-years-making-opinion-1509790.

Frank, P. 1949/1917: "Die Bedeutung der Physikalischen Erkenntnistheorie Ernst Machs für das Geisteslebens Unserer Zeit," *Die Naturwissenschaften* 5: 65–80. Trans.: "The Importance for Our Times of Ernst Mach's Philosophy of Science," in *Modern Science and Its Philosophy*, P. Frank (ed.), 61–78. Cambridge, MA: Harvard University Press.

1957: *Philosophy of Science: The Link between Science and Philosophy*. Englewood Cliffs: Prentice-Hall.

2021: *The Humanistic Background of Science*, G. A. Reisch and A. T. Tuboly (eds.), Albany: SUNY Press.

Fraser, N. 2009: *Scales of Justice: Reimagining Political Space in a Globalizing World*. New York: Columbia University Press.

Fraser, N. and A. Honneth 2003: *Redistribution or Recognition? A Political–Philosophical Exchange*. London: Verso.

Friedman, M. and R. Creath (eds.) 2007: *The Cambridge Companion to Carnap*. Cambridge: Cambridge University Press.

Galison, P. 2008: "Removing Knowledge: The Logic of Modern Censorship," in *Agnotology: The Making and Unmaking of Ignorance*, R. Proctor and L. Schiebinger (eds.), 37–54. Stanford, CA: Stanford University Press.

Galison, P. L. and D. J. Stump (eds.) 1996: *The Disunity of Science: Boundaries, Contexts, and Power*. Stanford, CA: Stanford University Press.

Garrison, J. 1997: *Dewey and Eros: Wisdom and Desire in the Art of Teaching*. Charlotte: Information Age.

Gaukroger, S. 2006: *The Emergence of a Scientific Culture: Science and the Shaping of Modernity 1210–1685*. Oxford: Oxford University Press.

2010: *The Collapse of Mechanism and the Rise of Sensibility: Science and the Shaping of Modernity 1680–1760*. Oxford: Oxford University Press.

2016: *The Natural and the Human: Science and the Shaping of Modernity 1739–1841*. Oxford: Oxford University Press.

2020: *Civilization and the Culture of Science: Science and the Shaping of Modernity 1795–1935*. Oxford: Oxford University Press.

Gentilini, U. and P. Webb 2008: "How Are We Doing on Poverty and Hunger Reduction? A New Measure of Country Performance," *Food Policy* 33: 521–532.

Gibbons, M. 1999: "Science's New Social Contract with Society," *Nature* 402 (Supp.): C81–C84.

Gibbons, M., C. Limoges, H. Nowotny, S. Schwartzman, P. Scott, and M. Trow 1994: *The New Production of Knowledge: The Dynamics of Science and Research in Contemporary Societies*. Thousand Oaks, CA: Sage.

Gignac, G. E. 2015: "Raven's Is Not a Pure Measure of General Intelligence: Implications for g Factor Theory and the Brief Measurement of g," *Intelligence* 52: 71–79.

Gläser, J. and K. Velarde 2018: "Changing Funding Arrangements and the Production of Scientific Knowledge: Introduction to the Special Issue," *Minerva* 56: 1–10.

Glover, D. 2010: "Is Bt Cotton a Pro-Poor Technology? A Review and Critique of the Empirical Record," *Journal of Agrarian Change* 10: 482–509.

Godlee, F. 2016: "At Your Next Conference Ask Where the Patients Are," *British Medical Journal* 354: i5123.

Goldblatt, P. 2006: "How John Dewey's Theories Underpin Art and Art Education," *Education and Culture* 22: 17–34.

Goldie, P. 2012: *The Mess Inside: Narrative, Emotion, and the Mind.* Oxford: Oxford University Press.

Goldman, A. 2010: "Huckleberry Finn and Moral Motivation," *Philosophy and Literature* 34: 1–16.

Gould, S. J. 1999: *Rocks of Ages: Science and Religion in the Fullness of Life*. New York: Ballantine.

Gould, E. and V. Wilson 2020: "Black Workers Face Two of the Most Lethal Preexisting Conditions for Coronavirus: Racism and Economic Inequality," Economic Policy Institute. www.epi.org/publication/black-workers-covid/.

Grafton, A. 1990: "Humanism, Magic and Science," in *The Impact of Humanism on Western Europe*, A. Goodman and A. MacKay (eds.), 99–117. London: Longman.

1996: "The New Science and Traditions of Humanism," in *The Cambridge Companion to Renaissance Humanism*, J. Kraye (ed.), 203–223. Cambridge: Cambridge University Press.

Gray, J. 1995: *Enlightenment's Wake: Politics and Culture at the Close of the Modern Age*. London: Routledge.

Grinnell, F. 2009: *Everyday Practice of Science: Where Intuition and Passion Meet Objectivity and Logic*. Oxford: Oxford University Press.

Guston, D. H. 2000a: "Retiring the Social Contract for Science," *Issues in Science and Technology* 16: 32–36.

2000b: *Between Politics and Science*. Cambridge: Cambridge University Press.

Hagen, E. H. 1999: "The Functions of Postpartum Depression," *Evolution and Human Behavior* 20: 325–359.

Halpern, D. F., C. P. Benbow, D. C. Geary, R. C. Gur, J. S. Hyde, and M. A. Gernsbacher 2007: "The Science of Sex Differences in Science and Mathematics," *Psychological Science in the Public Interest* 8: 1–51.

Halpern, D. F. and J. Wai 2020: "Sex Differences in Intelligence," in *The Cambridge Handbook of Intelligence*, R. Sternberg (ed.), 317–345. Cambridge: Cambridge University Press.

Halpern, D. F. and M. L. LaMay 2000: "The Smarter Sex: A Critical Review of Sex Differences in Intelligence," *Educational Psychology Review* 12: 229–246.

Hamington, M. 2010: "Care Ethics, John Dewey's 'Dramatic Rehearsal,' and Moral Education," *Philosophy of Education* 66: 121–128.

Hankins, J. 2007: "Humanism and Modern Political Thought," in *The Cambridge Companion to Renaissance Philosophy*, J. Hankins (ed.), 118–141. Cambridge: Cambridge University Press.

Haraway, D. J. 1988: "Situated Knowledges: The Science Question in Feminism and the Privilege of Partial Perspective," *Feminist Studies* 14: 575–599.

Harding, S. 1986: *The Science Question in Feminism*. Ithaca, NY: Cornell University Press.

1991: *Whose Science? Whose Knowledge? Thinking from Women's Lives*. Ithaca, NY: Cornell University Press.

2010: *The Postcolonial Science and Technology Studies Reader*. Durham, NC: Duke University Press.

Hardy, B. and T. D. Logan 2020: "Racial Economic Inequality amid the COVID-19 Crisis," Brookings Institution. www.brookings.edu/research/racial-economic-inequality-amid-the-covid-19-crisis/.

Harman, G. 1977: *The Nature of Morality*. Oxford: Oxford University Press.

Harris, C. 2021: "'Science Must Fall' and the Call for Decolonization in South Africa," in *Global Epistemologies and Philosophies of Science*, D. Ludwig, I. Koskinen, Z. Mncube, L. Poliseli, and L. Reyes-Galindo (eds.), 106–114. Milton Park: Routledge.

Harris, C. R. 2004: "The Evolution of Jealousy: Did Men and Women, Facing Different Selective Pressures, Evolve Different 'Brands' of Jealousy? Recent Evidence Suggests Not," *American Scientist* 92: 62–71.

Harrison, P. 2007: *The Fall of Man and the Foundations of Science*. Cambridge: Cambridge University Press.

Hartle, A. 2005: "Montaigne and Scepticism," in *The Cambridge Companion to Montaigne*, U. Langer (ed.), 183–206. Cambridge: Cambridge University Press.

Harvey, L. 2020: "Research Fraud: A Long-term Problem Exacerbated by the Clamour for Research Grants," *Quality in Higher Education* 26: 243–261.

Hastings, C. 2017: "Are Replication Studies Unwelcome?," *Frontiers*. blog.frontiersin.org/2017/05/01/are-replication-studies-unwelcome/.

Havstad, J. C. 2021: "Sensational Science, Archaic Hominin Genetics, and Amplified Inductive Risk," *Canadian Journal of Philosophy* 52: 295–320.

Hawkes, K., J. F. O'Connell, N. G. Bluton Jones, H. Alvarez, and E. L. Charnov 1998: "Grandmothering, Menopause, and the Evolution of Human Life Histories," *Proceedings of the National Academy of Sciences* 95: 1336–1339.

Haynes, S. 2020: "As Protesters Shine a Spotlight on Racial Injustice in America, the Reckoning Is Going Global," *Time*. time.com/5851879/racial-injustice-prsts-europe/.

Heidegger, M. 1962: *Being and Time*, J. Macquarrie and E. S. Robinson (trans.). Oxford: Blackwell.

1977: *The Question concerning Technology (and Other Essays)*, W. Lovitt (trans.). New York: Harper & Row.

1982: *Basic Questions of Phenomenology*, A. Hofstadter (trans.). Bloomington: Indiana University Press.

1987: *Gesamtausgabe, Volume 56/57*. Frankfurt am Main: Klostermann.

Heidegger, M. 1993: "Letter on Humanism," in *Basic Writings: Martin Heidegger*, D. F. Krell (ed.), 189–242. London: Routledge.

Helm, B. 2010: *Love, Friendship, and the Self: Intimacy, Identification, and the Social Nature of Persons*. New York: Oxford University Press.

Hempel, C. G. 1950: "Problems and Changes in the Empiricist Criterion of Meaning," *Revue Internationale De Philosophie* 41: 41–63.

Hempel, C. G. 1951: "The Concept of Cognitive Significance: A Reconsideration," *Proceedings of the American Academy of Arts and Sciences* 80: 61–77.

Hensel, P. 1930/1921: "Neuhumanismus," in *Kleine Schriften und Vorträge*, 272–277. Tübingen: J. C. B. Mohr.

Hickman, L. A. 1990: *John Dewey's Pragmatic Technology*. Bloomington: Indiana University Press.

Hicks, D. J. 2015: "Epistemological Depth in a GM Crops Controversy," *Studies in History and Philosophy of Biological and Biomedical Sciences* 50: 1–12.

Hinton, G. E., S. Osindero, and Y. W. Teh 2006: "A Fast Learning Algorithm for Deep Belief Nets," *Neural Computation* 18: 1527–1554.

Hoeveler, J. D. 1977: *The New Humanism: A Critique of Modern America, 1900–1940*. Charlottesville: University of Virginia Press.

Hofweber, T. 2005: "Number Determiners, Numbers, and Arithmetic," *Philosophical Review* 114: 179–225.

2016: *Ontology and the Ambitions of Metaphysics*. Oxford: Oxford University Press.

Holman, B. and T. Wilholt 2022: "The New Demarcation Problem," *Studies in History and Philosophy of Science* 91: 211–220.

Hope, C. and E. McGrath 1996: "Artists and Humanists," in *The Cambridge Companion to Renaissance Humanism*, J. Kraye (ed.), 161–188. Cambridge: Cambridge University Press.

Hoquet, T. 2020: "Bateman (1948): Rise and Fall of a Paradigm?," *Animal Behaviour* 164: 223–231.

Horkheimer, M. and T. Adorno 1994/1947: *The Dialectics of Enlightenment*, J. Cumming (trans.). New York: Continuum.

Howard, D. 2003: "Two Left Turns Make a Right: On the Curious Political Career of North American Philosophy of Science at Mid-Century," in *Logical Empiricism in North America*, G. Hardcastle and A. Richardson (eds.), 25–93. Minneapolis: University of Minnesota Press.

2009. "Better Red than Dead: Putting an End to the Social Irrelevance of Postwar Philosophy of Science," *Science & Education* 18: 199–220.

2019. "Otto Neurath: The Philosopher in the Cave," in *Neurath Reconsidered: New Sources and Perspectives*, J. Cat and A. Tuboly (eds.), 45–66. Cham: Springer.

Hrdy, S. 1999: *Mother Nature: Maternal Instincts and How They Shape the Human Species*. New York: Ballantine.

2006: "Empathy, Polyandry, and the Myth of the Coy Female," in *Conceptual Issues in Evolutionary Biology*, 3rd edition, E. Sober (ed.), 131–159. Cambridge, MA: MIT Press.

2009: *Mothers and Others: The Evolutionary Origins of Mutual Understanding*. Cambridge, MA: Belknap/Harvard University Press.

Humanists International 2022: "The Amsterdam Declaration." humanists.international/what-is-humanism/the-amsterdam-declaration/.

2023: "What Is Humanism?" humanists.international/what-is-humanism/.

Humanists UK 2023: "Humanism." humanists.uk/humanism/?gad_source=1.

Humphrey, N. 1999: "Why Human Grandmothers May Need Large Brains," *Psycoloquy* 10: 1–3.

Husserl, E. 1970: *The Crisis of European Sciences and Transcendental Philosophy*, D. Carr (trans.). Evanston: Northwestern University Press.

Ingram, D. 1994: "Foucault and Habermas on the Subject of Reason," in *The Cambridge Companion to Foucault*, G. Gutting (ed.), 215–261. Cambridge: Cambridge University Press.

Intemann, K. 2010: "25 Years of Feminist Empiricism and Standpoint Theory: Where Are We Now?," *Hypatia* 25: 778–796.

2017: "Feminism, Values, and the Bias Paradox: Why Value Management Is Not Sufficient," in *Current Controversies in Value and Science*, K. C. Elliott and D. Steel (eds.), 130–144. Abingdon: Routledge.

Ioannidis, J. 2018: "Rethink Funding," *Scientific American* 319: 52–55.

ISAAA 2019: *Global Status of Commercialized Biotech/GM Crops: 2019* (ISAAA Brief No. 55). Ithaca, NY: ISAAA.

Jacob, J. R. 1978: *Boyle and the English Revolution: A Study in Social and Intellectual Change*. New York: Burt Franklin.

Jacob, M. 1981: *Radical Enlightenment*. Crows Nest: Allen & Unwin.

James, W. 1896: "The Will to Believe," *New World* 5: 327–347.

1902: *The Varieties of Religious Experience: A Study in Human Nature*. New York: Longman.

1995/1907: *Pragmatism*. New York.

Janack, M. 2002: "Dilemmas of Objectivity," *Social Epistemology* 16: 267–281.

Jewett, A. 2020: *Science Under Fire: Challenges to Scientific Authority in Modern America*. Cambridge, MA: Harvard University Press.

Johnson, L. B. 1965: "To Fulfill These Rights: Commencement Address at Howard University." teachingamericanhistory.org/library/document/commencement-address-at-howard-university-to-fulfill-these-rights/.

Johnston, J. S. 2002: "John Dewey and the Role of the Scientific Method in Aesthetic Experience," *Studies in Philosophy and Education* 21: 1–15.

Kahane D. and K. Lopston 2013: "Stakeholder and Citizen Roles in Public Deliberation," *Journal of Public Deliberation* 9: 1–37.

Kang, S. T. 1973: *Sumerian Economic Texts from the Umma Archive*. Urbana: University of Illinois Press.

Kant, I. 1996/1784: "An Answer to the Question: What Is Enlightenment?," in *What Is Enlightenment: Eighteenth-Century Answers and Twentieth-Century Questions*, J. Schmidt (ed.), 53–58. Berkeley: University of California Press.

1929/1781/1787: *Critique of Pure Reason*, K. Smith (trans.). London: Macmillan.

1991: *Political Writings.* H. S. Reiss (ed.), H. B. Nisbet (trans.). Cambridge: Cambridge University Press.

Kaplan, H., K. Hill, J. Lancaster, and A. M. Hurtado 2000: "A Theory of Human Life History Evolution: Diet, Intelligence, and Longevity," *Evolutionary Anthropology* 9: 149–186.

Kellert, S. H., H. E. Longino, and C. K. Waters (eds.) 2006: *Scientific Pluralism.* Minneapolis: University of Minnesota Press.

Kennedy, B. and A. Tyson 2023: "Americans' Trust in Scientists, Positive Views of Science Continue to Decline," Pew Research Center. www.pewresearch.org/science/2023/11/14/americans-trust-in-scientists-positive-views-of-science-continue-to-decline/.

Kidd, I. J. 2020: "Humility, Contingency, and Pluralism in the Sciences," in *The Routledge Handbook on the Philosophy of Humility,* M. Alfano, M. Lynch, and A. Tanesini (eds.), 346–358. New York: Routledge.

2021: "Creativity in Science and the 'Anthropological Turn' in Virtue Theory," *European Journal of the Philosophy of Science* 11: 1–16.

Kidder, L., M. Fagan, and E. Cohn 1981: "Giving and Receiving: Social Justice in Close Relationships," in *The Justice Motive in Social Behavior,* M. J. Lerner and S. Lerner (eds.), 235–259. New York: Plenum.

King, M. L. Jr. 1964: *Why We Can't Wait.* New York: New American Library.

Kitcher, P. 1983: *Believing Where We Cannot Prove. Abusing Science: The Case against Creationism.* Cambridge, MA: MIT Press.

2001: *Science, Truth and Democracy.* New York: Oxford University Press.

2011: *Science in a Democratic Society.* Amherst: Prometheus Books.

2015: "Pragmatism and Progress," *Transactions of the Charles S. Peirce Society* 51: 475–494.

2020: "Can We Sustain Democracy and the Planet Too?," in *Science and the Production of Ignorance: When the Quest for Knowledge Is Thwarted,* J. A. Kourany and M. Carrier (eds.), 89–120. Cambridge, MA: MIT Press.

Kleis, R., T. Louwerens, and A. Sikkema 2019: "Fresco's Syngenta Board Position Controversial," *Resource.* resource.wur.nl/en/show/Frescos-Syngenta-board-position-controversial.html.

Kofman, A. 2018: "Bruno Latour, the Post-Truth Philosopher, Mounts a Defense of Science," *New York Times Magazine.* www.nytimes.com/2018/10/25/magazine/bruno-latour-post-truth-philosopher-science.html.

Kolenda, K. 1995: "Humanism," in *The Cambridge Dictionary of Philosophy,* R. Audi (ed.), 340–341. Cambridge: Cambridge University Press.

Kolm, S. C. 2002: *Modern Theories of Justice.* Cambridge, MA: MIT Press.

Kornblith, H. 2012: *On Reflection.* Oxford: Oxford University Press.

Korte, A. 2019: "U.S. Scientific Community Needs a Mission Reboot to Focus on Public Good," *American Association for the Advancement of Science.* www.aaas.org/news/us-scientific-community-needs-mission-reboot-focus-public-good.

Kourany, J. A. (ed.) 2002: *The Gender of Science.* Upper Saddle River: Prentice Hall.

2010: *Philosophy of Science after Feminism*. Oxford: Oxford University Press.

2016: "Should Some Knowledge Be Forbidden? The Case of Cognitive Differences Research," *Philosophy of Science* 83: 779–790.

2020: "Might Scientific Ignorance Be Virtuous? The Case of Cognitive Differences Research," in *Science and the Production of Ignorance: When the Quest for Knowledge Is Thwarted*, J. A. Kourany and M. Carrier (eds.), 123–143. Cambridge, MA: MIT Press.

Kranthi, K. R. and G. D. Stone 2020: "Long-Term Impacts of Bt Cotton in India," *Nature Plants* 6: 188–196.

Kraye, J. 1996: "Philologists and Philosophers," in *The Cambridge Companion to Renaissance Humanism*, J. Kraye (ed.), 142–160. Cambridge: Cambridge University Press.

Krimsky, S. 2003: *Science in the Private Interest: Has the Lure of Profits Corrupted Biomedical Research?* New York: Rowman & Littlefield.

Krishna, V. V. 2014: "Changing Social Relations between Science and Society: Contemporary Challenges," *Science, Technology, and Society* 19: 133–159.

Krishtel, P. and R. Malpani 2021: "Suspend Intellectual Property Rights for Covid-19 Vaccines," *British Medical Journal* 373: n1344.

Kristeller, P. O. 1990: "Humanism," in *The Cambridge History of Renaissance Philosophy*, C. B. Schmitt, J. Kraye, E. Kessler, and Q. Skinner (eds.), 111–138. Cambridge: Cambridge University Press.

Kuhn, T. S. 1962: *The Structure of Scientific Revolutions*. Chicago: University of Chicago Press.

1970: "Notes on Lakatos," *PSA: Proceedings of the Biennial Meeting of the Philosophy of Science Association* 1970: 137–146.

1977: "Objectivity, Value Judgment, and Theory Choice," in *The Essential Tension: Selected Studies in Scientific Tradition and Change*, 320–339. Chicago: University of Chicago Press.

1978: *Black Body Theory and the Quantum Discontinuity, 1894–1912*. Oxford: Oxford University Press.

2012: *The Structure of Scientific Revolutions: 50th Anniversary Edition*. Chicago: University of Chicago Press.

Kurtz, P. 2001: *Skepticism and Humanism: The New Paradigm*. London: Transaction.

Kwa, C. 2011: *Styles of Knowing: A New History of Science from Ancient Times to the Present*, D. McKay (trans.). Pittsburgh: University of Pittsburgh Press.

La Via Campesina 2020: "Annual Report." viacampesina.org/en/wp-content/uploads/sites/2/2021/05/EN_Annual_Report_2020_rev.pdf.

Lacey, A. 1995: "Humanism," in *The Oxford Companion to Philosophy*, T. Honderich (ed.), 375–376. Oxford: Oxford University Press.

Lacey, H. 2015: "Food and Agricultural Systems for the Future: Science, Emancipation and Human Flourishing," *Journal of Critical Realism* 14: 272–286.

2017: "Distinguishing between Cognitive and Social Values," in *Current Controversies in Value and Science*, K. C. Elliott and D. Steel (eds.), 15–30. Abingdon: Routledge.

Lacoue-Labarthe, P. 1990: *Heidegger, Art and Politics*. Oxford: Blackwell.

Ladyman, J. 2018: "Scientism with a Humane Face," in *Scientism: Prospects and Problems,* J. de Ridder, R. Peels, and R. van Woudenberg (eds.), 106–126. Oxford: Oxford University Press.

Lakatos, I. 1970: "Falsification and the Methodology of Scientific Research Programmes," in *Criticism and the Growth of Knowledge*, I. Lakatos and A. Musgrave (eds.), 91–196. Cambridge: Cambridge University Press.

Latour, B. 1993: *We Have Never Been Modern*. Minneapolis: University of Minnesota Press.

2004a: *Politics of Nature*. Cambridge, MA: Harvard University Press.

2004b: "Why Has Critique Run Out of Steam? From Matters of Fact to Matters of Concern," *Critical Inquiry* 30: 225–248.

2018: *Down to Earth: Politics in the New Climatic Regime*. Hoboken: John Wiley & Sons.

Laudan, L. 1984: *Science and Values: The Aims of Science and Their Role in Scientific Debate*. Berkeley: University of California Press.

LaVine, M. 2020: *Race, Gender and the History of Early Analytic Philosophy*. Lanham: Rowman & Littlefield.

Law, S. 2011: *Humanism: A Very Short Introduction*. New York: Oxford University Press.

Lee, R. B. and I. DeVore (eds.) 1968: *Man the Hunter*. Chicago: Aldine Publishing Company.

Lerner, M. J. 1980: *The Belief in a Just World: A Fundamental Delusion*. New York: Plenum.

Levya, X. and S. Speed 2008: "Hacia la Investigación Descolonizada: Nuestra Experiencia de Co-labor," in *Gobernar (en) la Diversidad: Experiencias Indígenas desde América Latina*, X. Leyva, A. Burguete, and S. Speed (eds.), 65–108. Mexico: CIESAS.

Lewis, W. 2018: "Louis Althusser," in *Stanford Encyclopedia of Philosophy*, E. Zalta (ed.). plato.stanford.edu/archives/spr2018/entries/althusser/.

Lipton, P. 2007/2017: "Science and Religion: The Immersion Solution," in *Realism and Religion: Philosophical and Theological Perspectives*, A. Moore and M. Scott (eds.), 31–46. London: Routledge.

Lloyd, E. 2006: *The Case of the Female Orgasm: Bias in the Science of Evolution*. Cambridge, MA: Harvard University Press.

Long, P. O. 2016: "Humanism and Science," in *Renaissance Humanism: Foundations, Forms, and Legacy, Volume 3: Humanism and the Disciplines*, A. Rabil, Jr. (ed.), 486–512. Philadelphia: University of Pennsylvania Press.

Longino, H. 1990: *Science as Social Knowledge*. Princeton, NJ: Princeton University Press.

2006: "Philosophy of Science after the Social Turn," in *Cambridge and Vienna: Frank P. Ramsey and the Vienna Circle*, M. C. Galavotti (ed.), 167–177. Dordrecht: Kluwer.

2016: *Underdetermination: A Dirty Little Secret?* STS Occasional Papers 4. London: UCL Department of Science and Technology Studies.

Low, B. S., R. D. Alexander, and K. M. Noonan 1987: "Human Hips, Breasts and Buttocks: Is Fat Deceptive?," *Ethology and Sociobiology* 8: 249–257.

Ludwig, D. 2016: "Ontological Choices and the Value-Free Ideal," *Erkenntnis* 81: 1253–1272.

Ludwig, D. and B. Boogaard 2021: "Making Transdisciplinarity Work," in *The Politics of Knowledge in Inclusive Development and Innovation*, D. Ludwig, B. Boogaard, P. Macnaghten, and C. Leeuwis 2022 (eds.), 19–33. London: Routledge.

Ludwig, D., B. Boogaard, P. Macnaghten, and C. Leeuwis (eds.) 2022: *The Politics of Knowledge in Inclusive Development and Innovation.* London: Routledge.

Luna, J. K. and B. Dowd-Uribe 2020: "Knowledge Politics and the Bt Cotton Success Narrative in Burkina Faso," *World Development* 136: 105–127.

Lynn, R. 1994: "Sex Differences in Intelligence and Brain Size: A Paradox Resolved," *Personality and Individual Differences* 17: 257–271.

2017: "Sex Differences in Intelligence: The Developmental Theory," *Mankind Quarterly* 58: 9–42.

Lynn, R. and P. Irwing 2004: "Sex Differences on the Progressive Matrices: A Meta-Analysis," *Intelligence* 32: 481–498.

MacIntyre, A. 1988: *Whose Justice? Which Rationality?* Notre Dame, IN: University of Notre Dame Press.

Magnus, P. D. 2013: "What Scientists Know Is Not a Function of What Scientists Know," *Philosophy of Science* 80: 840–849.

Maiman, T. H. 1960: "Optical and Microwave-Optical Experiments in Ruby," *Physical Review Letters* 4: 564–566.

Marcel, G. 1949: *Being and Having*, K. Farrar (trans.). London: Dacre.

March for Science 2017: "The Science behind the March for Science Crowd Estimates." medium.com/marchforscience-blog/the-science-behind-the-march-for-science-crowd-estimates-f337adf2d665.

Marx, K. 1986: *Karl Marx: A Reader*, J. Elster (ed.). Cambridge: Cambridge University Press.

1994: *Early Political Writing*, J. O'Malley (ed.). Cambridge: Cambridge University Press.

2009: *Economic and Philosophical Manuscripts of 1844 and The Communist Manifesto*, M. Mulligan (trans.). New York: Prometheus.

Mason, O. T. 1929: *Women's Share in Primitive Culture.* New York: Appleton.

Matthews, F. H. 1977: *Quest for an American Sociology: Robert E. Park and the Chicago School.* Montreal: McGill-Queen's University Press.

McAleer, G. J. and A. S. Rosenthal-Pubul 2023: *The Wisdom of Our Ancestors: Conservative Humanism and the Western Tradition.* Notre Dame, IN: University of Notre Dame Press.

McCarthy, C. 1999: "Dewey's Ethics: Philosophy or Science?," *Educational Theory* 49: 339–359.

McClelland, K. A. 2005: "John Dewey: Aesthetic Experience and Artful Conduct," *Education and Culture* 21: 42–62.

McDowell, J. 1979: "Virtue and Reason," *Monist* 62: 331–350.

1985: "Values and Secondary Qualities," in *Morality and Objectivity*, T. Honderich (ed.), 110–129. Boston, MA: Routledge.

2006: "Reply to Bilgrami," in *McDowell and His Critics*, C. Macdonald and G. Macdonald (eds.), 66–72. Hoboken: Wiley-Blackwell.

McElwain, G. 2019: *Mary Midgley: An Introduction*. London: Bloomsbury.

McGrew, W. C. 1981: "The Female Chimpanzee as a Human Evolutionary Prototype," in *Woman the Gatherer*, F. Dahlberg (ed.), 35–74. New Haven: Yale University Press.

McKnight, S. A. 2005: "Francis Bacon's God," *The New Atlantis* 10: 73–100.

Melogno, P., H. Miguel, and L. Giri (eds.) 2023: *Perspectives on Kuhn: Contemporary Approaches to the Philosophy of Thomas Kuhn*. Berlin: Springer.

Merleau-Ponty, M. 1962: *Phenomenology of Perception*, C. Morris (trans.). London: Routledge.

Michaels, D. 2008: *Doubt Is Their Product: How Industry's Assault on Science Threatens Your Health*. New York: Oxford University Press.

2020: *The Triumph of Doubt: Dark Money and the Science of Deception*. New York: Oxford University Press.

Midgley, M. 1994: *The Ethical Primate: Humans, Freedom, and Morality*. London: Routledge.

2002: *Beast and Man: The Roots of Human Nature*. London: Routledge.

Miller, B. 2014: "Science, Values, and Pragmatic Encroachment on Knowledge," *European Journal for Philosophy of Science* 4: 253–270.

Miller, G. 2000: *The Mating Mind*. New York: Random House.

Millstein, R. L. 2015: "GMOs? Not So Fast," *Common Reader: A Journal of the Essay* 1: 33–46.

Mirandolla, G. P. della 2012/1486: *Oration on the Dignity of Man: A New Translation and Commentary*, F. Borghesi, M. Papio, and M. Riva (eds.) (trans.). Cambridge: Cambridge University Press.

Mirowski, P. and E. Sent 2002: *Science Bought and Sold: Essays in the Economics of Science*. Chicago: University of Chicago Press.

Misak, C. J. 2004: *Truth and the End of Inquiry: A Peircean Account of Truth*. Oxford: Clarendon.

Mishima, K. 2020: "The 'Disenchantment of the World' or Why We Can No Longer Use the Formula as Max Weber Might Have Intended," in *The Oxford Handbook of Max Weber*, E. Hanke, L. Scaff, and S. Whimster (eds.), 352–373. Oxford: Oxford University Press.

Mizrahi, M. 2022: *For and Against Scientism: Science, Methodology, and the Future of Philosophy*. Lanham, MD: Rowman & Littlefield.

Monfasani, J. 1998: "Humanism, Renaissance," in *Routledge Encyclopedia of Philosophy*, E. Craig (ed.), 533–541. London: Taylor & Francis.

Montaigne, M. de 1991/1580–1595: *The Complete Essays*, M. A. Screech (trans.). Harmondsworth: Penguin.

Mooney, C. 2005: *The Republican War on Science*. New York: Basic Books.

More, L. T. 1915: *The Limitations of Science*. New York: Henry Holt.

1925: *The Dogma of Evolution*. Princeton, NJ: Princeton University Press.

1930: "The Pretensions of Science," in *Humanism and America: Essays on the Outlook of Modern Civilization*, N. Foerster (ed.), 2–24. New York: Farrar/Rinehart.

More, P. E. 1928. *Demon of the Absolute: Shelborne Essays, Volume 1*. Princeton, NJ: Princeton University Press.

Morgan, M. 2016: *Pragmatic Humanism: On the Nature and Value of Sociological Knowledge*. London: Routledge.

Moss, M. 2013: *Salt Sugar Fat: How the Food Giants Hooked Us*. New York: Random House.

Motta, R. 2014: "Social Disputes over GMOs: An Overview," *Sociology Compass* 8: 1360–1376.

Munafò, M. R., B. A. Nosek, D. V. M. Bishop, K. S. Button, C. D. Chambers, N. P. du Sert, U. Simonsohn, E. Wagenmakers, J. J. Ware, and J. P. A. Ioannidis 2017: "A Manifesto for Reproducible Science," *Nature Human Behaviour* 1. https://doi.org/10.1038/s41562-016-0021.

Murray, C. 2005a: "The Inequality Taboo," *Commentary* 120: 13.

2005b: "Where Are the Female Einsteins?," American Enterprise Institute. www.aei.org/articles/where-are-the-female-einsteins.

Nagel, E. 1960: *The Structure of Science*. New York: Harcourt.

Nagel, T. 1979: "Subjective and Objective," in *Mortal Questions*, T. Nagel, 196–213. Cambridge: Cambridge University Press.

National Science Board 2020: "Science and Technology: Public Attitudes, Knowledge, and Interest." ncses.nsf.gov/pubs/nsb20207/.

Nature 2017: "Researchers Should Reach beyond the Science Bubble," *Nature* 542: 391.

Nauert, C. G. 1995: *Humanism and the Culture of Renaissance Europe*. Cambridge: Cambridge University Press.

Nestle, M. 2018: *Unsavory Truth: How Food Companies Skew the Science of What We Eat*. New York: Basic Books.

Neurath, O. 1973/1912: "Das Problem des Lustmaximums," *Jahrbuch der Philosophischen Gesellschaft an der Universität Wien 1912*. Trans.: "The Problem of the Pleasure Maximum," in O. Neurath, *Empiricism and Sociology*, M. Neurath and R. S. Cohen (eds.), 113–122. Dordrecht: Reidel.

1973/1928: *Lebensgestaltung und Klassenkampf*. Berlin: Laub. Excerpts. Trans.: "Personal Life and Class Struggle," in O. Neurath, *Empiricism and Sociology*, M. Neurath and R. S. Cohen (eds.), 249–298. Dordrecht: Reidel.

1973/1931: *Empirische Soziologie: Der wissenschaftliche Gehat der Geschichte undf Nationalökonomie*. Vienna: Springer. Trans.: "Empirical Sociology," in O. Neurath, *Empiricism and Sociology*, M. Neurath and R. S. Cohen (eds.), 391–421. Dordrecht: Reidel.

1983/1913: "Die Verirrten des Cartesius und das Auxiliarmotiv (Zur Psychologie des Entschlusses)," *Jahrbuch der Philosophischen Gesellschaft an der Universität zu Wien 1913*, 45–59. Trans.: "The Lost Wanderers and the Auxiliary Motive (On the Psychology of Decision)," in O. Neurath,

Philosophical Papers 1913–1946, R. S. Cohen and M. Neurath (eds.), 1–12. Dordrecht: Reidel.

1983/1932a: "Soziologie im Physikalismus," *Erkenntnis* 2: 393–431. Trans.: "Sociology in the Framework of Physicalism," in O. Neurath, *Philosophical Papers 1913–1946*, R. S. Cohen and M. Neurath (eds.), 58–97. Dordrecht: Reidel.

1983/1932b: "Protokollsätze," *Erkenntnis* 3: 204–214. Trans.: "Protocol Statements," in O. Neurath, *Philosophical Papers 1913–1946*, R. S. Cohen and M. Neurath (eds.), 91–99. Dordrecht: Reidel.

1983/1934: "Radikaler Physikalismus und 'wirkliche Welt'," *Erkenntnis* 4: 346–362. Trans.: "Radical Physicalism and 'the Real World'," in O. Neurath, *Philosophical Papers 1913–1946*, R. S. Cohen and M. Neurath (eds.), 100–114. Dordrecht: Reidel.

2004/1913: "Zur Stellung des Sittlichen Werturteils in der Wissenschaftlichen Nationalökonomie," in *Verein für Sozialpolitik 1913*, 31–32. Trans.: "On the Role of Moral Value Judgements in Economic Science," in O. Neurath, *Economic Writings: Selections 1904–1945*, T. Uebel and R. S. Cohen (eds.), 297–298. Dordrecht: Kluwer.

2004/1920a: "Ein System der Sozialisierung," *Archiv für Sozialwissenschaft und Sozialpolitik* 48: 44–73. Trans.: "A System of Socialisation," in O. Neurath, *Economic Writings: Selections 1904–1945*, T. Uebel and R. S. Cohen (eds.), 345–370. Dordrecht: Kluwer.

2004/1920b: *Vollsozialisierung*. Jena: Diederichs. Trans.: "Total Socialisation," in O. Neurath, *Economic Writings: Selections 1904–1945*, T. Uebel and R. S. Cohen (eds.), 371–404. Dordrecht: Kluwer.

Neurath, O., H. Hahn, and R. Carnap 1973/1929: "Wissenschaftliche Weltauffassung: Der Wiener Kreis," in *Otto Neurath: Empiricism and Sociology*, M. Neurath, and R. S. Cohen (eds.), 299–318. Dordrecht: Reidel.

Nietzsche, F. 2001/1882: *The Gay Science: With a Prelude in German Rhymes and an Appendix of Songs*, B. Williams (ed.), J. Nauckhoff (trans., text), and A. Cel Caro (trans., poems). Cambridge: Cambridge University Press.

Niiniluoto, I. 2024: "Scientific Progress," in *Stanford Encyclopedia of Philosophy*, E. Zalta and U. Nodelman (eds.). plato.stanford.edu/archives/spr2024/entries/scientific-progress/.

Nola, R. 2018: "The Enlightenment: Truths behind a Misleading Abstraction," in *History, Philosophy, and Science Teaching: New Perspectives*, M. R. Matthews (ed.), 43–66. Dordrecht: Springer.

Noll, S. and E. G. Murdock 2020: "Whose Justice Is It Anyway? Mitigating the Tensions between Food Security and Food Sovereignty," *Journal of Agricultural and Environmental Ethics* 33: 1–14.

Noonan, J. 2022: *Embodied Humanism: Toward Solidarity and Sensuous Enjoyment*. London: Lexington.

Norman, R. 2004: *On Humanism*. London: Routledge.

Normore, C. 2010: "Fool's Good and Other Issues: Comment on Akeel Bilgrami's *Self-Knowledge and Resentment*," *Philosophy and Phenomenological Research* 81: 766–772.

Nosek, B. A. et al. 2018: "Evaluating the Replicability of Social Science Experiments in *Nature* and *Science* between 2010 and 2015," *Nature Human Behaviour* 2: 637–644.

NowThis Impact 2020: "Trump's Widespread and Well-Coordinated War on Science." www.youtube.com/watch?v=NFPa_aJ9X28.

Numbers, R. L. 2009: *Galileo Goes to Jail and Other Myths about Science and Religion.* Cambridge, MA: Harvard University Press.

Nussbaum, M. C. 1995: "Aristotle on Human Nature and the Foundations of Ethics," in *World, Mind, and Ethics: Essays on the Ethical Philosophy of Bernard Williams,* J. E. J. Altham and R. Harrison (eds.), 86–131. Cambridge: Cambridge University Press.

2016: *Not for Profit: Why Democracy Needs the Humanities.* Princeton, NJ: Princeton University Press.

O'Connell, H. E., K. V. Sanjeevan, and J. M. Hutson 2005: "Anatomy of the Clitoris," *The Journal of Urology* 174: 1189–1195.

Oh, S. S., et al. 2015: "Diversity in Clinical and Biomedical Research: A Promise Yet to Be Fulfilled," *PLOS Medicine* 12: e1001918.

Okruhlik, K. 2004: "Logical Empiricism, Feminism, and Neurath's Auxiliary Motive," *Hypatia* 19: 48–72.

Open Science Collaboration 2015: "Estimating the Reproducibility of Psychological Science," *Science* 349: aac4716-1–aac4716-8.

Oreskes, N. 2014: "Why We Should Trust Scientists." www.youtube.com/watch?v=RxyQNEVOElU&t=38s.

2021: *Why Trust Science?* Princeton, NJ: Princeton University Press.

Oreskes, N. and E. M. Conway 2010: *Merchants of Doubt: How a Handful of Scientists Obscured the Truth on Issues from Tobacco Smoke to Global Warming.* New York: Bloomsbury.

Osborne, N. G. and M. D. Feit 1992: "The Use of Race in Medical Research," *Journal of the American Medical Association* 267: 275–279.

Otto, S. 2016: *The War on Science: Who's Waging It, Why It Matters, What We Can Do about It.* Minneapolis: Milkweed.

Owen, D. 1994: *Maturity and Modernity: Nietzsche, Weber, Foucault, and the Ambivalence of Reason.* London: Routledge.

Page, A. E., et al. 2019: "Testing Adaptive Hypotheses of Alloparenting in Agta Foragers," *Nature Human Behavior* 3: 1154–1163.

Pappas, G. F. 2008: *John Dewey's Ethics: Democracy as Experience.* Bloomington: Indiana University Press.

Pascal, B. 1980/1670: *Pensées*, A. J. Krailsheimer (trans.). London: Penguin.

Peacocke, A. 1993: *Theology for a Scientific Age: Being and Becoming – Natural and Divine.* Oxford: Blackwell.

Pedwell, T. 2012: "Scientists Take Aim at Harper Cuts with 'Death of Evidence' Protest on Parliament Hill," *The Globe and Mail.* www.theglobeandmail

.com/news/politics/scientists-take-aim-at-harper-cuts-with-death-of-evidence-protest-on-parliament-hill/article4403233/.

Peels, R. 2018: "A Conceptual Map of Scientism", in *Scientism: Prospects and Problems*, J. de Ridder, R. Peels, and R. van Woudenberg (eds.), 28–56. Oxford: Oxford University Press.

Persson, R. S. 2007: "The Myth of the Antisocial Genius: A Survey Study of the Socio-Emotional Aspects of High-IQ Individuals," *Gifted and Talented International* 22: 19–34.

Pertry, I., I. R. E. Sanou, S. Speelman, and I. Ingelbrecht 2016: "The Success Story of Bt Cotton in Burkina Faso: A Role Model for Sustainable Cotton Production in Other Cotton-Growing Countries?," in *Innovative Farming and Forestry across the Emerging World: The Role of Genetically Modified Crops and Trees*, S. De Buck, I. Ingelbrecht, M. Heijde, and M. Van Montagu (eds.), 81–93. Ghent: International Industrial Biotechnology Network.

Pew Research Center 2021: "Amid a Series of Mass Shootings in the U.S., Gun Policy Remains Deeply Divisive," Pew Research Center. www.pewresearch.org/politics/2021/04/20/amid-a-series-of-mass-shootings-in-the-u-s-gun-policy-remains-deeply-divisive/.

Pielke, R. A. 2007: *The Honest Broker*. Cambridge: Cambridge University Press.

Pierce, A. J. 2020: "Whose Lives Matter? The Black Lives Matter Movement and the Contested Legacy of Philosophical Humanism," *Journal of Social Philosophy* 51: 261–282.

Pinker, S. 1999: *How the Mind Works*. New York: Norton.

2002: *The Blank Slate: The Modern Denial of Human Nature*. New York: Viking.

2011: *The Better Angels of Our Nature: Why Violence Has Declined*. New York: Viking.

2018: *Enlightenment Now: The Case for Reason, Science, Humanism, and Progress*. New York: Viking.

Pippin, R. 1998: *Idealism as Modernism: Hegelian Variations*. Cambridge: Cambridge University Press.

Plewis, I. 2019: "Adopting Hybrid Bt Cotton: Using Interrupted Time-Series Analysis to Assess Its Effects on Farmers in Northern India," *Review of Agrarian Studies* 9: 4–23.

Poellner, P. 1995: *Nietzsche and Metaphysics*. Oxford: Clarendon Press.

Polanyi, M. 1962: "The Republic of Science: Its Political and Economical Theory," *Minerva* 1: 54–74.

Popkin, R. 2003: *The History of Scepticism: From Savonarola to Bayle*. Oxford: Oxford University Press.

Porter, E. 2021: "Black Workers Stopped Making Progress on Pay. Is It Racism?," *New York Times*. www.nytimes.com/2021/06/28/business/economy/black-workers-racial-pay-gap.html.

Potochnik, A. 2017: *Idealization and the Aims of Science*. Chicago: University of Chicago Press.

Price, M. 2011: "To Replicate or Not to Replicate?," *Science*, December 2. www.sciencemag.org/careers/2011/12/replicate-or-not-replicate.
Proctor, R. 1995: *Cancer Wars: How Politics Shapes What We Know and Don't Know about Cancer*. New York: Basic Books.
Putnam, H. 1975: "What Is Mathematical Truth?," in *Mathematics, Matter and Method*, 60–78. Cambridge: Cambridge University Press.
1981: *Reason, Truth, and History*. Cambridge: Cambridge University Press.
2002: *The Collapse of the Fact/Value Dichotomy and Other Essays*. Cambridge, MA: Harvard University Press.
Quine, W. V. 1960: *Word and Object*. Cambridge, MA: MIT Press.
Radder, H. 2012: "What Prospects for a General Philosophy of Science?," *Journal for General Philosophy of Science* 43: 89–92.
(ed.) 2010: *The Commodification of Academic Research: Science and the Modern University*. Pittsburgh: University of Pittsburgh Press.
2019: *From Commodification to the Common Good: Reconstructing Science, Technology, and Society*. Pittsburgh: University of Pittsburgh Press.
Radnitzky, G. 1978: "The Boundaries of Science and Technology," in *The Search for Absolute Values in a Changing World, Volume 2: Proceedings of the Sixth International Conference on the Unity of the Sciences*, 1007–1036. New York: International Cultural Foundation Press.
Ratcliffe, M. 2013: "Phenomenology, Naturalism and the Sense of Reality," *Royal Institute of Philosophy Supplement* 72: 67–88.
Rattan, G. 2016: "Truth Incorporated," *Noûs* 50: 227–258.
Reichenbach, H. 1932/1930: *Atom and Cosmos*. London: Allen and Unwin.
1942/1927: *From Copernicus to Einstein*. New York: Philosophical Library.
1951: *The Rise of Scientific Philosophy*. Los Angeles: University of California Press.
1978/1928a: "Philosophy of the Natural Sciences," in *Hans Reichenbach: Selected Writings, 1909–1953, Volume 1*, M. Reichenbach and R. S. Cohen (eds.), 228–231. Dordrecht: Reidel.
1978/1928b: "The World View of the Exact Sciences," in *Hans Reichenbach: Selected Writings, 1909–1953, Volume 1*, M. Reichenbach and R. S. Cohen (eds.), 241–244. Dordrecht: Reidel.
1978/1930: "The Philosophical Significance of Modern Physics," in *Hans Reichenbach: Selected Writings, 1909–1953, Volume 1*, M. Reichenbach and R. S. Cohen (eds.), 304–323. Dordrecht: Reidel.
1978/1931: "Aims and Methods of Modern Philosophy of Nature," in *Hans Reichenbach: Selected Writings, 1909–1953, Volume 1*, M. Reichenbach and R. S. Cohen (eds.), 359–388. Dordrecht: Reidel.
Reichenberg, A. et al. 2006: "Advancing Paternal Age and Autism," *Archives for General Psychiatry* 63: 1026–1032.
Reilly, D., D. L. Neumann, and G. Andrews 2022: "Gender Differences in Self-estimated Intelligence: Exploring the Male Hubris, Female Humility Problem," *Frontiers in Psychology* 13: doi.org/10.3389/fpsyg.2022.812483.

Reisch, G. A. 2005: *How the Cold War Transformed Philosophy of Science: To the Icy Slopes of Logic*. Cambridge: Cambridge University Press.

Remarque, E. M. 1982/1928: *All Quiet on the Western Front*. New York: Ballantine Books.

Reyes-Galindo, L. 2021: "Post-truth and Science," in *Global Epistemologies and Philosophies of Science*, D. Ludwig, I. Koskinen, Z. Mncube, L. Poliseli, and L. Reyes-Galindo (eds.), 183–195. London: Routledge.

Richardson, A. 2013: "Taking the Measure of Carnap's Philosophical Engineering: Metalogic as Metrology," in *The Historical Turn in Analytic Philosophy*, E. Reck (ed.), 60–77. Basingstoke: Palgrave Macmillan.

2017: "'Neither a Confession nor an Accusation': Michael Polanyi, Hans Reichenbach, and Philosophical Modernity after World War One," *Historical Studies in the Natural Sciences* 47: 423–442.

2021: "Hans Reichenbach, Radio Philosopher: A Preliminary Report," *Synthese* 199: 12625–12641.

2022: "The Social Virtue of Science: Motivating Structural Objectivity in Logical Empiricism," in *The Socio-Ethical Dimension of Knowledge: The Mission of Logical Empiricism*, C. Damböck and A. T. Tuboly (eds.), 33–51. Vienna: Springer.

Richardson, S. S. 2009: "The Left Vienna Circle, Part 1: Carnap, Neurath, and the Left Vienna Circle Thesis" and "The Left Vienna Circle, Part 2: The Left Vienna Circle, Disciplinary History, and Feminist Philosophy of Science," *Studies in History and Philosophy of Science* 40: 14–24, 167–174.

Ridley, M. 1994: *The Red Queen: Sex and the Evolution of Human Nature*. Harmondsworth: Penguin.

Ritchie, S. J. et al. 2018: "Sex Differences in the Adult Human Brain: Evidence from 5216 UK Biobank Participants," *Cerebral Cortex* 28: 2959–2975.

Rivera Cusicanqui, S. 2010: *Ch'ixinakax Utxiwa: Una Reflexión Sobre Prácticas y Discursos Descolonizadores*. Buenos Aires: Tinta Limón.

Robbins, L. 1932: *On the Nature and Significance of Economic Science*. London: Macmillan.Rohe, W. 2017: "The Contract between Society and Science: Changes and Challenges," *Social Research: An International Quarterly* 84: 739–757.

Rollin, K. 2017: "Can Social Diversity Be Best Incorporated into Science by Adopting the Social Value Management Ideal?," in *Current Controversies in Value and Science*, K. C. Elliott and D. Steel (eds.), 113–129. Abingdon: Routledge.

Rømer, T. A. 2012: "Dewey's Philosophy: Intelligent Transactions in a Democratic Context," *Educational Philosophy and Theory* 44: 133–150.

Romizi, D. 2012. "The Vienna Circle's 'Scientific World-Conception': Philosophy of Science in the Political Arena," *HOPOS* 2: 205–242.

Rosenberg, A. 2018: "Philosophical Challenges for Scientism (and How to Meet Them?)," in *Scientism: Prospects and Problems*, J. de Ridder, R. Peels, and R. van Woudenberg (eds.), 83–105. Oxford: Oxford University Press.

Rossman, I. 1978: "Sexuality and Aging: An Internist's Perspective," in *Sexuality and Aging*, R. L. Solnick (ed.), 66–77. Los Angeles: University of Southern California Press.

Roughley, N. 2021: "Human Nature," in *Stanford Encyclopedia of Philosophy*, E. Zalta (ed.). plato.stanford.edu/entries/human-nature/.

Rouse, J. 2005: "Power/Knowledge," in *The Cambridge Companion to Foucault*, G. Gutting (ed.), 95–122. Cambridge: Cambridge University Press.

Rowell, T. E. 1974: "The Concept of Social Dominance," *Behavioural Biology* 11: 131–154.

Roy, S. and M. A. Edwards 2017a: "Academic Research in the 21st Century: Maintaining Scientific Integrity in a Climate of Perverse Incentives and Hypercompetition," *Environmental Engineering Science* 34: 51–61.

2017b: "Science Is Broken: Perverse Incentives and the Misuse of Quantitative Metrics Have Undermined the Integrity of Scientific Research," *Aeon*. aeon.co/essays/science-is-a-public-good-in-peril-heres-how-to-fix-it.

Rudner, R. 1953: "The Scientist qua Scientist Makes Value Judgments," *Philosophy of Science* 20: 1–6.

Ruphy, S. 2019: "Public Participation in the Setting of Research and Innovation Agenda: Virtues and Challenges from a Philosophical Perspective," in *Innovation beyond Technology: Science for Society and Interdisciplinary Approaches*, Y. Fujigaki, S. Laugier, and S. Lechevalier (eds.), 243–263. Berlin: Springer.

Said, E. W. 2004: *Humanism and Democratic Criticism*. New York: Columbia University Press.

Saini, A. 2019: *Superior: The Return of Race Science*. Boston: Beacon Press.

Sarewitz, D. 2004: "How Science Makes Environmental Controversies Worse," *Environmental Science & Policy* 7: 385–403.

2016: "Saving Science," *The New Atlantis* 49: 5–40.

Sargent, R. 2002: "Francis Bacon and the Humanistic Aspects of Modernity," *Midwest Studies in Philosophy* 26: 124–139.

2005: "Virtues and the Scientific Revolution," in *Scientific Values and Civic Virtues*, N. Koertge (ed.), 71–80. New York: Oxford University Press.

2012: "From Bacon to Banks: The Vision and the Realities of Pursuing Science for the Common Good," *Studies in History and Philosophy of Science* 43: 82–90.

Sarton, G. 1924: "The New Humanism," *Isis* 6: 9–42.

1930: "Review of *Humanism and America* by Norman Foerster," *Isis* 14: 446–449.

Sartre, J. 1957: *Being and Nothingness: An Essay on Phenomenological Ontology*, H. Barnes (trans.). London: Methuen.

1966/1946: *Existentialism and Humanism*, P. Mairet (trans.). London: Methuen.

Schaffer, S. 1997: "The Earth's Fertility as a Social Fact in Early Modern England," in *Nature and Society in Historical Context*, M. Teich, R. Porter, and B. Gustafsson (eds.), 124–147. Cambridge: Cambridge University Press.

Schlick, M. 1936: "Meaning and Verification," *Philosophical Review* 45: 339–369.

Schulenberg, U. 2021: "Pragmatism, Humanism, and Form," *European Journal of Pragmatism and American Philosophy* 13: 1–23.

Scimecca, J. A. and G. A. Goodwin 2003: "Jane Addams: The First Humanist Sociologist," *Humanity and Society* 27: 143–157.

Seigfried, C. H. 1999: "Socializing Democracy: Jane Addams and John Dewey," *Philosophy of the Social Sciences* 29: 207–230.

2002: "Introduction to the Illinois Edition," in *Democracy and Social Ethics*, C. H. Seigfried (ed.), ix–xxxviii. Chicago: University of Chicago Press.

Shah, E., D. Ludwig, and P. Macnaghten 2021: "The Complexity of the Gene and the Precision of CRISPR: What Is the Gene that Is Being Edited?," *Elementa: Science of Anthropocene* 9: 00072.

Shapin, S. 2010: "Lowering the Tone in the History of Science: A Noble Calling," in *Never Pure: Historical Studies of Science as If It Was Produced by People with Bodies, Situated in Time, Space, Culture, and Society, and Struggling for Credibility and Authority*, 1–14. Baltimore: Johns Hopkins University Press.

Sharar, S. 1983: *The Fourth Estate: A History of Women in the Middle Ages*. London: Routledge.

Shelby, T. 2016: *Dark Ghettos: Injustice, Dissent, and Reform*. Cambridge, MA: Harvard University Press.

Sheldrake, R. 2015. "The Replicability Crisis in Science." www.sheldrake.org/essays/the-replicability-crisis-in-science.

Shen, B. S. P. 1975: "Views – Science Literacy: Public Understanding of Science Is Becoming Vitally Needed in Developing and Industrialized Countries Alike," *American Scientist* 63: 265–268.

Sher, G. 2013: "Forms of Correspondence: The Intricate Route from Thought to Reality," in *Truth and Pluralism: Current Debates*, N. J. L. L. Pedersen and C. Wright (eds.), 157–179. Oxford: Oxford University Press.

Shipton, P. 1990: "African Famines and Food Security: Anthropological Perspectives," *Annual Review of Anthropology* 19: 353–394.

Shiva, V. 1991: *The Violence of the Green Revolution: Third World Agriculture, Ecology and Politics*. London: Zed Books.

Shiva, V., A. H. Jafri, A. Emani, and M. Pande 2000: *Seeds of Suicide: The Ecological and Human Costs of Globalisation of Agriculture*. New Delhi: Research Foundation for Science, Technology and Ecology.

Shusterman, R. 2012. *Thinking Through the Body: Essays in Somaesthetics*. Cambridge: Cambridge University Press.

Simba, P. 2021: *A Feminist Critique of Ubuntu: Implications for Citizenship Education in Zimbabwe,* doctoral dissertation, Stellenbosch University.

Simpson, L. C. 2001: *The Unfinished Project: Toward a Postmetaphysical Humanism*. London: Routledge.

Singh, D. 1993: "Adaptive Significance of Female Physical Attractiveness: Role of Waist-to-Hip Ratio," *Journal of Personality and Social Psychology* 65: 293–307.
Skoyles, J. R. 1999: "Human Evolution Expanded Brains to Increase Expertise Capacity, not IQ," *Psycoloquy* 10: 1–14.
Slater, M., J. Huxster, and J. Bresticker 2019: "Understanding and Trusting Science," *Journal for General Philosophy of Science* 50: 247–261.
Smilie, K. D. 2016: "Unthinkable Allies? John Dewey, Irving Babbitt and 'The Menace of the Specialized Narrowness'," *Journal of Curriculum Studies* 48: 113–135.
Smith, T. 2012: "Scientists Stage Mock Funeral to Protest Cuts to Research," *Canada.com*. www.canada.com/business/Scientists+stage+mock+funeral+protest+cuts+research/6913396/story.htm.
Smith-Spark, L. and J. Hanna 2017: "March for Science: Protesters Gather Worldwide to Support 'Evidence'," *CNN*. www.cnn.com/2017/04/22/health/global-march-for-science/index.html.
Snaza, N. 2017: "Is John Dewey's Thought 'Humanist'?," *Journal of Curriculum Theorizing* 32: 15–34.
Snow, C. P. 1959: *The Two Cultures and the Scientific Revolution*. Cambridge: Cambridge University Press.
Sokal, A. and J. Bricmont 1998: *Intellectual Impostures: Postmodern Philosophers' Abuse of Science*. London: Profile.
Soucheray, S. 2020: "US Blacks 3 Times More Likely than Whites to Get COVID-19," *CIDRAP News*. www.cidrap.umn.edu/news-perspective/2020/08/us-blacks-3-times-more-likely-whites-get-covid-19.
Sowell, E. R. et al. 2007: "Sex Differences in Cortical Thickness Mapped in 176 Healthy Individuals between 7 and 87 Years of Age," *Cerebral Cortex* 17: 1550–1560.
Specter, M. 2014: "Seeds of Doubt," *New Yorker*. www.newyorker.com/magazine/2014/08/25/seeds-of-doubt.
Spengler, O. 1991/1922/1918: *The Decline of the West*. Oxford: Oxford University Press.
Spinoza, B. 2018/1677: *Ethics*. Cambridge: Cambridge University Press.
Spivak, G. 1988: "Can the Subaltern Speak?," in *Marxism and the Interpretation of Culture*, C. Nelson and L. Grossberg (eds.), 271–313. Basingstoke: Macmillan.
Spriggs, W. 2020: "Is Now a Teachable Moment for Economists? An Open Letter to Economists from Bill Spriggs." www.minneapolisfed.org/~/media/assets/people/william-spriggs/spriggs-letter_0609_b.pdf.
Stadler, F. 2015: *The Vienna Circle: Studies in the Origins, Development, and Influence of Logical Empiricism*, 2nd edition. Dordrecht: Springer.
2018: "George Sarton, Ernst Mach, and the Unity of Science Movement: A Case Study in History and Philosophy of Science," *Sartoniana* 31: 63–121.

Steel, D. 2010: "Epistemic Values and the Argument from Inductive Risk," *Philosophy of Science* 77: 14–34.

Stenmark, M. 2001: *Scientism: Science, Ethics and Religion.* Aldershot: Ashgate.

2018: "Scientism and Its Rivals," in *Scientism: Prospects and Problems*, J. de Ridder, R. Peels, and R. van Woudenberg (eds.), 57–82. Oxford: Oxford University Press.

Stroud, S. R. 2011: *John Dewey and the Artful Life: Pragmatism, Aesthetics, Morality.* University Park: Pennsylvania State University Press.

Stuchlik, J. 2011: "Felicitology: Neurath's Naturalization of Ethics," *HOPOS* 1: 183–207.

Stuhr, J. J. 1997: *Genealogical Pragmatism: Philosophy, Experience, and Community.* Albany: University of New York Press.

Sumberg, J., J. Thompson, and P. Woodhouse 2012: *Contested Agronomy: Agricultural Research in a Changing World.* London: Routledge.

Swanson, H. 1989: *Medieval Artisans.* New York: Basil Blackwell.

Tabliabue, G. M. and H. Miller 2019: "Letter Regarding Dr Vandana Shiva's Anti-Scientific and Unethical Stances," *European Scientist.* www.europeanscientist.com/en/features/letter-regarding-dr-vandana-shivas-anti-scientific-and-unethical-stances/.

Tallis, R. 2011: *Aping Mankind: Neuromania, Darwinitis, and the Misrepresentation of Humanity.* Durham, NC: Acumen.

Taylor, C. 1989: *Sources of the Self: The Making of Modern Identity.* Cambridge: Cambridge University Press.

Taylor, P. J. and K. Patzke 2021: "From Radical Science to STS," *Science as Culture* 30: 1–10.

Taylor, R. and J. L. Jasinski 2011: "Femicide and the Feminist Perspective," *Homicide Studies* 15: 341–362.

Theocharis, T. and M. Psimopoulos 1987: "Where Science Has Gone Wrong," *Nature* 329: 595–598.

Toulmin, S. 1970: "Does the Distinction between Normal and Revolutionary Science Hold Water?," in *Criticism and the Growth of Knowledge*, I. Lakatos and A. Musgrave (eds.), 39–47. Cambridge: Cambridge University Press.

Touraille, P. and P. H. Gouyon 2008: "Why Are Women Smaller than Men? When Anthropology Meets Evolutionary Biology." *Nature Precedings.* https://doi.org/10.1038/npre.2008.1832.1.

Treves, A., R. B. Wallace, L. Naughton-Treves, and A. Morales 2006: "Co-managing Human–Wildlife Conflicts: A Review," *Human Dimensions of Wildlife* 11: 383–396.

Tuboly, A. T. and A. Richardson (eds.) 2024: *Interpreting Carnap.* Cambridge: Cambridge University Press.

Turner, C. 2013: *The War on Science: Muzzled Scientists and Wilful Blindness in Stephen Harper's Canada.* Vancouver: Greystone.

Twain, M. 2014/1884: *The Adventures of Huckleberry Finn*, R. K. Rasmussen (ed.). New York: Penguin Classics.

Uebel, T. 1996: "On Neurath's Boat," in *Otto Neurath: Philosophy between Science and Politics*, N. Cartwright, J. Cat, L. Fleck, and T. Uebel, 89–166. Cambridge: Cambridge University Press.

1998: "Enlightenment and the Vienna Circle's Scientific World-Conception," in *Philosophers on Education: New Historical Perspectives*, O. Rorty (ed.), 418–438. London: Routledge.

2005: "Political Philosophy of Science in Logical Empiricism: The Left Vienna Circle," *Studies in History and Philosophy of Science* 36: 754–773.

2009: "Neurath's Protocol Statements Revisited: Sketch of a Theory of Scientific Testimony," *Studies in History and Philosophy of Science* 40: 4–13.

2010: "What's Right about the Left Vienna Circle Thesis: A Refutation," *Studies in History and Philosophy of Science* 41: 214–221.

2012: "Carnap, Philosophy and 'Politics in Its Broadest Sense'," in *Carnap and the Legacy of Logical Empiricism*, R. Creath (ed.), 133–148. Dordrecht: Springer.

2018: "Calculation in Kind and Substantive Rationality: Neurath, Weber, Kapp," *History of Political Economy* 50: 289–320.

2019: "Neurath on Verstehen," *European Journal of Philosophy* 27: 912–938.

2020: "Intersubjective Accountability: Politics and Philosophy in the Left Vienna Circle," *Perspectives on Science* 28: 35–62.

2022: "More on Neurath on *Verstehen*: The Rejection of Weber's Ideal-Type Methodology," in *The History of Understanding in Analytic Philosophy: Around Logical Empiricism*, A. T. Tuboly (ed.), 103–134. London: Bloomsbury.

Vaesen, K. and J. Katzav 2019: "The National Science Foundation and Philosophy of Science's Withdrawal from Social Concerns," *Studies in History and Philosophy of Science* 78: 73–82.

Valentine, F., D. Allainé, J. Gaillard, and A. Cohas 2020: "Evolutionary Pathways to Communal and Cooperative Breeding in Carnivores," *The American Naturalist* 195: 1037–1055.

Van der Ploeg, J. D. 2018: *The New Peasantries: Struggles for Autonomy and Sustainability in an Era of Empire and Globalization*. Milton Park: Routledge.

Van der Vegt, R. 2018: "A Literature Review on the Relationships between Risk Governance and Public Engagement in Relation to Complex Environmental Issues," *Journal of Risk Research* 21: 1.

Van Fraassen, B. C. 2002: *The Empirical Stance*. New Haven: Yale University Press.

Varea, S. and S. Zaragocin (eds.) 2017: *Feminismo y Buen Vivir: Utopías Decoloniales*. Cuenca: Pydlos Ediciones.

Verein Ernst Mach (ed.) 2012/1929: *Wissenschaftliche Weltauffassung: Der Wiener Kreis*, Vienna: Wolf. Trans.: "The Scientific World-Conception: The Vienna Circle," in *Wissenschaftliche Weltauffassung: Der Wiener Kreis. Hrsg. vom Verein Ernst Mach (1929)*," F. Stadler and T. Uebel (eds.), 75–116. Vienna: Springer.

Verein für Sozialpolitik 1913: *Äusserungen zur Werturteilsdiskussion im Ausschuss des Vereins für Sozialpolitik*. Private printing for members. Reprinted 1996: *Der Werturteilsstreit*, H. H. Nau (ed.), 65–200. Marburg: Metropolis.

Vickers, B. 2000: "The Myth of Bacon's 'Anti-Humanism'," in *Humanism and Early Modern Philosophy*, J. Kraye and M. W. F. Stone (eds.), 135–158. London: Routledge.

Vijayan, D., D. Ludwig, C. Rybak, H. Kaechele, H. Hoffmann, H. C. Schönfeldt, and K. Löhr 2022: "Indigenous Knowledge in Food System Transformations," *Communications Earth & Environment* 3: 1–3.

Vives, J. L. 1948/1518: "A Fable about Man," N. Lenkeith (trans.), in *The Renaissance Philosophy of Man*, E. Cassirer, P. O. Kristeller, and J. H. Randall, Jr. (eds.), 387–393. Chicago: Chicago University Press.

Von Grebmer, K. et al. 2020: *2020 Global Hunger Index – One Decade to Zero Hunger: Linking Health and Sustainable Food Systems*. Bonn: Welthungerhilfe; and Dublin: Concern Worldwide. www.globalhungerindex.org/pdf/en/2020.pdf.

Vrieze, J. D. 2017: "Bruno Latour, A Veteran of the 'Science Wars,' Has a New Mission," *Science Magazine*. www.science.org/content/article/bruno-latour-veteran-science-wars-has-new-mission.

Wai, J., M. Cacchio, M. Putallaz, and M. C. Makel 2010: "Sex Differences in the Right Tail of Cognitive Abilities: A 30-Year Examination," *Intelligence* 38: 412–423.

Waks, L. J. 1998: "Experimentalism and the Flow of Experience," *Educational Theory* 1: 1–19.

Walker, R. S., K. R. Hill, M. V. Flinn, and R. M. Ellsworth 2011: "Evolutionary History of Hunter-Gatherer Marriage Practices," *PLOS One* 6: e19066–e19066.

Wallace, A. R. 1889: *Darwinism: An Exposition of the Theory of Natural Selection with Some of Its Applications*. London: Macmillan.

Watene, K. 2016: "Valuing Nature: Māori Philosophy and the Capability Approach," *Oxford Development Studies* 44: 287–296.

Weber, M. 2004/1917: "Science as a Vocation," in *The Vocation Lectures*, D. Owen and T. B. Strong (eds.), R. Livingstone (trans.), 1–31. Indianapolis: Hackett.

2012/1904: "Die 'Objektivität' Sozialwissenschaftlicher und Sozialpolitischer Erkenntnis," *Archiv für Sozialwissenschaft und Sozialpolitik* 19: 22–87. Trans.: "The 'Objectivity' of Knowledge in Social Science and Social Policy," in M. Weber, *Collected Methodological Writings*, H. H. Brun and S. Whimster (eds.), 100–138. London: Routledge.

2012/1917: "Der Sinn der 'Wertfreiheit' der Soziologischen und Ökonomischen Wissenschaften," revised edition, *Logos* 7: 40–88. Trans.: "The Meaning of 'Value Freedom' in the Sociological and Economic Sciences," in M. Weber, *Collected Methodological Writings*, H. H. Brun and S. Whimster (eds.), 304–334. London: Routledge.

2012/1919: *Wissenschaft als Beruf.* Munich: Duncker & Humblot. Trans.: "Science as Profession and Vocation," in M. Weber, *Collected Methodological Writings*, H. H. Brun and S. Whimster (eds.), 335–353. London: Routledge.

Welch, H. G., L. M. Schwartz, and S. Woloshin 2011: *Overdiagnosed: Making People Sick in the Pursuit of Health.* Boston, MA: Beacon Press.

Weldon, S. P. 2020: *The Scientific Spirit of American Humanism.* Baltimore: Johns Hopkins University Press.

Westbrook, R. B. 1991: *John Dewey and American Democracy.* Ithaca, NY: Cornell University Press.

Whiting, D. 2019: "Whither Higher-Order Evidence?," in *Higher-Order Evidence: New Essays*, A. Steglich-Petersen and M. Skipper (eds.), 246–264. Oxford: Oxford University Press.

Wikipedia Contributors 2023: "R/K selection theory," *Wikipedia.* en.wikipedia.org/w/index.php?title=R/K_selection_theory&oldid=1145864079.

Wilholt, T. and H. Glimell 2011: "Conditions of Science: The Three-Way Tension of Freedom, Accountability and Utility," in *Science in the Context of Application*, M. Carrier and A. Nordmann (eds.), 351–370. Berlin: Springer.

Williams, B. 1978: *Descartes: The Project of Pure Enquiry.* Harmondsworth: Penguin.

1985: *Ethics and the Limits of Philosophy.* Cambridge, MA: Harvard University Press.

2008: *Philosophy as a Humanistic Discipline.* A. W. Moore (ed.). Princeton, NJ: Princeton University Press.

Williams, W. C. 1923: *Spring and All.* Paris: Robert McAlmon.

Wilson, C. 2004: "The Preferences of Women," in *Moral Psychology: Feminist Ethics and Social Theory*, M. U. Walker and P. des Autels (eds.), 99–117. Lanham: Rowman & Littlefield.

Wilson, E. O. 1978: *On Human Nature.* Cambridge, MA: Harvard University Press.

2004: *On Human Nature*, revised edition. Cambridge, MA: Harvard University Press.

Wittgenstein, L. 1953: *Philosophical Investigations*, G. E. M. Anscombe (ed.). New York: Wiley-Blackwell.

Wittrock, C., E. M. Forsberg, A. Pols, P. Macnaghten, and D. Ludwig, 2021: *Implementing Responsible Research and Innovation: Organisational and National Conditions.* Cham: Springer Nature.

World Bank 2021a: "Employment in Agriculture (% of Total Employment) (Modelled ILO Estimate." data.worldbank.org/indicator/SL.AGR.EMPL.ZS.

2021b: "Responding to a Stark Rise in Food Insecurity across the Poorest Countries," *World Blogs.* blogs.worldbank.org/en/voices/responding-stark-rise-food-insecurity-across-poorest-countries.

Worthy, G. and C. Yestrebsky 2018: "Break Down Silos," *Scientific American* 319: 64–67.

Wray, K. B. 2021a: *Kuhn's Intellectual Path: Charting the Structure of Scientific Revolutions*. Cambridge: Cambridge University Press.
(ed.) 2021b: *Interpreting Kuhn: Critical Essays*. Cambridge: Cambridge University Press.
Wright, R. 1996: "Technology, Gender and Class: Worlds of Difference in Ur III Mesopotamia," in *Gender and Archaeology*, R. Wright (ed.), 79–110. Philadelphia: University of Pennsylvania Press.
WUR 2021: "University Rankings." www.wur.nl/en/education-programmes/wageningen-university/rankings.htm.
Wylie, A. 2012: "Feminist Philosophy of Science: Standpoint Matters," *Proceedings and Addresses of the American Philosophical Association* 86: 47–76.
2015: "A Plurality of Pluralisms: Collaborative Practice in Archaeology," in *Objectivity in Science*, F. Padovani, A. Richardson, and J. Y. Tsou, 189–210. Cham: Springer.
2022: "Humanizing Science and Philosophy of Science: George Sarton, Contextualist Philosophies of Science, and the Indigenous/Science Project," *Canadian Journal of Philosophy* 52: 256–278.
Yamamoto, Y. 2012: "Values, Objectivity and Credibility of Scientists in a Contentious Natural Resource Debate," *Public Understanding of Science* 21: 101–125.
Yap, A. 2010: "Feminism and Carnap's Principle of Tolerance," *Hypatia* 25: 437–454.
Yi, B.-U. 1999: "Is Two a Property?," *Journal of Philosophy* 96: 163–190.
Young, I. M. 1990: *Justice and the Politics of Difference*. Princeton, NJ: Princeton University Press.
Zimmerman, M. E. 1990: *Heidegger's Confrontation with Modernity*. Bloomington: Indiana University Press.
Zuckmayer, C. 1994/1929: "Erich Maria Remarque's All Quiet on the Western Front," in *The Weimar Republic Sourcebook*, A. Kaes, M. Jay, and E. Dimendberg (eds.), 23–24. Berkeley: University of California Press.

Index

Printed by Integrated Books International,
United States of America